Lecture Notes in Computer Sci

Edited by G. Goos, J. Hartmanis, and J. va

Springer
Berlin
Heidelberg
New York
Barcelona
Hong Kong
London
Milan
Paris
Tokyo

Holger Hermanns Roberto Segala (Eds.)

Process Algebra and Probabilistic Methods

Performance Modeling and Verification

Second Joint International Workshop PAPM-PROBMIV 2002
Copenhagen, Denmark, July 25-26, 2002
Proceedings

Springer

Series Editors

Gerhard Goos, Karlsruhe University, Germany
Juris Hartmanis, Cornell University, NY, USA
Jan van Leeuwen, Utrecht University, The Netherlands

Volume Editors

Holger Hermanns
University of Twente, Faculty of Computer Science
Formal Methods and Tools Group
P.O. Box 217, 7500 AE Enschede, The Netherlands
E-mail: hermanns@cs.utwente.nl

Roberto Segala
University of Verona, Department of Computer Science
Strada Le Grazie 15, 37134 Verona, Italy
E-mail: segala@sci.univr.it

Cataloging-in-Publication Data applied for

Die Deutsche Bibliothek - CIP-Einheitsaufnahme

Process algebra and probabilistic methods : performance modeling and verification ; second joint international workshop ; proceedings / PAPM PROBMIV 2002, Copenhagen, Denmark, July 25 - 26, 2002. Holger Hermanns ; Roberto Segala (ed.). - Berlin ; Heidelberg ; New York ; Barcelona ; Hong Kong ; London ; Milan ; Paris ; Tokyo : Springer, 2002
(Lecture notes in computer science ; Vol. 2399)
ISBN 3-540-43913-7

CR Subject Classification (1998): F.3.1, F.3, D.2.4, D.3.1, C.4

ISSN 0302-9743
ISBN 3-540-43913-7 Springer-Verlag Berlin Heidelberg New York

Springer-Verlag Berlin Heidelberg New York
a member of BertelsmannSpringer Science+Business Media GmbH

http://www.springer.de

Typesetting: Camera-ready by author, data conversion by Christian Grosche, Hamburg
Printed on acid-free paper SPIN 10873586 06/3142 5 4 3 2 1 0

Preface

This volume contains the proceedings of the second joint PAPM-PROBMIV Workshop, held at the University of Copenhagen, Denmark, July 25–26, 2002 as part of the *Federated Logic Conference* (FLoC 2002).

The PAPM-PROBMIV workshop results from the combination of two workshops: PAPM (Process Algebras and Performance Modeling) and PROBMIV (Probabilistic Methods in Verification). The aim of the joint workshop is to bring together the researchers working across the whole spectrum of techniques for the modeling, specification, analysis, and verification of probabilistic systems. Probability is widely used in the design and analysis of software and hardware systems, as a means to derive efficient algorithms (e.g. randomization), as a model for unreliable or unpredictable behavior (as in the study of fault-tolerant systems and computer networks), and as a tool to study performance and dependability properties. The topics of the workshop include specification, models, and semantics of probabilistic systems, analysis and verification techniques, probabilistic methods for the verification of non-probabilistic systems, and tools and case studies.

The first PAPM workshop was held in Edinburgh in 1993; the following ones were held in Regensberg (1994), Edinburgh (1995), Turin (1996), Enschede (1997), Nice (1998), Zaragoza (1999), and Geneva (2000). The first PROBMIV workshop was held in Indianapolis, Indiana (1998); the next one took place in Eindhoven (1999). In 2000, PROBMIV was replaced by a Dagstuhl seminar on Probabilistic Methods in Verification.

The first joint PAPM-PROBMIV was held as part of the *2001 Aachen Multiconference on Measurement, Modeling, and Evaluation of Computer-Communication Systems*. The proceedings were published in Springer's LNCS series as Volume 2165. Joining both research areas in a single workshop was considered very fruitful, and led to the second joint event, and to this volume.

Of the 19 regular papers submitted, the program committee accepted 10 for presentation at the workshop. They are included in the present volume as *selected papers*. The volume also contains four *short abstracts* which were selected on the basis of their innovative potential for the workshop. The workshop included two invited presentations, by David Sands (Chalmers University of Technology and Göteborg University), and André Schiper (Ecole Polytechnique Fédérale de Lausanne).

We thank all the members of the program committee, and their sub-referees, for selecting the papers to be presented. Our thanks go to the Centre for Telematics and Information Technology (CTIT) of the University of Twente for generously sponsoring the workshop, and to our FLoC sponsoring conference CAV. We also thank all the authors for their help in meeting the tight deadlines which we had to set.

May 2002 Holger Hermanns and Roberto Segala

PROBMIV Steering Committee

Marta Kwiatkowska	(Chair, University of Birmingham)
Luca de Alfaro	(University of California, Berkeley)
Rajeev Alur	(University of Pennsylvania)
Christel Baier	(University of Mannheim)
Michael Huth	(Kansas State University)
Joost-Pieter Katoen	(University of Twente)
Prakash Panangaden	(McGill University)
Roberto Segala	(University of Verona)

PAPM Steering Committee

Ed Brinksma	(University of Twente)
Roberto Gorrieri	(University of Bologna)
Ulrich Herzog	(University of Erlangen)
Jane Hillston	(University of Edinburgh)

Program Committee

Luca de Alfaro	University of California at Santa Cruz, USA
Christel Baier	University of Bonn, D
Gianfranco Balbo	University of Turin, I
Marco Bernardo	University of Urbino, I
Pedro R. D'Argenio	University of Cordoba, ARG
Stephen Gilmore	University of Edinburgh, UK
Holger Hermanns (co-chair)	University of Twente, co-chair
Michael Huth	Imperial College, UK
Marta Kwiatkowska	University of Birmingham, UK
Prakash Panangaden	Mc Gill University, CAN
William H. Sanders	University of Illinois at Urbana-Champaign, USA
Roberto Segala (co-chair)	University of Verona, co-chair
Markus Siegle	University of Erlangen-Nürnberg, D
Manuel Silva	University of Zaragoza, ES
Scott Smolka	SUNY at Stony Brook, USA

Referees

Alessandro Aldini
Suzana Andova
Christel Baier
Gianfranco Balbo
Simonetta Balsamo
Marco Bernardo
Henrik Bohnenkamp
Mario Bravetti
Lucia Cloth
Pedro R. D'Argenio
Massimiliano De Pierro
Susanna Donatelli
Stephen Gilmore
Holger Hermanns
Jane Hillston
Andras Horvath
Michael Huth
S. Purushothaman Iyer
Bertrand Jeannet
Joost-Pieter Katoen
Ulrich Klehmet
William Knottenbelt
Matthias Kuntz
Marcos E. Kurbán
Marta Kwiatkowska
Kai Lampka
Rom Langerak
Sophie Laplante
Olivier Markowitch
Joachim Meyer-Kayser
David Monniaux
Gethin Norman
Alfredo Olivero
Prakash Panangaden
Dave Parker
Sylvain Peyronnet
Tamara Rezk
Eike Ritter
Andrei Sabelfield
William H. Sanders
Roberto Segala
Markus Siegle
Manuel Silva
Scott Smolka
Jeremy Sproston
Eugene W. Stark
Vanessa Teague
Herbert Wiklicky

Table of Contents

Invited Contributions

Selected Papers

Short Abstracts

Failure Detection vs Group Membership in Fault-Tolerant Distributed Systems: Hidden Trade-Offs

André Schiper

Ecole Polytechnique Fédérale de Lausanne (EPFL)
1015 Lausanne, Switzerland
andre.schiper@epfl.ch

Abstract. Failure detection and group membership are two important components of fault-tolerant distributed systems. Understanding their role is essential when developing efficient solutions, not only in failure-free runs, but also in runs in which processes do crash. While group membership provides consistent information about the status of processes in the system, failure detectors provide inconsistent information. This paper discusses the trade-offs related to the use of these two components, and clarifies their roles using three examples. The first example shows a case where group membership may favourably be replaced by a failure detection mechanism. The second example illustrates a case where group membership is mandatory. Finally, the third example shows a case where neither group membership nor failure detectors are needed (they may be replaced by *weak ordering* oracles).

1 Introduction

Fault-tolerance in distributed systems may be achieved by replicating critical components. Although this idea is easily understood, the implementation of replication leads to difficult algorithmic problems. A distributed algorithm requires the specification of a system model. Two main models have been proposed: the *synchronous* model and the *asynchronous* model. The synchronous model assumes (1) a known bound on the transmission delay of messages, and (2) a known bound on the relative speed of processes — while the slowest process performs one step, the fastest process performs at most k steps where k is known). The asynchronous system does not assume any bound on the transmission delay of messages, and on the relative speed of processes. Obviously, the asynchronous model is more general. If some algorithm $\mathcal{A}$ is proven correct in the most general model (e.g., in the asynchronous model), $\mathcal{A}$ is also correct in a more restricted model (e.g. in the synchronous model). Clearly, it is advantageous to develop algorithms for the most general system model.

Unfortunately, it has been proven that a very basic fault-tolerant problem, the *consensus* problem, cannot be solved by a deterministic algorithm in the asynchronous model when a single process may crash [15]. The same problem may be solved by a deterministic algorithm in the synchronous system model. However, the synchronous system model requires that bounds be defined, which leads to a dilemma. If the bounds

H. Hermanns and R. Segala (Eds.): PAPM-PROBMIV 2002, LNCS 2399, pp. 1–15, 2002.

are chosen too small, they may be violated: in this case, the algorithm might behave erroneously. If the bound is too large, this has a negative impact on the performance of the algorithm if there is a crash: the crash detection time will be long, and the algorithm will be blocked in the meantime.

Two other system models have been defined, which are between the asynchronous and the synchronous system models: the partially synchronous system model [11,14] and the asynchronous model augmented with failure detectors (which we will simply refer to as the *failure detector* model) [5].[1] Consensus is solvable in these two system models. The partially synchronous model assumes that bounds exist, but they are not known and hold only eventually. The failure detector model specifies the properties with regard to failure detection in terms of two properties: *completeness* and *accuracy*. Completeness specifies the behaviour of the failure detectors with respect to a crashed process. Accuracy specifies the behaviour of failure detectors with respect to correct processes. For example, the $\Diamond\mathcal{S}$ failure detector is defined (1) by *strong completeness* — which requires that each faulty process is eventually suspected forever by each correct process — and (2) by *eventual weak accuracy* — which requires that eventually there exists some correct process that is no longer suspected by any correct process. The failure detector model has allowed a very important result to be established: $\Diamond\mathcal{S}$ is the weakest failure detector that allows us to solve consensus [4].

The results for failure detectors, and other work performed over the last 10 years, have contributed to providing a good understanding of the algorithms related to replication, e.g., consensus, atomic broadcast, group membership. The main open problem that remains is understanding the various algorithms from a quantitative point of view. This means not only comparing the cost of these algorithms in failure-free runs, but also in runs with process crashes. For crash detection, most existing infrastructures rely on a group membership service, whereas algorithmic papers rely on failure detectors. As shown below, this has an important impact on performance, and leads to the following questions: when is a membership service really needed, and when is a failure detection mechanism preferable?

Before addressing these questions, Section 2 introduces the group membership problem, and discusses solutions to this problem. Section 3 illustrates a case where group membership can favourably be replaced by a failure detection mechanism. However, failure detection alone is not enough: Section 4 gives an example where membership is necessary. Finally, Section 5 shows that it is sometimes possible to do without failure detection and group membership. Section 6 concludes the paper.

2 The Group Membership Problem

2.1 Specification

Roughly speaking, a *group membership service* manages the formation and maintenance of a set of processes called a *group*. The successive memberships of a group are called *views*, and the event by which a new is provided to a process is called the *install* event. A process may *leave* the group as a result of an explicit leave request

[1] Other system models have been defined, e.g., [10,18].

or because it failed. Similarly, a process may *join* the group, for example to replace a process that has left the group. One distinguishes two types of group membership services: *primary-partition* and *partitionable*. Primary-partition group membership services attempt to maintain a *single* agreed view of the current membership of the group. On the contrary, partitionable group membership services allow *multiple* views of the group to coexist in order to model network partitions. In the paper we only consider the primary-partition membership service.

2.2 Solving Group Membership

Many algorithms have been proposed to solve the group membership problem. These algorithms have in common to be complex. This is the case of the protocol in [24], but there are two more recent examples. In [17] Lotem et al. describe a membership protocol that requires the introduction of notions such as quorums, sub quorums, ambiguous sessions, last formed sessions, resolution rules, learning rules. In [7], and in its recent version [20], an Atomic Broadcast algorithm is described, which is based on a ring of processes. The protocol requires a *reformation* phase if one of the processes in the ring is suspected. The reformation phase decides on the processes that form the new ring, i.e., it solves the membership problem. The authors of [20] propose a complex protocol, based here on a three-phase commit protocol.

Understanding these protocols is not easy and takes time. However, the membership problem becomes trivial using consensus, which is a well understood problem [5,25]. Consider the current membership (also called *view*) v_i and the problem of defining the next membership v_{i+1}. This is can be seen as a consensus problem to be solved among processes in v_i, where the initial value of each process is a proposal for the next view (e.g., the set of processes not suspected), and the decision is the next view v_{i+1} (see Algorithm 1) [19].[2] Algorithm 1 completely hides the complexity of group membership in the consensus black box. In [17] the authors claim that the solution based on consensus is more costly in terms of communication rounds. However, the figures given (i.e., five communication rounds) is not correct: the right number is one plus the cost of consensus, i.e., the protocol can terminate in three communication rounds.[3] It is doubtful that [17] requires less that three communication rounds.

3 Failure Suspicions Instead of Membership Exclusion

Developing complex membership protocols — instead of reducing membership to consensus — had an indirect consequence. It has hidden the benefit of decoupling "failure detection" from "membership exclusion". While a group membership service gives a consistent information about the state of processes (correct or not), failure detection

[2] One part of the algorithm is missing here. If one correct process starts the protocol, all other processes have also to start the protocol (otherwise consensus might not terminate). This can be done using Reliable Broadcast [16].

[3] Consensus can be solved in two rounds, e.g., [25].

Algorithm 1 Solving group membership among *current-view* by reduction to consensus (code of process p)

$v_p \leftarrow \mathit{current\text{-}view} \setminus \mathit{suspected\text{-}processes}$;
$\mathit{decision}_p \leftarrow \mathit{consensus}5v_p0$;
    {execute consensus among *current-view*; v_p is the initial value for consensus}

$\mathit{new\text{-}view} \leftarrow \mathit{decision}_p$;

provides an inconsistent information. It may sometimes be sufficient and less costly to rely on inconsistent failure detection information rather than on consistent group membership information. Consider the following example. Let $v_i = \{p, q, r\}$ be the current view of a group (information known to p, q and r) and let p wait for a message from q. If only membership information are accessible to p, then p waits for the message until a new view v' is installed from which q is excluded. If failure detection information is accessible to p, than p waits until it suspects q to have crashed: the view is still v, other processes not necessarily suspect q, process p might later change its mind about q, and it is possible that q is never excluded from the membership. Failure detection is a lightweight service, compared to a membership service — which relies on a failure detection service. We show below a concrete example of the benefit of relying on failure suspicions instead of membership exclusion.

3.1 Replication Techniques

There exists two main classes of replication techniques that ensure strong consistency: *active* and *passive* replication (Fig. 1). Both replication techniques are useful since they have complementary features. With active replication [26], each request is processed by all replicas. This ensures a fast reaction to failures, and sometimes makes it easier to replicate legacy systems. However, active replication uses processing resources heavily and requires processing of requests to be deterministic.[4] With passive replication (also called *primary-backup* replication) [3] only one replica (the primary) processes the request, and sends update messages to the other replicas (the backups). This uses less resources than active replication does, without the requirement of operation determinism. However, passive replication is known to have a slow reaction to failures. The reason is related to failure detection. Passive replication is usually based on a group membership, which excludes the primary whenever it is suspected to have crashed [2,21].

Excluding a process from the membership has a high cost, which leads a group membership to avoid excluding processes that have not crashed. This requires a high failure detection timeout value, which leads to a slow reaction to the crash of the primary, and a high response time for the client. High response time can be prevented by decoupling failure suspicions from membership exclusion, as shown by the semi-passive replication technique.

[4] Determinism means that the result of an operation depends only on the initial state of a replica and the sequence of operations it has already performed.

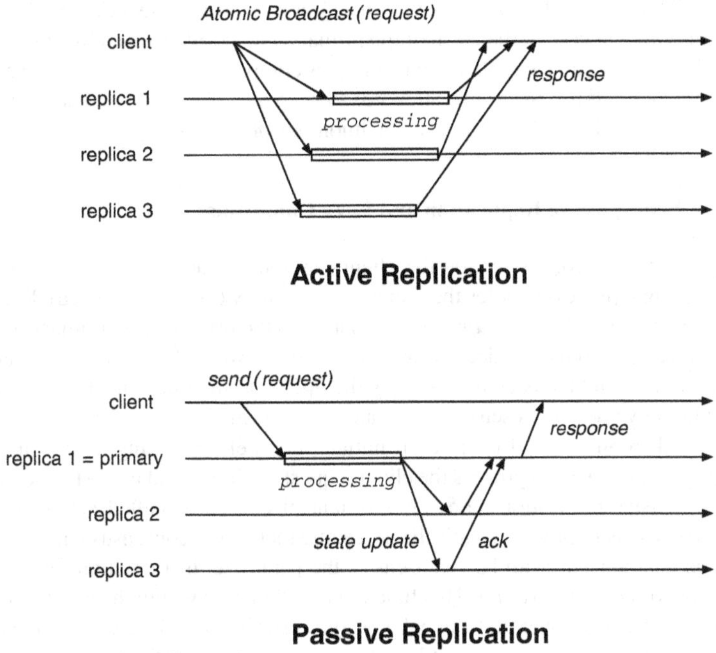

Fig. 1. Principle of active *vs* passive replication

3.2 Semi-passive Replication

Semi-passive replication [13,12] is a variant of passive replication: it retains its major characteristics (e.g., allows for non-deterministic processing). The main difference between passive and semi-passive replication is the selection of the primary. In semi-passive replication the selection of the primary is based on the rotating coordinator paradigm [7,14]; in passive replication the selection of the primary is based on a group membership service. The rotating coordinator paradigm allows the primary to be suspected without being excluded. This has a big advantage: it reduces the overhead of an incorrect suspicions. Consider the two cases: (1) the correct primary has been suspected and excluded from the membership, and (2) the correct primary has been suspected but not excluded from the membership. In case (1), in order to keep the same degree of replication, the excluded process needs to join again the membership, which leads to an new execution of the membership protocol (*join* operation), followed by the costly state transfer.[5] In case (2) no special action needs to be taken. In other words, an incorrect failure detection is costly in case (1), while it costs almost nothing in case (2). This allows in case (2) the failure detection mechanism to be much more aggressive, while in

[5] A correct process that is excluded from the membership is forced to commit suicide, and has to take a fresh copy of the state shared among the members.

case (1) it needs to be conservative. An aggressive failure detection time reduces the response time in case of the crash of the primary, i.e., corrects one of the major limitation of the class of passive replication techniques compared to the class of active replication techniques. Semi-passive also allows us to keep one of the major advantage of the class of passive replication techniques: parsimonious processing.

3.3 Semi-passive Replication and Lazy Consensus

With semi-passive replication the client sends its request to all replicas (see Fig. 2), but a single process handles the request, the primary (unless suspected). Processing the request provides the state *update* information to the primary. The primary then starts an instance of consensus to decide on the *state update value*. Upon decision, all replicas — the primary and the backups — apply the update to their current state. In other words, the initial value for consensus is a "state update value".

If all replicas need to have an initial value before starting consensus, then each replica would have to process the client request, which would be costly. To prevent this, semi-passive replication relies on a variant of consensus called *lazy consensus* [12]. With consensus, process p calls the procedure that solves consensus with its initial value v_p as a parameter. With lazy consensus, the parameter is a function called giv (which stands for *get initial value*). This function is called by p within the consensus algorithm whenever p needs an initial value. Lazy consensus is solved by a variant of the Chandra-Toueg $\Diamond S$ consensus algorithm based on the rotating coordinator paradigm [5]. If the first coordinator c is not suspected, then only c calls the giv function to get the update value. In other words:

- Semi-passive replication technique leads to a sequence of lazy consensus.
- Lazy consensus is based on the rotating coordinator paradigm, where the coordinator for consensus is the primary from the point of view of the replication algorithm.
- The initial value for consensus is obtained by calling the giv function, which processes the client request.
- If the first coordinator is suspected (Fig. 2, Scenario 2), a new process takes over the coordinator role for consensus, i.e., becomes the primary from the point of view of the replication algorithm. Changing the primary does not exclude the previous primary from the membership!

To summarise, group membership is a nice abstraction, but needs be used with care. Group membership transforms failure suspicions into process exclusion. There are cases where failure suspicions should not lead to process exclusion.

4 Failure Suspicion with Membership Exclusion

The previous section has illustrated a case where failure suspicion should not lead to process exclusion. In this section we give an example of failure suspicion that requires process exclusion.

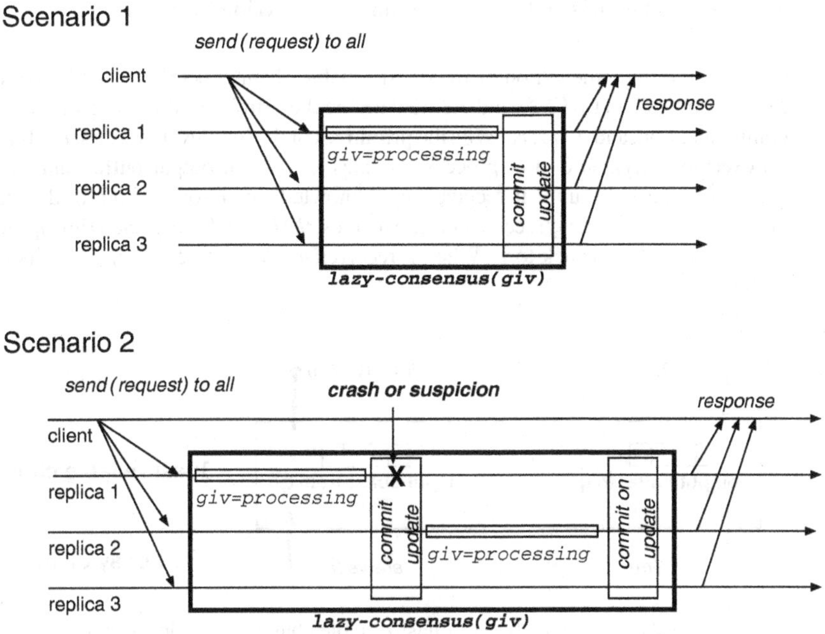

Fig. 2. Semi-passive replication. In Scenario 1, only the first coordinator (replica 1) calls the *get initial value (giv)* function. In Scenario 2, the first coordinator crashes or is wrongly suspected, replica 2 takes over the role of the coordinator, and calls the *giv* function.

4.1 Reliable Channels?

In the context of fault-tolerance, theoretical papers usually assume that channels are *reliable* — if p sends a message to q, and q is correct,[6] then q eventually receives m — or *quasi-reliable* — if p sends a message to q, and the two processes are correct, then q eventually receives m. However, real channels are neither reliable nor quasi-reliable. Lossy channels (and finite memory) lead to the exclusion of processes from the membership (see below). This explains that group membership is always considered in real systems, but is mostly absent from theoretical papers (apart from papers solving the group membership problem).[7]

---

[6] A correct process is a process that never crashes.

[7] This explains also the difficulty to come to a convincing specification for the group membership problem: a convincing specification requires to understand exactly when membership is needed, and when membership is not needed.

4.2 Lossy Channels and the Time-Bounded Buffering Problem

Consider the implementation of the quasi-reliable channel between p and q over fair-lossy channels.[8] Let $SEND$ and $RECEIVE$ be the primitives providing quasi-reliable communication, and $send$, $receive$ the primitives of the low-level lossy channel (Fig. 3). To execute $SEND5m0$ to q, process p copies m into an output buffer and executes $send5m0$ repeatedly until it receives an acknowledgement of m from q, denoted by $ack5m0$. The first time q receives m, it executes $RECEIVE5m0$. Each time q receives m, it sends $ack5m0$ back to p. When p receives $ack5m0$, it deletes m from its output buffer.

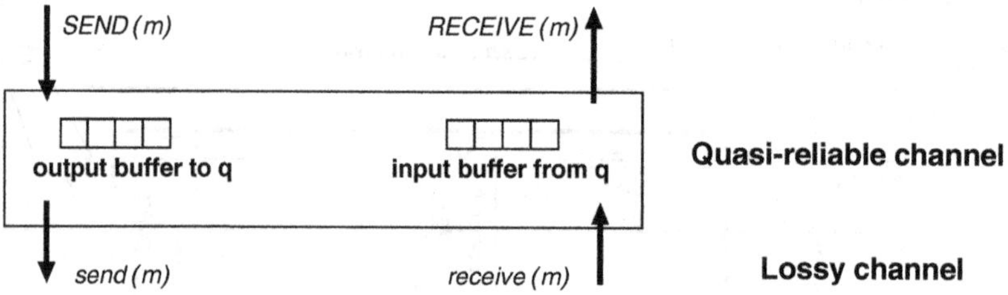

Fig. 3. Implementation of quasi-reliable channels over lossy channels

In this implementation, if q crashes, p might never receive $ack5m0$, and so might never delete m from its output buffer. This issue can be formalised by the *time-bounded buffering* problem [8]. Let m be a message in the output buffer of process p that must be sent to process q: time-bounded buffering ensures that p eventually deletes m from its output buffer. The problem cannot be solved in an asynchronous system model, neither in an asynchronous system model augmented with failure detectors of class $\mathcal{S}$ or class $\Diamond\mathcal{P}$ [5,8]. The same holds for Reliable Broadcast over fair-lossy channels. Real system overcome this impossibility by relying on *program-controlled crash* [6], which gives processes the ability to kill other processes. Consider process p with message m in its output buffer to q. If after some duration p has not received $ack5m0$ from q, it decides (1) to exclude q from the membership (i.e., to kill q), and (2) to discard m from its output buffer: as q eventually crashes, there is no obligation for q to deliver m.

There is a better solution than using timeouts to kill q. Process p kills q if, upon execution of $SEND5m0$ to q, *p's output buffer to q is full.* The murder of q is here the consequence of lack of resources, and not time-related. This is the best solution: it makes sense for p to kill q iff p has not enough space to buffer messages for q.

[8] A fair-lossy channels do not create, duplicate and garble messages, and ensure that if p sends an infinite number of messages to q, and q is correct, then q receives and infinite number of messages from p.

4.3 Process Suspicion vs Process Exclusion

In the above example, the exclusion of q from the membership is different from the exclusion of the primary r in the context of passive replication. With passive replication, if the primary r crashes at time t, then the replication service is immediately blocked. In the example of Section 4.2, if q crashes at time t, then p is blocked much later (depending on the size of its output buffer to q). This shows that when the blocking time an issue, suspecting the primary r fast is important, whereas suspecting q fast in the example of Section 4.2 is not important. In one case (primary r) the suspicion is *input-triggered*, whereas in the other case the suspicion is *output-triggered* [8]:

- *Input-triggered suspicions:* p suspects q because p waits a message from q, and its input buffer from q is empty. Process p is blocked until it suspects q.
- *Output-triggered suspicions:* p suspects q because some message remains for a long time in its output buffer to q, or because its output buffer to q is full.

There is no reason for input-triggered suspicions to lead to process exclusion: semi-passive replication is a good example. On the other hand, if the output buffer of p to q is full, p does not have many options: block or exclude q. In other words, input triggered suspicions *should never* lead to exclusion, while output-triggered suspicions *should always* lead to exclusion. This gives a clean picture. A group membership service should not react to input-triggered suspicions, but only to output-triggered suspicions.

Most existing systems handle input-triggered and output-triggered suspicions in the same way, i.e., by excluding processes. This is a poor choice from a performance point of view. If p waits a message from q, the timeout to detect the crash of q should be short (in order to reduce the blocking period). If p sends a message to q, there is no need to detect the crash of q quickly. A single failure detection mechanism, requires a compromise: the suspicion should not be too fast (do avoid too many wrong suspicions) and not too slow (do reduce the blocking time). Having input-triggered suspicions (without exclusions) on one hand side, and output-triggered exclusions on the other hand side allows to escape from the dilemma.

4.4 Input/Output-Triggered Suspicions vs Partitionable Membership

The distinction between input and output triggered suspicions is orthogonal to the distinction between primary partition and partitionable membership. Partitionable membership [9] does not distinguish between input-triggered and output-triggered suspicions: it relies on one single failure detection mechanism.

The difference between (1) input and output-triggered suspicions in the context of primary partition membership, and (2) partitionable membership can be clarified on the following example. Consider a system of five processes (Fig. 4): two client processes c_1, c_2, and three server processes s_1, s_2, s_3 (which implement semi-passive replication). Consider the following scenario, where the exclusion of the servers s_i is output-triggered:

- At time t_0 all processes are reachable from all processes. The server membership is $\{s_1, s_2, s_3\}$. No process is suspected, and all client requests are handled by the s_1. The server s_1 broadcasts the "update" message (Sect. 3) to s_2 and s_3.

- At time t_1, a link failure occurs, which partitions the system in two components: $\Pi_1 = \{c_2, s_1\}$ and $\Pi_2 = \{c_1, s_2, s_3\}$: processes s_2, s_3 suspect s_1 (and s_1 suspects probably s_2 and s_3). The requests of c_1 are handled by another server, say s_2. The server s_2 broadcasts the update message to s_1 and s_3. If the output buffer of s_2 to s_1 is large enough, s_1 is not excluded.
- At time t_2, the link failure is repaired. No special action needs to be taken. The update sent by s_2 to s_1 during the link failure have been buffered, and can now reach s_1. The occurrence of the partition is transparent to the servers.

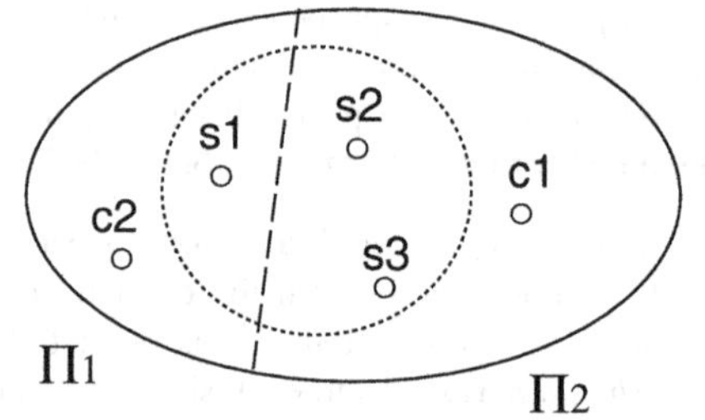

Fig. 4. Partition of the processes in two components Π_1 and Π_2

With partitionable membership, the link failure is not transparent. In the interval $[t_1, t_2]$ a partitionable membership defines two concurrent views. Once the link failure is repaired, the states of the two partitions would have to be merged (at the application level).

5 Doing without Failure Suspicion

In Section 3 we have shown the benefit of decoupling failure suspicion from membership exclusion. In Section 4 we have introduced the distinction between input and output-triggered suspicions, and explained the relationship between output-triggered suspicions and membership exclusion. We address now the following question: can input-triggered suspicions be avoided? Indeed the following dilemma remains: what timeout value should be chosen for input-triggered suspicions? On one hand the timeout value should be large in order to avoid wrong suspicions (they still have a negative impact on performance[9]). On the other hand timeout values should be small in order to ensure fast reaction to failures. Avoiding input-triggered suspicions avoids the dilemma of fine tuning the failure detection mechanism.

Randomisation is one solution, since it allows us to solve consensus in asynchronous systems [1,23]. However, randomisation leads to slow algorithms. Another solution is to augment the asynchronous system with *weak ordering oracles* [22], which order

[9] Even if wrong suspicions have less impact on performance than process exclusion, they still have a negative impact.

messages frequently, but might also deliver messages out of order. Weak ordering oracles capture the behaviour of network multicast in state-of-the-art local area networks: if messages are multicast in a local area network, there is a good chance that some of them will be received by all processors in the same order. Experiments have shown that if the interval between broadcast is around 0.15 ms, then very few messages are received out of order (about 5%) [22].

5.1   Weak Ordering Oracles

Weak ordering oracles are defined by two primitives, *W-ABroadcast(r,m)* and *W-ADeliver(r,m)*. The first primitive asks the oracle to broadcast m. The second primitive corresponds to the delivery of m by the oracle. The parameter r groups messages with the same r value (they can be seen as *round* numbers). The weak ordering property holds for round r if there exists some message m such that all processes deliver m before the other messages of round r. To illustrate this property, consider three processes p_1, p_2, p_3 executing the following queries to the oracle:

- p_1 executes W-ABroadcast$5D, m_1 0$; W-ABroadcast$5L, m_2 0$; W-ABroadcast$5z, m_3 0$
- p_2 executes W-ABroadcast$5D, m_4 0$; W-ABroadcast$5L, m_5 0$; W-ABroadcast$5z, m_6 0$
- p_3 executes W-ABroadcast$5D, m_7 0$; W-ABroadcast$5L, m_8 0$; W-ABroadcast$5z, m_9 0$

and assume the following sequences of W-ADeliver$5r, m0$ (for brevity, we denote next W-ADeliver$5r, m0$ by $D(r, m)$):

- on p_1: D$5D, m_1 0$; D$5L, m_2 0$; D$5D, m_4 0$; D$5z, m_3 0$; D$5D, m_7 0$; ...
- on p_2: D$5D, m_4 0$; D$5D, m_1 0$; D$5L, m_5 0$; D$5D, m_7 0$; D$5z, m_3 0$; ...
- on p_3: D$5D, m_4 0$; D$5D, m_7 0$; D$5z, m_3 0$; D$5L, m_8 0$; ...

The weak ordering property holds for round $r = z$ (m_3 is the first message with $r = z$ delivered by p_1, p_2, p_3), but does not hold for either $r = D$ or $r = L$. The oracle can make mistake: it does not have to satisfy the weak ordering property for all rounds.

The definition of the oracle assumes that each process executes W-ABroadcast$5r, m0$ sequentially for rounds $r = L, z, \ldots$. Let $first_p 5r0$ denote the first message of round r delivered by p. The *k-Weak Atomic Broadcast (or k-WAB)* Oracle is defined by the following properties:

- **Validity:** If a correct process executes W-ABroadcast$5r, m0$, then all correct processes eventually execute W-ADeliver$5r, m0$.
- **Uniform Integrity:** For every par $5r, m0$, W-ADeliver$5r, m0$ is executed at most once, and only if W-ABroadcast$5r, m0$ was previously executed.
- **Eventual Uniform k-Order:** If all processes execute an infinite sequence of W-ABroadcast$5r, m0$, for $r = L, z, \ldots$, then there exist k values $r_1, \ldots, r_k$ such that, for all $i \in [L, k]$ and all processes p, q, we have $first_p 5r_i 0 = first_q 5r_i 0$.

Algorithm 2 Ben-Or binary consensus algorithm: code of process p $(f < n/z)$

```
 1: Consensus (initVal):

 2:   estimate_p ← initVal
 3:   decided ← false
 4:   r_p ← 0

 5: while true do

 6:   send (FIRST, r_p, estimate_p) to all
 7:   wait until received (FIRST, r_p, v) from n − f processes
 8:   if ∃ v s.t. received (FIRST, r_p, v) from n − f processes then
 9:     estimate_p ← v
10:   else
11:     estimate_p ← ⊥

12:   send (SECOND, r_p, estimate_p) to all
13:   wait until received (SECOND, r_p, v) from n − f processes
14:   if not decided_p and (∃ v ≠ ⊥ s.t. received (second, r_p, v) from f + 1 processes) then
15:     decide v          {continue the algorithm after the decision}
16:     decided_p ← true
17:   if ∃ v ≠ ⊥ s.t. received (SECOND, r_p, v) then
18:     estimate_p ← v
19:   else
20:     estimate_p ← coin()          {toss the coin}

21:   r_p ← r_p + 1
```

5.2 Solving Consensus with 1-WAB Oracles

Consensus can be solved in an asynchronous system augmented with a 1-WAB oracle [22]. The algorithm is inspired by Ben-Or's randomised binary consensus algorithm [1] (Algorithm 2). Ben-Or's algorithm executes a sequence of rounds, where each round has two phases (n is the number of processes, $f < n/z$ is the maximum number of processes that can crash):

- *Phase I, lines 6-11*: process p sends its current $estimate_p$ (0 or 1) of the decision value to all, and waits to receive the estimate value from $n - f$ processes. If the same value v is received from $n - f$ processes, then $estimate_p$ is updated to the value received, else $estimate_p$ is set to $\perp$.
- *Phase II, lines 12-20*: process p sends again its $estimate_p$ to all, and waits to receive the estimate value from $n - f$ processes. If the same value v is received from $f +$ L processes, then p decides v. If some value different from $\perp$ is received, then $estimate_p$ is set to v, otherwise $estimate_p$ is updated with random value (0 or 1) .

Algorithm 3 is the 1-WAB consensus algorithm. Lines 6-22 of Algorithm 2 are identical to lines 6-19 of Algorithm 3. Line 20 in Algorithm 2 (random coin toss) is replaces with line 23 in Algorithm 3. Lines 6-8 in Algorithm 3 are new (query and response of the oracle): the weak ordering oracle replaces the coin toss. It is interesting to note that, contrary to Ben-Or's algorithm which solves only the binary consensus problem, the 1-WAB consensus algorithm solves the non-binary consensus problem. Proving that the Algorithm 3 satisfies the safety properties of consensus is not very different from the proof of Ben-Or's algorithm. Proving termination relies on the eventual uniform 1-WAB property of the 1-WAB oracle.

Algorithm 3 Consensus alg. using 1-WAB oracles: code of process p $(f < n/z)$

1: Consensus ($initVal$):

2: $estimate_p \leftarrow initVal$
3: $decided \leftarrow false$
4: $r_p \leftarrow 0$

5: **while** $true$ **do**

6: W-ABroadcast($r_p, estimate_p$)
7: **wait until** W-ADeliver of the first message (r_p, v)
8: $estimate_p \leftarrow v$

9: send (FIRST, $r_p, estimate_p$) to all
10: **wait until** received (FIRST, r_p, v) from $n - f$ processes
11: **if** $\exists\, v$ s.t. received (FIRST, r_p, v) from $n - f$ processes **then**
12: $estimate_p \leftarrow v$
13: **else**
14: $estimate_p \leftarrow \perp$

15: send (SECOND, $r_p, estimate_p$) to all
16: **wait until** received (SECOND, r_p, v) from $n - f$ processes
17: **if not** $decided_p$ **and** ($\exists\, \overline{v} \neq \perp$ s.t. received ($second,\ r_p, \overline{v}$) from $f + 1$ processes) **then**
18: decide $\overline{v}$ {*continue the algorithm after the decision*}
19: $decided_p \leftarrow true$
20: **if** $\exists\, \overline{v} \neq \perp$ s.t. received (SECOND, $r_p, \overline{v}$) **then**
21: $estimate_p \leftarrow \overline{v}$
22: **else**
23: $estimate_p \leftarrow initVal$

24: $r_p \leftarrow r_p + 1$

5.3 Solving Atomic Broadcast with WAB Oracles

A Weak Atomic Oracle (or WAB Oracle) is a k-WAB Oracle where $k = \infty$. The 1-WAB consensus algorithm can be extended to an atomic broadcast algorithm, which requires a WAB oracle [22]. Contrary to the classical solution, in which atomic broadcast is reduced to consensus [5], the solution directly relies on the oracle. This shows that consensus can sometimes be bypassed. Moreover, contrary to failure detection based solution, the algorithm does not suffer from the failure detection dilemma. There is no timeout to tune, and no notion of *reaction time to failures*. The performance is as good in the presence of failures as in the absence of failures.

6 Conclusion

While fault-tolerant distributed algorithms in the context of replication are nowadays well understood, the important trade-off related to the reaction to process failures has not attracted the attention that it deserves. The trade-off is between (1) fast reaction to crashes, and (2) infrequent wrong failure suspicions. We have seen how to escape from this trade-off using semi-passive replication (which relies on consensus) rather than passive replication (which relies on group membership). More generally, we have seen how to escape from this trade-off by distinguishing on one hand input-triggered suspicions that do not lead to process exclusions, and on the other hand output-triggered suspicions that lead to process exclusion. Finally, while the timeout trade-off remains for input-triggered suspicions, we have seen that it can be avoided by using weak ordering oracles instead of failure detectors.

These trade-offs are important in the context of quantitative evaluation of consensus and atomic broadcast algorithms, and more generally of group communication algorithms. Such evaluations represent an important challenge. Some preliminary results have been obtained, but much more needs to be done. Understanding group communication algorithms from a quantitative point of view is mandatory, before considering that group communication is a solved problem.

Acknowledgements

I would like to thank Bernadette Charron-Bost, Xavier Défago, Fernando Pedone, Péter Urbán, Matthias Wiesmann, David Cavin and Luiz Barchet Estefanel for their contributions to the results presented in the paper.

References

1. M. Ben-Or. Another Advantage of Free Choice: Completely Asynchronous Agreement Protocols. In *proc. 2nd annual ACM Symposium on Principles of Distributed Computing*, pages 27–30, 1983.
2. K.P. Birman and R. van Renesse. *Reliable Distributed Computing with the Isis Toolkit*. IEEE Computer Society Press, 1994.
3. N. Budhiraja, K. Marzullo, F.B. Schneider, and S. Toueg. The Primary-Backup Approach. In Sape Mullender, editor, *Distributed Systems*, pages 199–216. ACM Press, 1993.
4. T.D. Chandra, V. Hadzilacos, and S. Toueg. The Weakest Failure Detector for Solving Consensus. *Journal of ACM*, 43(4):685–722, 1996.
5. T.D. Chandra and S. Toueg. Unreliable failure detectors for reliable distributed systems. *Journal of ACM*, 43(2):225–267, 1996.

6. Tushar Deepak Chandra, Vassos Hadzilacos, Sam Toueg, and Bernadette Charron-Bost. On the impossibility of group membership. In *Proc. of the 15th ACM Symposium on Principles of Distributed Computing*, pages 322–330, Philadelphia, Pennsylvania, USA, May 1996.
7. J.M. Chang and N. Maxemchuck. Reliable Broadcast Protocols. *ACM Trans. on Computer Systems*, 2(3):251–273, August 1984.
8. B. Charron-Bost, X. Défago, and A. Schiper. Broadcasting Messages in Fault-Tolerant Distributed Systems: the benefit of handling input-triggered and output-triggered suspicions differently. TR IC/2002/020, EPFL, May 2002.
9. G.V. Chockler, I. Keidar, and R. Vitenberg. Group Communication Specifications: A Comprehensive Study. *Computing Surveys*, 4(33):1–43, December 2001.
10. F. Cristian and C. Fetzer. The timed asynchronous distributed system model. *IEEE Transactions on Parallel & Distributed Systems*, 10(6):642–657, June 1999.
11. D.Dolev, C. Dwork, and L. Stockmeyer. On the minimal synchrony needed for distributed consensus. *Journal of ACM*, 34(1):77–97, January 1987.
12. X. Défago and A. Schiper. Specification of Replication Techniques, Semi-Passive Replication and Lazy Consensus. TR IC/2002/007, EPFL, February 2002.
13. X. Défago, A. Schiper, and N. Sergent. Semi-passive Replication. In *17th IEEE Symp. on Reliable Distributed Systems (SRDS-17)*, pages 43–58, West Lafayette, USA, October 1998.
14. C. Dwork, N. Lynch, and L. Stockmeyer. Consensus in the presence of partial synchrony. *Journal of ACM*, 35(2):288–323, April 1988.
15. M. Fischer, N. Lynch, and M. Paterson. Impossibility of Distributed Consensus with One Faulty Process. *Journal of ACM*, 32:374–382, April 1985.
16. V. Hadzilacos and S. Toueg. Fault-Tolerant Broadcasts and Related Problems. Technical Report 94-1425, Department of Computer Science, Cornell University, May 1994.
17. E.Y. Lotem, I. Keidar, and D. Dolev. Dynamic Voting for Consistent Components. In *Proc. 17th Annual ACM Symposium on Principles of Distributed Computing (PODC-97)*, 1997.
18. N.A. Lynch. *Distributed Algorithms*. Morgan Kaufmann, 1996.
19. C. Malloth and A. Schiper. View Synchronous Communication in Large Scale Networks. In *ESPRIT Basic Research BROADCAST, Third Year Report, Vol 4*, July 1995.
20. N.F. Maxemchuk and D.H. Shur. An Internet multicast system for the stock market. *ACM Trans. on Computer Systems*, 19(3):384–412, August 2001.
21. L.E. Moser, P.M. Melliar-Smith, D.A. Agarwal, C.A. Lingley-Papadopoulis, and T.P. Archambaud. The Totem system. In *IEEE 25th Int Symp on Fault-Tolerant Computing (FTCS-25)*, pages 61–66, 1995.
22. F. Pedone, A. Schiper, P. Urban, and D. Cavin. Solving Agreement Problems with Weak Ordering Oracles. TR IC/2002/010, EPFL, March 2002. Appears also as Technical Report HPL-2002-44, Hewlett-Packard Laboratories, March 2002.
23. M. Rabin. Randomized Byzantine Generals. In *Proc. 24th Annual ACM Symposium on Foundations of Computer Science*, pages 403–409, 1983.
24. A.M. Ricciardi and K. P. Birman. Using Process Groups to Implement Failure Detection in Asynchronous Environments. In *Proc. of the 10th ACM Symposium on Principles of Distributed Computing*, pages 341–352, August 1991.
25. A. Schiper. Early consensus in an asynchronous system with a weak failure detector. *Distributed Computing*, 10(3):149–157, April 1997.
26. F.B. Schneider. Replication Management using the State-Machine Approach. In Sape Mullender, editor, *Distributed Systems*, pages 169–197. ACM Press, 1993.

Probability and Timing: Challenges for Secure Programming
(Invited Talk)

David Sands

Chalmers University of Technology and Göteborg University
http://www.cs.chalmers.se/~dave

Summary

When can a program be trusted with your secret data? The setting which motivates this work is that of confidentiality and privacy in mobile code. Assume that some user wants to run a program that originates from an untrusted source. For example, the program can have been downloaded from an untrusted site on the Internet. When the program is run, it has to be given access to some data that the user regards as confidential in order to compute the desired results. While running, the program also needs to have access to the Internet in order to fetch various kinds of information from databases etc. This setting has been the motivation behind a recent resurgence of interest in the analysis and certification of confidentiality properties of programs.

In this talk we will provide an overview of our recent work on the specification and verification of secure information flow in such programs. We highlight how probabilistic considerations enter in two quite different ways. In the first instance, probabilities are an additional security threat. Concurrent systems might exhibit probabilistic behaviour which an attacker could exploit to leak information. In this case we look at the modelling and verification of secure information flow using probabilistic bisimulations. In the second case, we look at how probabilistic behaviour can come to our aide when trying to eliminate the information flows which arise through the *timing behaviour* of programs.

This talk is based on joint work with Andrei Sabelfeld and Johan Agat [2,1].

References

1. J. Agat and D. Sands. On confidentiality and algorithms. In Francis M. Titsworth, editor, *Proceedings of the 2001 IEEE Symposium on Security and Privacy (S&P-01)*, pages 64–77. IEEE Computer Society, May 2001.
2. A. Sabelfeld and D. Sands. Probabilistic noninterference for multi-threaded programs. In *Proceedings of the 13th IEEE Computer Security Foundations Workshop*, pages 200–214, Cambridge, England, July 2000. IEEE Computer Society Press.

H. Hermanns and R. Segala (Eds.): PAPM-PROBMIV 2002, LNCS 2399, pp. 16–16, 2002.

Security Analysis of a Probabilistic Non-repudiation Protocol

Alessandro Aldini and Roberto Gorrieri

Università di Bologna, Dipartimento di Scienze dell'Informazione
Mura Anteo Zamboni 7, 40127 Bologna, Italy
{aldini,gorrieri}@cs.unibo.it

Abstract. Non-interference is a definition of security introduced for the analysis of confidential information flow in computer systems. In this paper, a probabilistic notion of non-interference is used to reveal information leakage which derives from the probabilistic behavior of systems. In particular, as a case study, we model and analyze a non-repudiation protocol which employs a probabilistic algorithm to achieve a fairness property. The analysis, conducted by resorting to a definition of probabilistic non-interference in the context of process algebras, confirms that a solely nondeterministic approach to the information flow theory is not enough to study the security guarantees of cryptographic protocols.

1 Introduction

The original notion of non-interference [14] has been proposed to analyze the kind of information flowing through computer systems. More precisely, assuming a classification which divides the information into (high level) confidential data and (low level) public data, the goal of such a property consists of verifying the presence of an insecure information flow from high level to low level. The formalization of non-interference based security properties is a topic that has received a lot of interest in the framework of nondeterministic process algebras (see, e.g., [19, 10, 11]). However, as emphasized in the literature (see, e.g., [15, 20, 1, 9, 22]), the security analysis of real systems should take into consideration further aspects of the system behavior, such as time and probabilities, that may cause undesirable information flows which are not observable in a solely nondeterministic context. In [1, 2] we exploited the well-established theory developed in process algebras to extend the non-interference theory based on nondeterminism to a probabilistic setting, by emphasizing the relation between the two frameworks.

In this paper, we present a simple but effective case study showing that a purely nondeterministic approach to the analysis of information flow is not enough to derive significant results, because the probabilistic behavior of the system we consider is the most important aspect affecting the security guarantees. In particular, we model and analyze a non-repudiation protocol [17] based on a probabilistic algorithm which allows two participants to exchange information in a fair way without resorting to a trusted third party. In the context

H. Hermanns and R. Segala (Eds.): PAPM-PROBMIV 2002, LNCS 2399, pp. 17–36, 2002.

of cryptographic protocols, fairness means that during the protocol, either both parties obtain their expected information, or the probability that one of them gains information while the other one gains nothing is less than a negligible threshold ε. Therefore, in order to evaluate the security level of the protocol, we have to estimate ε, which is the most important parameter affecting the satisfaction of the fairness property of the system. With this in view, by employing a qualitative, nondeterministic approach based on the classical, binary view according to which the system is (or is not) secure, we cannot obtain a *quantitative estimate* of potential, unwanted information flows so that an approach which takes into consideration probabilities is needed.

In the following, we describe in Sect. 2 the probabilistic non-repudiation protocol [17] and we introduce in Sect. 3 a process algebraic approach to the probabilistic non-interference theory [1, 2, 4]. In Sect. 4 we define the security property we intend to check and then we analyze the fairness condition for two different models of the probabilistic algorithm employed by the non-repudiation protocol. In Sect. 5 some conclusions terminate the paper.

2 A Probabilistic Non-repudiation Protocol

In a message exchange based protocol, repudiation consists of the denial by one of the entities involved in the communication of having participated in all or part of the protocol (see, e.g., [16, 23]). In this context, the *non-repudiation of origin* is intended to prevent the *originator* of a message from denying having sent the message, while the *non-repudiation of receipt* is intended to prevent the *recipient* of a message from denying having received the message. Especially in electronic commerce, non-repudiation services are mandatory if we want to guarantee the validity of a transaction, against any attempt to repudiate all or part of the transaction (e.g., if a payment for a service is represented by an acknowledgment, the client could refuse to send the ack message when the service has been delivered). Usually, to achieve non-repudiation services, a third party is involved by the participants as a trusted authority that comes into play in case of future disputes. Here, we consider an interesting alternative proposal that offers the non-repudiation service, obtained with a certain probability, without resorting to a trusted third party [17]. Such a protocol guarantees a fair exchange of a message, sent by the originator O which offers a service, for an acknowledgment, sent by the recipient R which is expected to confirm the received service. In particular, the probabilistic protocol is *ε-fair*, i.e. at each step of the protocol run, either both parties receive their expected information, or the probability that a cheating party gains any valuable information, while the other party gains nothing, is less than ε.

Before describing the protocol, we introduce some notation and assumption. We suppose that an authentication phase precedes the non-repudiation protocol, during which the involved parties exchange their public keys of a public key cryptosystem. We denote by $Sign_E(M)$ the message M encrypted with the private key of the entity E. The number n of steps for the protocol is decided

Table 1. Probabilistic non-repudiation protocol

1.	$R \rightarrow O$	: $Sign_R(\text{request}, R, O, t)$
2.	$O \rightarrow R$	: $Sign_O(f_n(M), O, R, t)$
3.	$R \rightarrow O$	: $Sign_R(\text{ack}_1)$
⋮		
2n.	$O \rightarrow R$	: $Sign_O(f_1(M), O, R, t)$
2n + 1.	$R \rightarrow O$	: $Sign_R(\text{ack}_n)$

by the originator, that also computes n functions $f_1, \ldots, f_n$ which are parts of a function composition. In particular, we have $f_n(M) \circ f_{n-1}(M) \circ \ldots \circ f_1(M) = M$. Finally, we use t to denote a timestamp which each message is enriched with.

A description of the protocol is as shown in Table 1, where with the notation $R \rightarrow O : Msg$ we express a message Msg sent by R and received by O. The recipient R starts the protocol by sending a signed, timestamped request for a service to the originator O, that in turns secretly decides the number n of steps and then sends the first signed, timestamped message containing $f_n(M)$. For each received message containing $f_i(M)$, the recipient sends a related signed, timestamped acknowledgment message containing $\text{ack}_i = (i, R, O, t)$. Since the recipient does not know n, he cannot determine when the protocol will end and, as a consequence, when he will receive the final message which allows him to compute M. In particular, since the functions f_i are sent in reverse order and since we assume that the composition operator does not permit the commutativity, the recipient cannot use intermediate results to speed up the computations in order to guess M before the reception of the last message. Upon the reception of the ack related to the last message containing $f_1(M)$, the originator correctly terminates the protocol. Note that an *end of protocol* message sent by the originator is not mandatory. Indeed, after not receiving further messages, the recipient is aware of the protocol state and is able to compute M.

It is worth noting that each message conveys a timestamp, which can be considered as a random value used to guarantee the freshness of the message and to protect the parties against replication attacks. Intuitively, the non-repudiation of origin is guaranteed by the sequence of messages $Sign_O(f_i(M), O, R, t)$, with $i \in \{1, \ldots, n\}$, while the non-repudiation of receipt is given by the last message $Sign_R(\text{ack}_n)$. If the protocol terminates after the delivery of the nth ack message, both parties obtain their expected information and the protocol is fair. If the protocol terminates before sending the message containing $f_1(M)$, then neither the originator nor the recipient obtain any valuable information, so that the fairness is preserved. Since the number n of steps is not revealed by the originator during the protocol run, a strategy for a dishonest recipient can consists of (i) guessing the last message containing $f_1(M)$ by computing the composition of the received functions f_i, and (ii) violating the fairness of the protocol by blocking the transmission of the last ack message. Therefore, the key to success of the protocol is the immediacy in sending back the ack messages, so that if a malicious recipient delays each ack message in order to compute the composition of the received functions f_i to see if he has obtained the message M, then the originator realizes this unfair strategy and can stop the protocol. To this end, the choice of the functions f_i must be in such a way that the composition computation takes more time than the transmission of an ack message. In this way, the originator decides a deadline for the reception of each ack, after which, if the ack message is not received, the protocol is stopped. Anyway, a malicious recipient can try to randomly guess the number of steps of the protocol. We point out that the probability for the recipient of guessing the last message depends on the size of the sampling space and the probability distribution chosen by the originator to compute n. In the next section, we propose a formal approach which includes all the ingredients needed to evaluate such a potential unwanted behavior of the recipient.

We conclude this section by showing a concrete proposal [17] based on the above generic non-repudiation algorithm, which noticeably simplifies the choice of the functions f_i. We take f_n as a ciphering function which uses a secret key k, namely $f_n(M) = \{M\}_k$, f_{n-1} until f_2 as random garbage, and $f_1(M)$ as the key k in clear. In this way, the recipient immediately obtains M encrypted with an unknown key and, before the last transmission, he does not receive any valuable information. Only upon the reception of the last message, the recipient obtains the key needed to compute M. Against a cheating recipient, the cryptosystem must be adequately chosen, in such a way that the time needed to verify a key, by deciphering the message, is to be too long with respect to the transmission time for an ack message. For instance, the most efficient implementations of RSA guarantee a ciphering of 600 Kbit per second; therefore the greater the message size, the longer the transmission delay for the ack messages which can be tolerated by the originator.

3 A Process Algebraic Approach to the Probabilistic Non-interference Theory

In this section, we describe an adequate technique for the formal description and the security analysis of cryptographic protocols like the one described in Sect. 2. In particular, we employ the process algebra proposed in [3] and used in [1, 2, 4] to extend the classical nondeterministic approach to non-interference theory with probabilities. Such a calculus derives from a simple nondeterministic CSP-like process algebra where actions are syntactically divided into input actions and output actions. The extension is obtained by adding probabilistic information to the algebraic operators and by employing an appropriate model of probabilities. In particular, the model of probabilities adopted in our calculus is an integration of the generative and reactive approaches in the line of [13], where we assume the output actions behaving as *generative* actions (a generative process autonomously decides, on the basis of a probability distribution, which action will be executed and how to behave after such an event) and the input actions behaving as *reactive* actions (a reactive process reacts internally to an external action of type a, chosen by the environment, on the basis of a probability distribution associated with the reactive actions of type a it can perform).

The labeled transition system associated with a term of our probabilistic calculus, called generative-reactive transition system (*GRTS*), consists of a bundle of generative transitions representing all the output actions that the system can perform, and several bundles of reactive actions (one for each action type) representing the input actions enabled by the system. The choice within a bundle is purely probabilistic, while the choice among bundles is nondeterministic.

Formally, *AType* denotes the set of action types, ranged over by $a, b, \ldots$, including also the special type τ denoting an internal action. We denote the set of reactive actions by $RAct = \{a_* \mid a \in AType - \{\tau\}\}$ and the set of generative actions by $GAct = AType$ (note that τ is a generative action, because it expresses an autonomous internal move that does not react to external stimuli). The set of actions is denoted by $Act = RAct \cup GAct$, ranged over by $\pi, \pi', \ldots$. The set $\mathcal{L}$ of process terms is generated by the syntax:

$$P ::= \underline{0} \mid \pi.P \mid P +^p P \mid P \,\|_S^p\, P \mid P \backslash L \mid P/_a^p \mid A$$

where $S, L \subseteq AType - \{\tau\}$, $a \in AType - \{\tau\}$, and $p \in]0,1[$. The set $\mathcal{L}$ is ranged over by $P, Q, \ldots$. As usual in security models, we distinguish among high level visible actions and low level visible actions by defining two disjoint sets $AType_H$ of high level types and $AType_L$ of low level types that form a covering of $AType - \{\tau\}$, such that $a \in GAct$ and $a_* \in RAct$ are high (low) level actions if $a \in AType_H$ ($a \in AType_L$). Now, we give an informal intuition of the operators, while we delay a complete presentation of their semantics to the appendix.

- $\underline{0}$ represents the terminated or deadlocked term.
- $\pi.P$ performs the action π with probability 1 and then behaves like P.
- $P +^p Q$ is a CCS-like alternative choice operator, where a mixed nondeterministic and probabilistic choice among the actions of P and Q is performed. More precisely, $P +^p Q$ executes a generative (reactive of type a) action of P with probability p and a generative (reactive of type a) action of Q with probability $1 - p$. If one process P or Q cannot execute generative (reactive of type a) actions, $P +^p Q$ chooses a generative (reactive of type a) action of the other process with probability 1. The choice among generative and reactive actions and among reactive actions of different type is purely nondeterministic.
- $P \,\|_S^p\, Q$ is a CSP-like parallel composition operator, where the actions not belonging to the type set S are locally executed, while the actions belonging to S are constrained to synchronize. In particular, a synchronization between two actions may occur only if either they are both input actions of the same type a (and the result is an input action of type a), or one of them is an output action of type a and the other one is an input action of type a (and the result is an output action of type a). The probabilistic choice mechanism among the actions of P and Q is the same as that described for the choice operator.

- $P \backslash L$ prevents the execution of the actions of type in L.
- $P/^p_a$ turns generative and reactive actions of type a into generative actions τ. Parameter p expresses the probability that actions τ obtained by hiding reactive actions a_* of P are executed with respect to the generative actions previously enabled by P. Parameter p is, instead, not used when hiding generative actions, because the choice among generative actions is already probabilistic. As an example, consider process $P \stackrel{\Delta}{=} a_* +^q b$, where the choice is purely nondeterministic (parameter q is not considered), and hide the action a_*. The semantics of $P/^p_a \stackrel{\Delta}{=} \tau +^p b$ is a probabilistic choice between the action τ obtained by hiding the action a_* and the action b, performed according to probabilities p and $1-p$, respectively. We turn reactive actions into generative internal actions τ, because the hiding operator is used to obtain fully specified systems from open systems (i.e. systems enabling reactive choices). In particular, when fully specifying a system, the nondeterministic choices due to the possible interactions with the environment have to be resolved, and parameter p turns such choices into probabilistic choices.
- Constants A are used to specify recursive systems. In general, when defining an algebraic specification, we assume a set of constants defining equations of the form $A \stackrel{\Delta}{=} P$ (with P a guarded term [18]) to be given.

In the following, we denote by $\mathcal{G}$ the set of guarded and closed terms of $\mathcal{L}$ [18] and by $sort(P)$, with $P \in \mathcal{G}$, the set of actions which syntactically occur in the action prefix operators within P. Moreover, let $\mathcal{G}_H = \{P \in \mathcal{G} \mid sort(P) \subseteq AType_H\}$ be the set of high level terms. For the sake of simplicity, we use the following abbreviations. We assume the parameter p to be equal to $\frac{1}{2}$ whenever it is omitted from any probabilistic operator. Moreover, when it is clear from the context, we use the abbreviation P/S, with $S = \{a_1, \ldots, a_n\}$, to denote any expression of the form $P/^{p_1}_{a_1} \ldots /^{p_n}_{a_n}$, $\forall p_i \in]0,1[, i \in \{1 \ldots n\}$, which hides the actions of type in $S \subseteq AType - \{\tau\}$. This notation simplifies the description of those terms where, e.g., we hide generative actions only and the probabilistic parameters of the hiding operators are not meaningful.

To conclude the presentation of the calculus, we report the definition of weak bisimulation for *GRTS*s [1], based on which we define the security properties. Such an equivalence relation is inspired by [5], where the classical relation $\approx$ of [18] is replaced by a function *Prob* that computes the probability of reaching classes of equivalent states. In particular, we recall that each transition of a *GRTS* is labeled with an action in the set *Act* and a probability, and that the probability associated to a sequence of transitions is given by the product of the probabilities of the transitions. Then, by $Prob(P, \tau^* \hat{a}, C)$ [1] we denote the overall probability associated to the set of transition sequences with initial state P which lead to a system state in the set C via a sequence of transitions, whose ordered labels generate a sequence of the form $\tau^* \hat{a}$.

[1] $\tau^* \hat{a}$ stands for the sequence $\tau^* a$ if $a \in GAct - \{\tau\}$ and for the sequence τ^* if $a = \tau$. Note that τ^* means zero or more τ occurrences.

Definition 1. *An equivalence relation $R \subseteq \mathcal{G} \times \mathcal{G}$ is a probabilistic weak bisimulation if and only if, whenever $(P, Q) \in R$ then for all $C \in \mathcal{G}/R$ we have that $Prob(P, \tau^*\hat{a}, C) = Prob(Q, \tau^*\hat{a}, C)$ for all $a \in GAct$, and $Prob(P, a_*, C) = Prob(Q, a_*, C)$ for all $a_* \in RAct$.*

Two terms $P, Q \in \mathcal{G}$ are weakly probabilistic bisimulation equivalent, denoted $P \approx_{PB} Q$, if there exists a probabilistic weak bisimulation R containing the pair (P, Q).

Based on the above process algebra and the related notion of equivalence, several security properties can be defined. The reader interested in a classification of security properties for processes of our calculus should refer to [2, 4]. As an example, here we propose the probabilistic version of the bisimulation based Non Deducibility on Composition (*PBNDC*, for short), which informally says that, from the low level standpoint, the probability distribution of the low level events as observed in a system P in isolation is not to be altered when considering the potential interactions of P with any high level user.

Definition 2. *Let $P \in \mathcal{G}$. We say that $P \in PBNDC$ if and only if*

$$P \backslash AType_H \approx_{PB} ((P \parallel_S^p \Pi)/S) \backslash AType_H \quad \forall p \in]0,1[, \Pi \in \mathcal{G}_H, S \subseteq AType_H.$$

$P \backslash AType_H$, where the execution of the high level actions is prevented, represents the low behavior of P in isolation, i.e. when every high level activity is absent, while $((P \parallel_S^p \Pi)/S) \backslash AType_H$ models the low behavior of P which possibly interacts with any high level user Π and such interactions are unobservable by the low level user. Hence, the *PBNDC* property checks the equivalence between these two different views of the low behavior of P, which turns out to be secure if the low level user is not able to reveal the potential interference of the high level user on the system behavior. Such a definition extends the original nondeterministic *BNDC* property [10], in that it also considers the influence of the probabilistic high behavior upon the probabilistic low view. We point out that in [5] an algorithm is described that computes probabilistic weak bisimulation equivalence classes in time $\mathcal{O}(n^3)$ and space $\mathcal{O}(n^2)$, where n is the number of states. Hence, by defining a process Π which affects the low level view of the system, we are able to formally capture, through the automatic verification provided by probabilistic weak bisimulation laws, those interferences which reveal confidential information to low level users.

Example 1. Let us consider the system $P \stackrel{\Delta}{=} l.\underline{0} +^p h.h.l.\underline{0}$, where $l \in AType_L$ and $h \in AType_H$. The high level part alters the low view of the system in the case the high level user first executes the first action h and then blocks the execution of the second action h. In such a case a deadlock is caused by the high level user that prevents the action l from being observed by a low level user. This information flow is captured by the *PBNDC* property if we take $S = \{h\}$ and $\Pi \stackrel{\Delta}{=} h_*.\underline{0}$. Indeed, in such a case we have $P \backslash AType_H \approx_{PB} l.\underline{0} \not\approx_{PB} l.\underline{0} +^p \tau.\underline{0} \approx_{PB} (((l.\underline{0} +^p h.h.l.\underline{0}) \parallel_S (h_*.\underline{0}))/S) \backslash AType_H$. □

By modeling the probabilistic behavior a quantitative estimate of the information flowing through the system can be obtained. More precisely, with respect to the nondeterministic case, where we can just verify the presence of an information leakage, in our setting we can measure the probability of observing an unwanted information flow from high level to low level. To this end, we need a quantitative notion of behavioral equivalence for deciding if two probabilistic processes behave almost (up to small ε-fluctuations) the same or, more formally, for measuring the distance between probabilistic transition systems (see, e.g., [8, 7]). In [4], we formalize the notion of bisimulation equivalence which takes into consideration ε-perturbations in the context of *GRTS*s.

Definition 3. *An equivalence relation $R \subseteq \mathcal{G} \times \mathcal{G}$ is a probabilistic weak bisimulation with ε-precision, where $\varepsilon \in]0,1[$, if and only if, whenever $(P,Q) \in R$, then for all $C \in \mathcal{G}/R$*

- $|\, Prob(P, \tau^* \hat{a}, C) - Prob(Q, \tau^* \hat{a}, C) \,| \leq \varepsilon \;\; \forall a \in GAct$
- $|\, Prob(P, a_*, C) - Prob(Q, a_*, C) \,| \leq \varepsilon \;\; \forall a_* \in RAct$

Two terms $P, Q \in \mathcal{G}$ are ε-bisimulation equivalent, denoted $P \approx_{PB\varepsilon} Q$, if there exists a probabilistic weak bisimulation with ε-precision R containing the pair (P,Q).

By replacing in the definition of any security property, say SP, the relation $\approx_{PB}$ by the relation $\approx_{PB\varepsilon}$, we make the condition verified by SP less restrictive in the sense that it can tolerate those small fluctuations which the observer considers scarcely significant for the security level of the system.

Example 2. Let us consider the system $P \stackrel{\Delta}{=} h.l'.\underline{0} +^p \tau.(l.\underline{0} +^q h.l.\underline{0})$, where $l, l' \in AType_L$ and $h \in AType_H$. P is clearly not secure because the execution of l' reveals to a low level observer that h has been executed. In particular, if we employ the probabilistic weak bisimulation $\approx_{PB}$, then P is not *PBNDC* secure, because $P \backslash AType_H \approx_{PB} \tau.l.\underline{0} \not\approx_{PB} \tau.l'.\underline{0} +^p \tau.(l.\underline{0} +^q \tau.l.\underline{0}) \approx_{PB} ((P \,\|_{\{h\}} (h_*.\underline{0}))/\{h\}) \backslash AType_H$. Anyway, we also have that P is *PBNDC* secure if the equivalence relation we employ is the probabilistic weak bisimulation with p-precision $\approx_{PBp}$. To this end, we just observe that $\tau.l.\underline{0} \approx_{PBp} \tau.l'.\underline{0} +^p \tau.l.\underline{0} \approx_{PB} \tau.l'.\underline{0} +^p \tau.(l.\underline{0} +^q \tau.l.\underline{0})$. Hence, depending on a tolerance ε, we can decide if P is to be considered secure enough. □

4 Security Analysis of the Non-repudiation Protocol

In this section, we employ our process algebraic approach to model the concrete proposal of the protocol described in Sect. 2 in order to verify the non-repudiation property. To this end, we resort to the following assumptions. Since the non-repudiation property we intend to check depends on the order and the number of packets which are exchanged, we abstract from the cryptosystem used within the protocol and we simply model the packet exchange between the two involved parties. Moreover, we also abstract from the channel and the transmission delays,

Table 2. Probabilistic non-repudiation protocol - *Originator* model

$$\begin{aligned}
Orig &\triangleq accept_request_*.snd_first_msg.ack_*.O' \\
O' &\triangleq O_1 +^{1/\eta} (O_2 +^{1/(\eta-1)} \ldots (O_{\eta-1} +^{1/2} O_\eta) \ldots) \\
O_i &\triangleq snd_msg.(ack_*.O_{i-1} + stop_*.\underline{0}) \quad \text{if } 2 \leq i \leq \eta \\
O_1 &\triangleq snd_msg.(ack_*.stop.\underline{0} + stop_*.\mathbf{unfair}.\underline{0})
\end{aligned}$$

by assuming that a message which is delayed (not sent) by a participant is not delivered to the other participant. As far as the probabilistic algorithm is concerned, we model two different scenarios, depending on the probability distribution used by the originator to sample the number n of protocol steps. In the first scenario we assume an equally distributed probability, while in the second one we assume a Bernoulli distribution. In both cases, we show that the protocol is ε-fair, where the parameter ε is decided by the honest participant.

We start by introducing the models of an originator and of a recipient which behave correctly. The algebraic specification of the originator is shown in Table 2. Process *Orig* is always ready to start the message exchange by (*i*) accepting a request (action $accept\_request_*$), (*ii*) sending the first message containing the information M encrypted with a secret key k (action *snd_first_msg*), and (*iii*) receiving the first ack message (action ack_*). Then, by assuming that the number n of steps of the protocol is chosen in the set $\{1, \ldots, \eta\}$ with an equally distributed probability, in term O' a value for n is sampled by performing a probabilistic choice among η generative actions of type *snd_msg*, where the probability associated with each action is $1/\eta$. If the action *snd_msg* of term O_n is the winning action, then the originator sends the first message (of n messages) to the recipient and waits for the related ack message. Then, if the ack is received through the reactive action ack_*, term O_{n-1} is reached and the next step is scheduled. On the other hand, if the ack is not received before the expected deadline the protocol is prematurely stopped. Since we abstract from time, we perform a nondeterministic choice between the reactive action ack_* and a reactive action $stop_*$ to model the time-out. Moreover, for the sake of simplicity, we assume that whenever an ack message is sent by the recipient then the originator receives it before the deadline. The fair, correct termination of the protocol is reached when the originator receives the last ack message and performs the generative action *stop* (see term O_1). The protocol terminates in an unfair way if and only if the reactive action $stop_*$ is executed when the originator is in term O_1. Indeed, in such a case, the originator has already sent the last message containing the key k which reveals M, but he has not received the final ack message which completes the fair exchange of information. We make explicit such a situation by performing the special action **unfair**.

Table 3. Probabilistic non-repudiation protocol - Honest *Recipient* model

$$\begin{aligned} H\_Recip &\triangleq accept\_request.snd\_first\_msg_*.ack.HR \\ HR &\triangleq snd\_msg_*.ack.HR + stop_*.\underline{0} \end{aligned}$$

In Table 3, we show the algebraic specification of a recipient that behaves correctly. Process *H_Recip* starts the protocol by sending a request (action *accept_request*), and then it waits for the first message (action $snd\_first\_msg_*$). Upon the reception of such a message, it sends the first ack (action *ack*) and then behaves like term *HR*, where each time a message is received through the action snd_msg_*, a related ack message is sent by executing the action *ack*. The protocol terminates when the reactive action $stop_*$ is executed to denote that the last received message contains the key k in clear needed to decrypt M.

To complete the algebraic description of the protocol, we specify the communication interface between the two parties by composing in parallel process *Orig* and process *H_Recip* in the system $Orig \parallel^p_S H\_Recip$, where the two processes interact by synchronizing on the actions in the set $S = \{accept\_request, snd\_first\_msg, ack, snd\_msg, stop\}$ containing all the actions except for the special action **unfair**. Note that parameter p is not considered (in the following we omit it), because each state of the composed system is such that a probabilistic choice among actions of process *Orig* and of process *H_Recip* is never to be performed.

In order to verify a security property in the line of the method described in Sect. 3, we have to single out the high level actions and the low level ones. Here, we follow an idea proposed in the general case for the analysis of cryptographic protocols in [12]. Since the recipient, which can be considered a potential malicious party, has the complete control of the network, it is reasonable to assume that the names used for message exchange (i.e., all the actions in S) are the high level actions. Instead, the low level actions are extra observable actions that we include into the model to observe the non-repudiation property. With this in view, it is natural to consider the extra action **unfair**, put in the algebraic specification of the originator, as the unique low level action of the system which is executed when the protocol terminates without preserving the fairness condition.

Now, we have to formalize the non-repudiation property that will be checked. To this end, we intend to employ the *PBNDC* property described in Sect. 3. With respect to the original definition of *PBNDC*, where the system behavior is observed when put in parallel with the external environment, in the case of the non-repudiation property a malicious behavior is due to an internal component of the system itself which tries to obtain a valuable information violating the fairness condition. Therefore, potential attacks by third parties which are not

Table 4. Probabilistic non-repudiation protocol - Malicious *Recipient* model

$$
\begin{aligned}
M\_Recip &\triangleq accept\_request.snd\_first\_msg_*.ack.MR_1 \\
MR_i &\triangleq snd\_msg_*.(stop.\underline{0} +^{1/(\eta-i+1)} ack.MR_{i+1}) \quad \text{if } 1 \le i < \eta \\
MR_\eta &\triangleq snd\_msg_*.stop.\underline{0}
\end{aligned}
$$

involved in the protocol are not meaningful. In particular, forgery and authentication attacks are prevented by the signed, exchanged messages (we assume that the initial authentication phase is secure so that both parties know the public keys to decrypt the messages), and replication attacks are prevented by the timestamps. Since the originator has not interest in stopping the protocol before the end, the hostile environment is represented by the recipient that possibly tries to obtain its expected information without sending the last ack message. According to this, to verify the non-repudiation property we check the semantic equivalence between the model of the protocol where both parties behave honestly and each possible model of the protocol involving a potentially malicious recipient. More formally, the property to be checked is as follows.

The protocol satisfies the non-repudiation property if and only if:

$$((Orig \parallel_S H\_Recip)/S)\backslash AType_H \approx_{PB} ((Orig \parallel^p_{S'} Recip)/S')\backslash AType_H$$

for all $S' \subseteq AType_H, p \in]0,1[$, *and for any* $Recip \in \mathcal{G}_H$.

Intuitively, we observe that the protocol does not satisfy the property above. In particular, if both participants behave honestly, then the unfair behavior cannot be executed; instead, it is possible to find a malicious recipient that receives the expected information and denies sending the final non-repudiation of receipt message.

Formally, process *H_Recip* never stops the protocol prematurely by executing the generative action *stop*, so the reactive action $stop_*$ in term O_1 is never enabled and the low level action **unfair** cannot be executed. Hence, if both participants are honest, the protocol is fair with probability 1. In particular, it is easy to see that $Prob(((Orig \parallel_S H\_Recip)/S)\backslash AType_H, \tau^*\mathbf{unfair}, C) = 0 \ \forall C \in \mathcal{G}/\approx_{PB}$ and $\sum_{C \in \mathcal{G}/\approx_{PB}} Prob(((Orig \parallel_S H\_Recip)/S)\backslash AType_H, \tau^*, C) = 1$.

On the other hand, we now describe a malicious recipient model *Recip* which allows the low action **unfair** to be executed in $((Orig \parallel^p_{S'} Recip)/S')\backslash AType_H$. To this aim, we take $S' = S$, any $p \in]0,1[$, and as term *Recip* the process *M_Recip* shown in Table 4, which models the malicious recipient with the maximum knowledge of the protocol variables, i.e. he knows both the sampling space size and the probability distribution associated with each possible event of the sampling space chosen by the originator to compute the number n of steps of the protocol. Based on such a knowledge, process *M_Recip* is able to maximize the probability of guessing the last message of the protocol by adopting the following strategy.

For each step of the protocol, the probability of receiving the last message is 1 over the maximum number of steps which can remain before the end of the protocol. Therefore, upon the reception of the first message represented by the execution of the action of type *snd_msg*, the recipient stops the protocol by executing the generative action *stop* with probability $\frac{1}{\eta}$, where η is the size of the sampling space, and sends the ack message by executing the generative action *ack* with probability $1 - \frac{1}{\eta}$. If the recipient decides to send the ack and the protocol is not terminated yet, upon the reception of the second message, the recipient stops the protocol with probability $\frac{1}{\eta-1}$, because the probability of guessing the last step is redistributed among the $\eta - 1$ possible events of the reduced sampling space (see terms MR_i). Finally, if after $\eta - 1$ steps the protocol is not terminated yet, the recipient knows that the next step must be the final one, so that he decides to stop the protocol with probability 1 (see term MR_η). Intuitively, if we want to compute by hand the probability for the recipient of cheating with success, we perform the following steps:

- the probability that the originator samples the value 1 is $\frac{1}{\eta}$ and the probability that the recipient stops the protocol after a single step is $\frac{1}{\eta}$, therefore the probability of guessing the event $n = 1$ is $\frac{1}{\eta} \cdot \frac{1}{\eta} = \frac{1}{\eta^2}$;
- the probability that the originator samples the value 2 is $\frac{1}{\eta}$ and the probability that the recipient stops the protocol after two steps is $\frac{\eta-1}{\eta} \cdot \frac{1}{\eta-1} = \frac{1}{\eta}$, where $\frac{\eta-1}{\eta}$ is the probability of sending the first ack and $\frac{1}{\eta-1}$ is the probability of stopping the protocol at the second step. Therefore the probability of guessing the event $n = 2$ is $\frac{1}{\eta} \cdot \frac{1}{\eta} = \frac{1}{\eta^2}$;
- the probability that the originator samples the value $i > 2$ is $\frac{1}{\eta}$ and the probability that the recipient stops the protocol after i steps is $\frac{\eta-1}{\eta} \cdot \frac{\eta-2}{\eta-1} \cdot \ldots \cdot \frac{1}{\eta-i+1} = \frac{1}{\eta}$, therefore the probability of guessing the event $n = i$ is $\frac{1}{\eta} \cdot \frac{1}{\eta} = \frac{1}{\eta^2}$.

Given that the sampling space is composed of η possible events, the overall probability of guessing the last message is $\eta \cdot \frac{1}{\eta^2} = \frac{1}{\eta}$, exactly the value we expected.

Formally, by analyzing the *GRTS* associated to the composed system, we immediately obtain that $Prob(((Orig \parallel_S M\_Recip)/S)\backslash AType_H, \tau^* \mathbf{unfair}, C) = \frac{1}{\eta}$ (where C is the equivalence class of the terminated term). Hence, the non-repudiation protocol is not *PBNDC* secure if we employ as the equivalence relation the weak probabilistic bisimulation $\approx_{PB}$. Moreover, if we resort to the definition of ε-bisimilarity, we can add that the protocol is secure with $\frac{1}{\eta}$-tolerance. Indeed, as we can formally verify, we have that $((Orig \parallel_S H\_Recip)/S)\backslash AType_H \approx_{PB\frac{1}{\eta}} ((Orig \parallel_S M\_Recip)/S)\backslash AType_H$. Therefore, the level of security of the protocol is parameterized by the size of the sampling space η.

By calculating probabilistic weak bisimulation equivalence classes, we have formally derived the same result obtained computationally in [17]. On the other hand, we think that our technique can be automated and generalized, by following an approach inspired by [12], for the analysis of not only non-repudiation,

Table 5. Probabilistic Non-repudiation Protocol - *Originator* alternative model

$$\begin{aligned}
OrigB &\stackrel{\Delta}{=} accept_request_{*}.snd_first_msg.ack_{*}.OB' \\
OB' &\stackrel{\Delta}{=} snd_msg.Last_msg +^{p} snd_msg.OB'' \\
OB'' &\stackrel{\Delta}{=} ack_{*}.OB' + stop_{*}.\underline{0} \\
Last_msg &\stackrel{\Delta}{=} ack_{*}.stop.\underline{0} + stop_{*}.\mathbf{unfair}.\underline{0}
\end{aligned}$$

but also other security properties related to the algebraic specification of cryptographic protocols. Moreover, it is worth noting that in a purely nondeterministic setting [2], we could only conclude that the protocol of [17] does not satisfy the non-repudiation of receipt, since an insecure behavior is observable. Instead, in the probabilistic setting we have seen that the insecure behavior caused by the dishonest recipient *M_Recip* can be estimated and can be acceptable for the honest participant, that is in charge to decide the risk factor ε affecting the security condition.

As another example of the relation between the behavior of a cheating recipient and the behavior of a honest originator, we now present an alternative model of the system where the originator follows a Bernoulli distribution with parameter p as the particular probability distribution chosen to sample the duration of the protocol. With such an example we emphasize that the probability of cheating for a dishonest recipient depends on a parameter which is under the control of the honest participant.

The originator model is depicted in Table 5. After the initial set-up phase, at each step of the protocol (see term OB') the originator sends either the last message with probability p and then behaves as term *Last_msg*, or the next garbage message with probability $1-p$ and then behaves as term OB'', where the reception of the related ack message leads to term OB' again. Such a version of the protocol is *time-bounded* (i.e. the protocol will be completed before a finite amount of time), because, even if the protocol is never stopped in term OB'' via the execution of the action $stop_{*}$, the probability of reaching state *Last_msg* (where the protocol terminates with probability 1) from the initial state is given by:

$$\sum_{i=0}^{\infty} p \cdot (1-p)^i = p \cdot \sum_{i=0}^{\infty} (1-p)^i = p \cdot \frac{1}{1-(1-p)} = 1$$

In Table 6, we show the model of a cheating recipient that follows a Bernoulli distribution with parameter q to decide either to send the ack message or to stop the protocol. In particular, after the initial message exchange (see term *RecipB*),

[2] We can omit the probabilistic information from the operators of the probabilistic calculus, thus obtaining a process algebra on which the nondeterministic non-interference theory can be rephrased [2].

Table 6. Probabilistic Non-repudiation Protocol - *Recipient* alternative model

$$RecipB \stackrel{\Delta}{=} accept\_request.snd\_first\_msg_*.ack.RB$$
$$RB \stackrel{\Delta}{=} snd\_msg_*.(stop.\underline{0} +^q ack.RB) + stop_*.\underline{0}$$

in term RB the recipient accepts each arriving message (action snd_msg_*) and then either sends the related ack message (the action ack is executed with probability $1 - q$) or tries to compute M by employing the last received key (the action $stop$ is executed with probability q). In Fig. 1 we show the $GRTS$ associated to the composed system $((OrigB \parallel_S RecipB)/S)\backslash AType_H$. State s_1 corresponds to term $((OB' \parallel_S RB)/S)\backslash AType_H$ from which the system evolves either into state $s_2 = ((OB'' \parallel_S (stop.\underline{0} +^q ack.RB))/S)\backslash AType_H$ with probability $1 - p$, denoting that the protocol is not terminated yet, or into state $s_3 = ((Last\_msg \parallel_S (stop.\underline{0} +^q ack.RB))/S)\backslash AType_H$ with probability p, denoting that the last message has been sent. Then, from state s_2 the system evolves either into state s_1 again if the two processes synchronize on the action of type ack with probability $1 - q$, or into the termination state if they synchronize on the action of type $stop$ with probability q. Note that from state s_3 the unfair state is reached only if the two processes synchronize on the action of type $stop$ with probability q.

By verifying the $PBNDC$, we obtain $Prob(((OrigB \parallel_S H\_Recip)/S)\backslash AType_H, \tau^*\mathbf{unfair}, C) = 0\ \forall C \in \mathcal{G}/\approx_{PB}$, while $Prob(((OrigB \parallel_S RecipB)/S)\backslash AType_H, \tau^*\mathbf{unfair}, C')$, where C' is the equivalence class of the terminated term $\underline{0}$, is equal to:

$$z = p \cdot q \cdot \sum_{i=0}^{\infty}((1-p)\cdot(1-q))^i = \frac{p \cdot q}{1 - (1-p)\cdot(1-q)}$$

Given $0 < p < 1$ chosen by the originator and by assuming the constraint $0 \le q \le 1$, the maximum value for z is p, obtained by taking $q = 1$. Therefore,

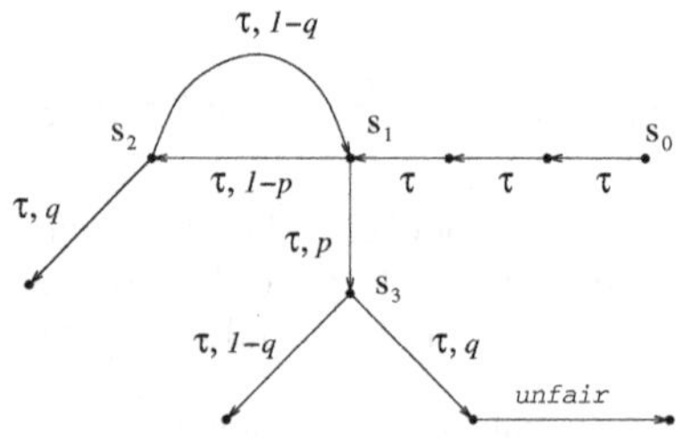

Fig. 1. *GRTS* underlying the non-repudiation protocol specification

it is easy to see that the recipient model which optimizes the probability of violating the fairness condition is obtained by assuming term RB of Table 6 to be $RB \stackrel{\Delta}{=} snd\_msg_*.stop.\underline{0}$. In general, even if the recipient knows the parameter p of the Bernoulli distribution chosen by the originator, process *RecipB* cannot adopt a strategy for which the probability of guessing the number of steps of the protocol is greater than p. Hence, such a system can be considered as a secure system if we can tolerate p-perturbations. In particular, independently of the behavior of the recipient, the upper bound p can be kept as low as will fit the security needs of the originator, by simply choosing adequate parameters for the sampling space size and the probability distribution.

5 Conclusion

As the literature has shown (see, e.g., [21, 6]), the role of the probabilistic aspect is important for the analysis of distributed systems. The well-known and established results in the theory of (probabilistic) process algebras allowed us to easily extend the nondeterministic, qualitative notion of non-interference to a probabilistic, quantitative definition of security. The need for such an extension has been emphasized in a simple but effective case study, a probabilistic non-repudiation protocol, whose analysis reveals the importance of quantifying the potential insecure behaviors of systems.

To the best of our knowledge, a formal quantitative estimate of probabilistic information flows is studied also in [22] in the context of a type system for secure information flow and in [9] in the framework of a declarative programming language. However, we think that the formalisms of [22] and [9] lack the expressivity needed to analyze cryptographic protocols like the one presented in this paper.

From a theoretical viewpoint, it would be useful to find a static (i.e., without universal quantification over all possible adversaries) or a stronger, easily verifiable characterization of Probabilistic *BNDC*. Indeed, while in this case study it seems reasonable to guess that *M_Recip* is the recipient model that optimizes the probability of cheating (but we have not proved it), in a more general setting it would be needed to characterize the most powerful adversary that can interact with the system to be analyzed. Then, the notion of probabilistic weak bisimulation which tolerates small fluctuations can be employed to verify different security properties of cryptographic protocols (not only non-repudiation, but also authentication, secrecy, integrity, and anonymity), depending on the particular observable low level actions we add to the system specification. This approach can help the modeler of complex systems to reveal unexpected, subtle information flows that otherwise are difficult to be captured because of the large number of possible behaviors.

We conclude with two observations concerning the analysis of the system specification. On the one hand, we could exchange the roles of the recipient and the originator, in order to study the behavior of a honest recipient for each potential malicious originator. As already seen, such an analysis would reveal that there does not exist a system state which is unfair from the recipient standpoint. On the other hand, we could explicitly add a component modeling a lossy channel. In such a case, we would obtain that the probability of observing an unfair behavior cannot be limited to any ε fixed by the originator. This is because the lower bound for the unfairness condition would be represented by the percentage of lost (delayed) packets, which is a parameter that is not under the control of the originator. Our probabilistic approach can be employed to study the trade-off between the security guarantees of the cryptographic protocol and the communication channel conditions, by revealing the adaptiveness of the protocol in tolerating hostile environment conditions.

Acknowledgment

The authors thank anonymous referees for their valuable comments. This research has been funded by Progetto MURST "Metodi Formali per la Sicurezza e il Tempo" (MEFISTO).

References

1. A. Aldini, *"Probabilistic Information Flow in a Process Algebra"*, in Proc. of the 12th Int. Conf. on Concurrency Theory, LNCS 2154, pp. 152-168, Springer-Verlag, 2001
2. A. Aldini, *"On the Extension of Non-interference with Probabilities"*, 2nd ACM SIGPLAN and IFIP WG 1.7 Workshop on Issues in the Theory of Security, 2002
3. A. Aldini, M. Bravetti, *"An Asynchronous Calculus for Generative-Reactive Probabilistic Systems"*, in Proc. of the 8th Int. Workshop on Process Algebra and Performance Modeling, pp. 591-605, Carleton Scientific, 2000
4. A. Aldini, M. Bravetti, R. Gorrieri, *"A Process Algebraic Approach for the Analysis of Probabilistic Non-interference"*, Technical Report UBLCS-2002-2, University of Bologna, Italy, 2002, ftp://ftp.cs.unibo.it/pub/techreports/
5. C. Baier, H. Hermanns, *"Weak Bisimulation for Fully Probabilistic Processes"*, in Proc. of the 9th Int. Conf. on Computer Aided Verification, LNCS 1254, pp. 119-130, Springer-Verlag, 1997
6. C. Baier, M. Kwiatkowska, *"Domain Equations for Probabilistic Processes"*, in Mathematical Structures in Computer Science 10(6), pp. 665-717, 2000
7. F. van Breugel, J. Worrell, *"Towards Quantitative Verification of Probabilistic Systems (extended abstract)"*, in Proc. of the 28th International Colloquium on Automata, Languages and Programming, LNCS 2076, pp. 421-432, Springer-Verlag, 2001
8. J. Desharnais, V. Gupta, R. Jagadeesan, P. Panangaden, *"Metrics for Labeled Markov Processes"*, in Proc. of the 10th Int. Conf. on Concurrency Theory, LNCS 1664, pp. 258-273, Springer-Verlag, 1999
9. A. Di Pierro, C. Hankin, H. Wiklicky, *"On Approximate Non-Interference"*, 2nd ACM SIGPLAN and IFIP WG 1.7 Workshop on Issues in the Theory of Security, 2002
10. R. Focardi, R. Gorrieri, *"A Classification of Security Properties"*, Journal of Computer Security, 3(1):5-33, 1995

11. R. Focardi, R. Gorrieri, *"Classification of Security Properties (Part I: Information Flow)"*, Foundations of Security Analysis and Design - Tutorial Lectures (R. Focardi and R. Gorrieri, Eds.), LNCS 2171, pp. 331-396, Springer-Verlag, 2001
12. R. Focardi, R. Gorrieri, F. Martinelli, *"Non Interference for the Analysis of Cryptographic Protocols"*, in Proc. of the 27th International Colloquium on Automata, Languages and Programming, LNCS 1853, pp. 354-372, Springer-Verlag, 2000
13. R.J. van Glabbeek, S.A. Smolka, B. Steffen, *"Reactive, Generative and Stratified Models of Probabilistic Processes"*, in Information and Computation 121:59-80, 1995
14. J.A. Goguen, J. Meseguer, *"Security Policy and Security Models"*, in Proc. of Symposium on Security and Privacy, pp. 11-20, IEEE CS Press, 1982
15. J.W. Gray III, *"Toward a Mathematical Foundation for Information Flow Security"*, Journal of Computer Security, 1:255-294, 1992
16. Y. Han, *"Investigation of Non-repudiation Protocols"*, in ACISP: Information Security and Privacy: Australasian Conference, LNCS 1172, pp. 38-47, Springer-Verlag, 1996
17. O. Markowitch, Y. Roggeman, *"Probabilistic Non-Repudiation without Trusted Third Party"*, 2nd Conference on Security in Communication Networks, Amalfi, Italy, 1999
18. R. Milner, *"Communication and Concurrency"*, Prentice Hall, 1989
19. P.Y.A. Ryan, *"A CSP Formulation of Non-Interference"*, Cipher, pp. 19-27, IEEE CS Press, 1991
20. A. Sabelfeld, D. Sands, *"Probabilistic Noninterference for Multi-threaded Programs"*, in Proc. of the 13th Computer Security Foundations Workshop, IEEE CS Press, 2000
21. R. Segala, N.A. Lynch, *"Probabilistic Simulations for Probabilistic Processes"*, in Proc. of the 5th Int. Conf. on Concurrency Theory, LNCS 836, pp. 481-496, Springer-Verlag, 1994
22. G. Smith, *"Weak Probabilistic Bisimulation for Secure Information Flow"*, 2nd ACM SIGPLAN and IFIP WG 1.7 Workshop on Issues in the Theory of Security, 2002
23. J. Zhou, D. Gollmann, *"An Efficient Non-repudiation Protocol"*, in Proc. of the 10th Computer Security Foundations Workshop, pp. 126-132, IEEE CS Press, 1997

A Operational Semantics of the Calculus

The formal semantics of our probabilistic process algebra is given by the *GRTS* $(\mathcal{G}, AType, T)$, whose states are terms of the calculus and the transition relation T is the minimal relation satisfying the operational rules reported in Table 7 and in Table 8, where in addition to rules undersigned with l, which refer to the local moves of the lefthand process P, we also consider the symmetric rules (undersigned with r) taking into account the local moves of the righthand process Q, obtained by exchanging the roles of terms P and Q in the premises and by replacing p with $1-p$ in the label of the derived transitions.

As far as the notation is concerned, we use $P \xrightarrow{\pi} P'$ to stand for $\exists p : P \xrightarrow{\pi, p} P'$, meaning that P can execute action π with probability p and then behaving as P'; $P \xrightarrow{\pi}$ to stand for $\exists P' : P \xrightarrow{\pi} P'$, meaning that P can execute action π; $P \xrightarrow{G}$ to stand for $\exists a \in G \subseteq AType : P \xrightarrow{a}$, meaning that P can execute a generative action of type belonging to set G.

It is easy to see that rules $(gr1)$, $(gr2)$, and $(r1)$ to $(g2)$ express exactly the informal description of the semantics of the operators given in Sect. 3.

Rules $(r3)$ to $(r5)$ are concerned with reactive transitions executable by $P \parallel^p_S Q$. In particular, rules $(r3)$ and $(r4)$ express the independent reactive transitions of P and Q (note that they follow the same probabilistic mechanism seen for the alternative choice operator). Moreover, as far as reactive actions of a given type $a \in S$ are concerned, if both P and Q may execute some reactive action a_*, the choice of the two actions a_* of P and Q forming the actions a_* executable by $P \parallel^p_S Q$ is made according to the probability they are *independently* chosen by P and Q (see rule $(r5)$).

Rules $(g3)$ to $(g6)$ are concerned with generative transitions executable by $P \parallel^p_S Q$. In particular, since we consider a restricted set of executable actions, we redistribute the probabilities of the generative transitions of P (Q) executable by $P \parallel^p_S Q$ so that their overall probability sums up to 1 (as standard when restricting actions in the generative model [13]). To this aim, in semantics rules we employ the function $\nu_P(G) : \mathcal{P}(AType) \longrightarrow]0,1]$, with $P \in \mathcal{G}$, defined as $\nu_P(G) = \sum\{\!| p \mid \exists P', a \in G : P \xrightarrow{a,p} P' |\!\}$ that computes the sum of the probabilities of the generative transitions executable by P whose type belongs to set G. Note that the set of types of the generative transitions of P executable by $P \parallel^p_S Q$ is composed by the action types not belonging to the synchronization set S and the action types belonging to S for which a synchronization between a generative action of P and a reactive action of Q is allowed to be performed. Therefore, if we take as set G in $\nu_P(G)$ the set $G_{S,Q} \subseteq AType$, with $S \subseteq AType - \{\tau\}$ and $Q \in \mathcal{G}$, to be defined as $G_{S,Q} = \{a \in AType \mid a \notin S \vee (a \in S \wedge Q \xrightarrow{a_*})\}$, we have that $\nu_P(G_{S,Q})$ computes the overall probability of the generative transitions of P executable by $P \parallel^p_S Q$ and can be used to normalize the probabilities of the generative transitions of P. Moreover, the probability of the generative transitions a (with $a \in S$) of P are further distributed among the reactive transitions of type a of Q. Finally, parameter p is used to redistribute the probabilities of the normalized generative transitions of P and Q executable by $P \parallel^p_S Q$.

As far as the restriction operator is concerned, rule $(r6)$ states that $P \backslash L$ does not inherit those reactive bundles executable by P whose type belongs to the restriction set L. Moreover, by following the same mechanism seen for the parallel operator, in rule $(g7)$ we use function $\nu_P(G_L)$, which computes the overall probability of the generative transitions of P executable by $P \backslash L$, to normalize the probabilities of the generative transitions of P. In particular, since we want G_L to be the set of action types not belonging to the restriction set L, we assume $G_L \subseteq AType$, with $L \subseteq AType - \{\tau\}$, to be defined as follows: $G_L = \{a \in AType \mid a \notin L\}$.

Table 7. Operational semantics: Part I (symmetric rules omitted)

$$(gr1) \quad \pi.P \xrightarrow{\pi,1} P$$

$$(r1_l) \quad \frac{P \xrightarrow{a_*,q} P' \quad Q \xrightarrow{a_*}}{P +^p Q \xrightarrow{a_*,p\cdot q} P'}$$

$$(r2_l) \quad \frac{P \xrightarrow{a_*,q} P' \quad Q \not\xrightarrow{a_*}}{P +^p Q \xrightarrow{a_*,q} P'}$$

$$(g1_l) \quad \frac{P \xrightarrow{a,q} P' \quad Q \xrightarrow{GAct}}{P +^p Q \xrightarrow{a,p\cdot q} P'}$$

$$(g2_l) \quad \frac{P \xrightarrow{a,q} P' \quad Q \not\xrightarrow{GAct}}{P +^p Q \xrightarrow{a,q} P'}$$

$$(r3_l) \quad \frac{P \xrightarrow{a_*,q} P' \quad Q \xrightarrow{a_*}}{P \,\|^p_S\, Q \xrightarrow{a_*,p\cdot q} P' \,\|^p_S\, Q} \quad a \notin S$$

$$(r4_l) \quad \frac{P \xrightarrow{a_*,q} P' \quad Q \not\xrightarrow{a_*}}{P \,\|^p_S\, Q \xrightarrow{a_*,q} P' \,\|^p_S\, Q} \quad a \notin S$$

$$(r5) \quad \frac{P \xrightarrow{a_*,q} P' \quad Q \xrightarrow{a_*,q'} Q'}{P \,\|^p_S\, Q \xrightarrow{a_*,q\cdot q'} P' \,\|^p_S\, Q'} \quad a \in S$$

$$(g3_l) \quad \frac{P \xrightarrow{a,q} P' \quad Q \xrightarrow{G_{S,P}}}{P \,\|^p_S\, Q \xrightarrow{a,p\cdot q/\nu_P(G_{S,Q})} P' \,\|^p_S\, Q} \quad a \notin S$$

$$(g4_l) \quad \frac{P \xrightarrow{a,q} P' \quad Q \not\xrightarrow{G_{S,P}}}{P \,\|^p_S\, Q \xrightarrow{a,q/\nu_P(G_{S,Q})} P' \,\|^p_S\, Q} \quad a \notin S$$

$$(g5_l) \quad \frac{P \xrightarrow{a,q} P' \quad Q \xrightarrow{a_*,q'} Q' \quad Q \xrightarrow{G_{S,P}}}{P \,\|^p_S\, Q \xrightarrow{a,p\cdot q'\cdot q/\nu_P(G_{S,Q})} P' \,\|^p_S\, Q'} \quad a \in S$$

$$(g6_l) \quad \frac{P \xrightarrow{a,q} P' \quad Q \xrightarrow{a_*,q'} Q' \quad Q \not\xrightarrow{G_{S,P}}}{P \,\|^p_S\, Q \xrightarrow{a,q'\cdot q/\nu_P(G_{S,Q})} P' \,\|^p_S\, Q'} \quad a \in S$$

$$(gr2) \quad \frac{P \xrightarrow{\pi,q} P'}{A \xrightarrow{\pi,q} P'} \quad \text{if } A \triangleq P$$

Table 8. Operational semantics: Part II

$$(r6)\quad \frac{P \xrightarrow{a_*,q} P'}{P\backslash L \xrightarrow{a_*,q} P'\backslash L}\ a \notin L \qquad (g7)\quad \frac{P \xrightarrow{a,q} P'}{P\backslash L \xrightarrow{a,q/\nu_P(G_L)} P'\backslash L}\ a \notin L$$

$$(r7)\quad \frac{P \xrightarrow{a_*,q} P' \quad P \xrightarrow{GAct}}{P/^{p}_{a} \xrightarrow{\tau,p\cdot q} P'/^{p}_{a}} \qquad (r8)\quad \frac{P \xrightarrow{a_*,q} P' \quad P \stackrel{GAct}{\not\longrightarrow}}{P/^{p}_{a} \xrightarrow{\tau,q} P'/^{p}_{a}}$$

$$(r9)\quad \frac{P \xrightarrow{b_*,q} P'}{P/^{p}_{a} \xrightarrow{b_*,q} P'/^{p}_{a}}\ a \neq b$$

$$(g8)\quad \frac{P \xrightarrow{b,q} P' \quad P \xrightarrow{a_*}}{P/^{p}_{a} \xrightarrow{b,(1-p)\cdot q} P'/^{p}_{a}}\ a \neq b \qquad (g9)\quad \frac{P \xrightarrow{b,q} P' \quad P \stackrel{a_*}{\not\longrightarrow}}{P/^{p}_{a} \xrightarrow{b,q} P'/^{p}_{a}}\ a \neq b$$

$$(g10)\quad \frac{P \xrightarrow{a,q} P' \quad P \xrightarrow{a_*}}{P/^{p}_{a} \xrightarrow{\tau,(1-p)\cdot q} P'/^{p}_{a}} \qquad (g11)\quad \frac{P \xrightarrow{a,q} P' \quad P \stackrel{a_*}{\not\longrightarrow}}{P/^{p}_{a} \xrightarrow{\tau,q} P'/^{p}_{a}}$$

Finally, as far as the hiding operator is concerned, the generative bundle executable by $P/^{p}_{a}$ is obtained from the generative bundle and the reactive bundle of type a executable by P. In particular, we distinguish the following cases:

- P enables both some reactive action a_* and some generative action. In such a case, in order to form the generative bundle of $P/^{p}_{a}$, the probabilities of the transitions a_* of P are multiplied by p when turned into the generative transitions τ (rule ($r7$)), while the probabilities of the generative transitions of P are multiplied by $1-p$ and transitions a are turned into transitions τ (rules ($g8$) and ($g10$));
- P does not enable generative actions, therefore transitions a_* of P are simply turned into generative transitions τ and parameter p is not used (rule ($r8$));
- P does not enable reactive actions a_*, therefore the probability distribution of the generative transitions does not change, transitions a are turned into transitions τ, and parameter p is not used (rules ($g9$) and ($g11$));
- for each $b \in AType$ such that $b \neq a$, $P/^{p}_{a}$ simply inherits the reactive bundle of type b executable by P (rule ($r9$)).

The Mean Value of the Maximum

Henrik Bohnenkamp[1] and Boudewijn Haverkort[2]

[1] University of Twente, The Netherlands
bohnenka@cs.utwente.nl
[2] Technical University of Aachen, Germany
haverkort@cs.rwth-aachen.de

Abstract. This paper treats a practical problem that arises in the area of stochastic process algebras. The problem is the efficient computation of the mean value of the maximum of phase-type distributed random variables. The maximum of phase-type distributed random variables is again phase-type distributed, however, its representation grows exponentially in the number of considered random variables. Although an efficient representation in terms of Kronecker sums is straightforward, the computation of the mean value requires still exponential time, if carried out by traditional means. In this paper, we describe an approximation method to compute the mean value in only polynomial time in the number of considered random variables and the size of the respective representations. We discuss complexity, numerical stability and convergence of the approach.

1 Introduction

Formal verification techniques and the evaluation of performance have proven to be important stages in the design of distributed systems. Stochastic process algebras (SPAs) attempt to integrate the qualitative (functional) and quantitative (temporal or stochastic) specifications within one formal modelling framework.

Markovian SPAs (MPAs) which are considered in this paper are a restricted class of SPA, where durations or delays of actions can only be exponentially distributed. The semantics of MPAs is usually given in in terms of ordinary labelled transition systems, and the restriction to exponential distributions allows us to describe the stochastic behaviour of an MPA specification by means of a continuous-time Markov chain (CTMC), which can be extracted easily from the labelled transition system. To evaluate an MPA model then boils down to the construction of the underlying CTMC and the subsequent computation of the steady-state probabilities. The latter is done by solving a system of linear equations.

Unfortunately, MPAs are, as any formalism referring to the global state space, subject to the *state space explosion problem*, i.e., the state space of the model (and thus that of the CTMC) increases exponentially with increasing degree of parallelism in the specified system. Therefore, performance evaluation of even small models becomes difficult, if not impossible, due to enormous memory requirements.

H. Hermanns and R. Segala (Eds.): PAPM-PROBMIV 2002, LNCS 2399, pp. 37–56, 2002.

In [2,3], we have identified a class of MPA processes for which a solution approach is feasible and that does not require the construction of the global state-space. The processes in this class have to meet several criteria, most importantly the following: processes have to be of the form $R = P_1||_S \cdots ||_S P_n$, *i.e.,* every component P_i, $i = 1, \ldots, n$, has to synchronise with the other components over the same synchronisation set. Processes of this class can therefore only do *barrier synchronisations.* On an abstract level, they describe semi-Markov chains, where the synchronisations are the points of regeneration. The idea for the solution of these models is to translate the components $P_1, \ldots, P_n$ of the original SPA process into these semi-Markov-chains, solve them in a compositional manner (which is possible in this case), and re-import the obtained stochastic measures *i.e.,* steady-state throughputs, into the original SPA components. The measures that can be obtained in this way are *local* steady-state measures of the respective components $P_1, \ldots, P_n$.

At the heart of the above compositional solution approach for the semi-Markov chains is the efficient computation of the mean value of the maximum of a number of phase-type distributed random-variables. A phase-type distribution can be described as the time to absorption of an absorbing Markov chain. The distribution of the maximum of two phase-type distributions can be described by the time to absorption of the *Kronecker sum* [5] of the respective absorbing Markov chains, and which, again, describes an absorbing Markov chain. The Kronecker sum describes actually the cross-product of the involved Markov chains. It is therefore not recommended to compute this sum explicitly, since it leads immediately to a state-space explosion.

In [3] we proposed to compute the mean value of the maximum of the considered random variables using so-called *exponomials*, *i.e.,* a semi-numerical representation of the phase-type distributions [13,14]. The computation of these exponomials, however, requires knowledge about the eigenstructure of the involved phase-type distributions (or, rather, the absorbing Markov chains representing them), which implies numerical difficulties. Most often, the involved distributions will not have a simple structure, so the eigenvalues can not be known in advance. Moreover, the time complexity of this approach is still exponential in the number of considered random variables (which is identical to the number of considered components of the model). This approach has therefore serious drawbacks.

Another, more straightforward and more promising approach is an *implicit representation* of the cross-product by means of a Kronecker sum. This approach is much better from a numerical point of view, but since the computations are still carried out on the cross-product of the involved absorbing Markov chains, even though it is not constructed explicitly, the time to compute a solution is still exponential in the number of considered random variables. So, the Kronecker approach is better from a numerical point of view, but it is still inefficient.

For our purposes, that we have sketched above, it is most desirable to have a method that avoids the problem of state-space explosion but is still time-efficient. In this paper we show that such an approach indeed does exist. We present a method (that we call MEANMAX throughout this paper) that allows us to compute the mean value of the maximum of a number of random variables

- *without the problem of state-space explosion*, and
- *in polynomial time* in the number of components.

MEANMAX is based on the observation that the maximum of positive random variables can be reformulated as a sum of some other, related random variables.

The remainder of this paper is organised as follows. In Section 2, we introduce notation and terminology used throughout the paper. As a motivation, in Section 3 we describe the class of SPA processes that we can solve efficiently with MEANMAX in more detail. In Section 4 we then develop the new approach to compute the mean value of the maximum of a number of random variables. In Section 5 we then focus on a particular aspect of the computation in case the random variables are all phase-type distributed. Section 6 addresses numerical and complexity issues when actually computing mean values of maxima. In Section 7 we compare our results with the Kronecker approach. In Section 8, we conclude the paper.

2 Notation and Terminology

2.1 Phase-Type Distributions

Phase-type distributions [12] can be represented by finite, absorbing CTMCs with starting probability distributions over their state space. We consider only CTMCs with one absorbing state. We assume an n-state absorbing Markov chain with generator matrix

$$\mathbf{Q} = \left(\begin{array}{c|c} \mathbf{T} & \underline{\mathbf{T}}_0 \\ \hline 0 \cdots 0 & 0 \end{array}\right) \tag{1}$$

and starting distribution $\underline{\pi} = (\pi_1, \ldots, \pi_n)$. The last row of the matrix represents the absorbing state, and $\underline{\mathbf{T}}_0$ all transition that go into this state. The *representation* of the phase-type distribution is defined to be the tuple $(\underline{\alpha}, \mathbf{T})$, where $\underline{\alpha} = (\pi_1, \ldots, \pi_{n-1})$. In this paper, we have to consider both the generator matrices as well as the representations of phase-type distributions. For ease of notation, we define $\dot{\mathbf{T}} = \mathbf{Q}$, and $\dot{\underline{\alpha}} = \underline{\pi}$. Note that $\underline{\mathbf{T}}_0 = -\mathbf{T}\underline{1}$, where, assuming that $\mathbf{T}$ has dimension $(n-1) \times (n-1)$, $\underline{1}$ is the column vector of length $n-1$ comprising ones only. If X is a phase-type distributed random variable with representation $(\alpha, \mathbf{T})$, then $\Pr\{X \leq t\} = F_X(t) = 1 - \underline{\alpha} e^{\mathbf{T}t} \underline{1}$. The density of X is defined as $f_X(t) = \underline{\alpha} e^{\mathbf{T}t} \underline{\mathbf{T}}_0$. The n-th moment of X can be derived as

$$\mathsf{E}[X^n] = (-1)^n \, n! \, \underline{\alpha} \, \mathbf{T}^{-n} \, \underline{1}. \tag{2}$$

2.2 Max-Plus Notation

For ease of notation we introduce a concise notation to express maxima as so-called max-plus expressions.

Let $x_1, x_2, \ldots$ be real-valued numbers, functions, or random variables. Then, we define $x_1 \boxplus x_2 := \max\{x_1, x_2\}$, $x_1 \odot x_2 := x_1 + x_2$, and $\frac{x_1}{x_2} := x_1 - x_2$. When the context allows, we omit $\odot$ and write $x_1 x_2$ instead.

In our case here, $\odot$ over $\boxplus$ are associative and commutative. Moreover, $\odot$ is distributive over $\boxplus$.

2.3 Kronecker Products and Sums

The Kronecker product $\otimes$ of two matrices $\mathbf{D} = (d_{ij})_{n,m}$ and $\mathbf{E} = (e_{ij})_{k,l}$ is defined to be the $(nk \times ml)$ matrix concisely written as

$$\mathbf{D} \otimes \mathbf{E} := \begin{pmatrix} d_{1,1}\mathbf{E} & \cdots & d_{1,m}\mathbf{E} \\ \vdots & \ddots & \vdots \\ d_{n,1}\mathbf{E} & \cdots & d_{n,m}\mathbf{E} \end{pmatrix}.$$

When $\mathbf{A}$ is an $n \times n$ matrix and $\mathbf{B}$ is an $m \times m$ matrix, then the *Kronecker sum* of $\mathbf{A}$ and $\mathbf{B}$ is a $(nm) \times (nm)$ matrix defined by

$$\mathbf{A} \oplus \mathbf{B} := (\mathbf{A} \otimes \mathbf{I}_B) + (\mathbf{I}_A \otimes \mathbf{B}), \tag{3}$$

where $\mathbf{I}_A$ and $\mathbf{I}_B$ are the $n \times n$ and $m \times m$ unit matrices, respectively.

If A is a phase-type distributed random variable with generator matrix $\mathbf{A}$ and B is a phase-type distributed random variable with generator matrix $\mathbf{B}$, then $\mathbf{A} \oplus \mathbf{B}$ is a generator matrix of the random variable $A \boxplus B$.

2.4 Semi-Markov Chains

In this section, we describe the most important definitions for semi-Markov chains (SMCs).

A semi-Markov chain is a stochastic process that can be described by a tuple $(\Sigma, \underline{p}, \mathbf{E}, J)$, where Σ is the state space of the stochastic process, $\mathbf{E}$ a stochastic matrix describing a DTMC with state space Σ, the *embedded Markov chain* (EMC), $\underline{p}$ is a starting probability distribution over $\mathbf{E}$ and $J : \Sigma \longrightarrow \mathcal{F}^+$ is a function that assigns non-negative distribution functions (the elements of $\mathcal{F}^+$) to states. $J(i)$ describes the distribution of the state sojourn time of state $i \in \Sigma$. An SMC is a CTMC, if for all $i \in \Sigma$, $J(i)$ is a negative-exponential distribution (with arbitrary rates $r_i \in \mathbb{R}^+$). An SMC is a DTMC, if the sojourn time distributions for all states are ignored.

The steady-state solution of a finite SMC with state space $\Sigma = \{1, \ldots, n\}$ is a probability vector $\underline{\pi} = (\pi_1, \ldots, \pi_n)$ that can be computed quite easily. Let $\underline{p} = (p_1, \ldots, p_n)$ be the steady-state solution of EMC $\mathbf{E}$, from which we assume that it exists. Let μ_i be the mean value of a random variable with distributions

$J(i)$, $i = 1, \ldots, n$. Then, the steady-state probabilities π_i of the SMC are defined as follows:

$$\pi_i = \frac{p_i \mu_i}{\sum_{j=1}^{n} p_j \mu_j}. \tag{4}$$

For a proof we refer to [9, Chapters 10 and 11]. Note that for a steady-state solution of a SMC only the mean sojourn times are needed.

In the following, we will only consider SMCs for which the sojourn time distributions are of phase-type. Moreover, we assume that the absorbing Markov chains describing the distributions have a unique starting state, *i.e.*, they start in a certain state with probability one. To keep the notation more simple, we assume that $J(i) = \mathbf{Q}_i$, where $\mathbf{Q}_i$ is a generator matrix describing the absorbing Markov chain of the sojourn time distribution of state i. Also, we silently assume that the starting distributions of the $\mathbf{Q}_i$ are vectors of the form $\underline{\pi}_0^{(i)} = (1, 0, \ldots, 0)$ with dimension compatible with $\mathbf{Q}_i$.

3 Efficient Solution of a Class of MPA Processes

The main motivation to develop MeanMax was a problem that occurred in in [2,3]. There, a special class of Markovian SPA models has been characterised, which describe semi-Markov processes on an abstract level.

In this section we define the most important features of this class, and explain why the mean value of the maximum must be computed efficiently.

3.1 Markovian Process Algebra

In this section we give a very concise overview over the syntactical and semantical properties of the MPA we consider.

Syntax and Semantics. Let *Act* be a set of actions, and $\mathbf{i} \in Act$ an *internal* action. The language $\mathcal{L}$ of our MPA is given by the following grammar:

$$\mathsf{P} \longrightarrow \mathsf{stop} \;\mid\; X \;\mid\; a.\mathsf{P} \;\mid\; [\lambda].\mathsf{P} \;\mid\; \mathsf{P} + \mathsf{P} \;\mid\; recX : \mathsf{P} \;\mid\; \mathsf{P} \setminus H \;\mid\; \mathsf{P} \,\|_S\, \mathsf{P},$$

where X is a recursion variable, $a \in Act$, $\lambda \in \mathbb{R}$, and $H \cup S \subseteq (Act \setminus \{\mathbf{i}\})$. The semantics of this SPA is given in terms of a labelled transition system that is defined by an SOS very similar to that of IMC [8], and which is described in detail in [2]; it will not be repeated here. The intuitive meaning of the operators is as follows. stop is the process that does nothing. $a.P$ is a process that can execute action a and then behaves like P. $[\lambda].P$ is a process that waits for an exponentially distributed time (rate λ) and then behaves like P. $P + Q$ behaves either as P or Q. $recX : \mathsf{P}$ behaves as the process P, in which each occurrence of variable X is textually replaced by $recX : \mathsf{P}$. $P\|_S Q$ denotes the parallel execution of P and Q, with synchronisation over the actions $a \in S$. The type of synchronisation is well known from process algebras like LOTOS [4].

3.2 The Class of Processes

We assume a set $\mathcal{P} = \{P_1, \ldots, P_n\} \subseteq \mathcal{L}^1$ of components for $n \in \mathbb{N}$, and a set $S \subseteq (Act \setminus \{\mathbf{i}\})$ of visible actions. Generally, the processes we consider are of the form

$$R \stackrel{\text{def}}{=} P_1 \|_S P_2 \|_S \cdots \|_S P_n \ . \tag{5}$$

Sometimes we abbreviate and write $R \stackrel{\text{def}}{=} !_S \mathcal{P}$. The set $\mathcal{P}$ as well as the individual $P \in \mathcal{P}$ have to fulfil the following requirements.

A: *Alternation.* For each $P \in \mathcal{P}$, the process P has to show an *alternating behaviour* between local, invisible or timed transitions and synchronising transitions. Formally, this means that, whenever $P \xrightarrow{a}$ for $P \in Act$, then there is no transition originating in P such that $P \xrightarrow{\lambda}$.

WC: *Weak (functional) congruence of components.* We define the (multi-)set $\mathcal{P}' = \{P'_1, \ldots, P'_n\}$ such that P'_i is defined as P_i, only that all occurrences of an action in P_i (be they timed or functional) that are *not* in S, are replaced by the internal action $\mathbf{i}$, for $i = 1, \ldots, n$. Then all processes $P \in \mathcal{P}'$ must be pairwise weakly congruent [8].

ID: *Irreducibility.* We require the irreducibility of the components $P \in \mathcal{P}$ and R, as well as the absence of non-determinism in R. The question how this property can be checked without generating the state-space of R is treated thoroughly in [2].

The three requirements **A**, **WC**, and **ID** are restricting the structure of the components $P \in \mathcal{P}$, and of R. Requirement **A** makes sure that the times between the occurrence of synchronising actions $a \in S$ are of phase-type. Requirements **WC** and **ID** ensure that process R is deadlock-free and irreducible. Then it can be shown that the times between the occurrence of synchronising actions $a \in S$ in R are also of phase-type. Moreover, each component $P \in \mathcal{P}$ of R takes part in all synchronisations over the actions $a \in S$. Therefore, whenever a synchronisation happens, the history of every component (and for R itself) becomes irrelevant for its future: a synchronisation is a regeneration point of the stochastic process underlying the considered R. In fact, if we only observe the instances of synchronisations, it can be shown that the resulting stochastic process is a semi-Markov chain, where the distributions assigned to the states describe the times between the synchronisations. We can construct this semi-Markov chain explicitly. Leaving out the details here, we assume that we have a function SMC($\cdot$) which maps a process $P_i \in \mathcal{P}$ on this semi-Markov chain $S_i = (\Sigma_i, \mathbf{E}_i, \underline{p}_i, J_i)$, *i.e.*, $\text{SMC}(P_i) \mapsto S_i$, for $i = 1, \ldots, n$.

[1] Actually, we must assume $\mathcal{P}$ to be a multi-set, since different components can be syntactically equal. However, for sake of simplicity we assume that we can distinguish all components from each other, so that we can assume that $\mathcal{P}$ is an ordinary set.

3.3 Combining SMCs

We have defined SMC(·) on components only, but we can define an operator $\diamond$ on SMCs such that

$$\mathcal{S} := \mathrm{SMC}(P_1\|_S \dots \|_S P_n) = \mathrm{SMC}(P_1) \diamond \dots \diamond \mathrm{SMC}(P_n). \tag{6}$$

The definition of $\diamond$ depends on the synchronisation set S. $\diamond$ is defined as follows. We assume two SMCs $S_i = (\Sigma_i, \mathbf{E}_i, \underline{p}_i, J_i)$ and functions $T_i : \Sigma \to S$, for, say, $i = 1, 2$. The functions T_i assign *tags*, *i.e.*, actions, $a \in S$ to the SMC states, which are needed to define the synchronisations between the SMCs properly. The tag functions are derived from the respective processes P_i. We define

$$S = S_1 \diamond S_2 := (\Sigma, \underline{p}, \mathbf{E}, J),$$

where $\Sigma \subseteq \Sigma_1 \times \Sigma_2$ with $(s_1, s_2) \in \Sigma \iff T_1(s_1) = T_2(s_2)$, $\underline{p} = (\underline{p}_1 \otimes \underline{p}_2)\big|_{\Sigma}$, $\mathbf{E} = (\mathbf{E}_1 \otimes \mathbf{E}_2)\big|_{\Sigma}$ and $J((s_1, s_2)) = J_1(s_1) \oplus J_2(s_2)$ for $(s_1, s_2) \in \Sigma)^2$.

3.4 Solving the Semi-Markov Chains

What we have done so far is to translate a set of MPA processes into a set of semi-Markov chains. We can combine these semi-Markov chains by means of $\diamond$ such that (6) holds. But what can we do with the representation of R in terms of semi-Markov chains? The idea is to obtain a steady-state solution for $\mathcal{S} = \mathrm{SMC}(P_1) \diamond \cdots \diamond \mathrm{SMC}(P_n)$. This solution provides us with stochastic measures for $\mathcal{S}$ *i.e.*, state throughputs, which can be re-imported into the process R. The same throughputs can also be interpreted in the components of R, *i.e.*,, the processes $P_i \in \mathcal{P}$. These throughputs can be used to compute *local* steady-state probabilities for the components, and then we are able to compute local steady-state measures for these components.

To summarise, once we have a solution for $\mathcal{S}$, we can compute local performance measures for the components $P_i \in \mathcal{P}$. The problem is now to solve $\mathcal{S}$ efficiently. With (6), we have essentially carried over the compositional structure of process R to the semi-Markov side. As shown in Section 2.4, to obtain a steady-state solution for an SMC, we need the steady-state solution of the EMC, and the mean values of the random residence times assigned to the SMC states. As shown in [2], the computation of the solution for the EMC can be done compositionally. But as we can see in the definition of $\diamond$ in Section 3.3, the distributions of $\mathcal{S}$ are in fact the Kronecker sum of some generator matrices of phase-type distributions, *i.e.*, as already pointed out in Section 2.3, the generator of the maximum of some random variables that are distributed according to these phase-type distributions.

We have now finally come to the main motivation of the rest of this paper. The common approach to derive the mean value of a phase-type distribution

[2] This definition is a bit sloppy, since Σ could contain states that are not reachable. This does, however, not play a role in the following, therefore, we leave the definition as it is.

is Equation (2). If the involved generator matrix is the Kronecker sum of some smaller generator matrices, then the computational effort usually grows exponentially with the number of Kronecker addends. For the efficient solution of the class of processes we have defined in this section it is therefore important to find a method that is more advanced. In this paper we will show that there is indeed an efficient method to compute the mean value of the maximum of some phase-type distributed random variables, that is, a method that exploits the compositional structure of (6). We will describe this approach in the rest of the paper.

4 A Bit of Random Variable Arithmetic

From now on, we will abstract from the concrete problem that we have described in Section 3 for our class of SPA processes and only assume phase-type distributed random variables.

In this section we show how we can express the maximum of a set of independent random variables as a sum of a number of other, different but related random variables. Let $X_1, X_2, \ldots, X_n$ be pairwise independent positive random variables with CDF F_i and pdf f_i for $i = 1, \ldots, n$. Let $Z = X_1 \boxplus \cdots \boxplus X_n$. Then

$$\mathsf{E}[Z] = \mathsf{E}[X_1 \boxplus \cdots \boxplus X_n]. \tag{7}$$

To begin with, we now pick an arbitrary random variable out of $\{X_1, \ldots, X_n\}$, say X_1. Then we can rewrite (7) as follows:

$$\begin{aligned}
\mathsf{E}[Z] = \mathsf{E}\left[\frac{(X_1 \boxplus \cdots \boxplus X_n)\,X_1}{X_1}\right] \quad &= \quad \mathsf{E}\left[\left(\frac{X_1}{X_1} \boxplus \frac{X_2}{X_1} \boxplus \cdots \boxplus \frac{X_n}{X_1}\right) X_1\right] \\
= \mathsf{E}\left[\left(0 \boxplus \frac{X_2}{X_1} \boxplus \cdots \boxplus \frac{X_n}{X_1}\right) X_1\right] & \\
= \mathsf{E}\left[\left(\left(0 \boxplus \frac{X_2}{X_1}\right) \boxplus \left(0 \boxplus \frac{X_3}{X_1}\right) \boxplus \cdots \boxplus \left(0 \boxplus \frac{X_n}{X_1}\right)\right) X_1\right] &
\end{aligned} \tag{8}$$

If we define

$$Y_i^{(2)} = 0 \boxplus \frac{X_i}{X_1}$$

for $i = 2, \ldots, n$, we can rewrite (8) as follows:

$$\mathsf{E}[Z] = \mathsf{E}[X_1] + \mathsf{E}\left[Y_2^{(2)} \boxplus Y_3^{(2)} \boxplus \cdots \boxplus Y_n^{(2)}\right]. \tag{9}$$

The random variables $Y_i^{(2)}$, for $i = 2, \ldots, n$, are all non-negative since they are all bounded by 0. Hence, we can repeat the above derivation towards (9), with the random variables $Y_j^{(2)}$, $j = 2, \ldots, n$, taking over the roles of the X_i, $i = 1, \ldots, n$. Again, we choose one of the random variable, say, $Y_2^{(2)}$, and derive an equation similar to (9):

$$\mathsf{E}[Z] = \mathsf{E}\left[X_1\right] + \mathsf{E}\left[Y_2^{(2)}\right] + \mathsf{E}\left[\left(Y_3^{(3)} \boxplus Y_4^{(3)} \boxplus \cdots \boxplus Y_n^{(3)}\right)\right], \tag{10}$$

where

$$Y_i^{(3)} = 0 \boxplus \frac{Y_i^{(2)}}{Y_2^{(2)}},$$

for $i = 3, \ldots, n$. Applying this procedure $n-1$ times, we eventually end up with the equation

$$\mathsf{E}[Z] = \mathsf{E}[X_1] + \mathsf{E}\left[Y_2^{(2)}\right] + \mathsf{E}\left[Y_3^{(3)}\right] + \cdots + \mathsf{E}\left[Y_n^{(n)}\right]. \tag{11}$$

If we set $Y_1^{(1)} = X_1$, then

$$\mathsf{E}[Z] = \mathsf{E}[X_1 \boxplus \cdots \boxplus X_n] = \sum_{i=1}^{n} \mathsf{E}\left[Y_i^{(i)}\right].$$

We have hence expressed the mean value of the maximum of a set of positive random variables by the sum of the mean values of some other, related, random variables. Of course, our derivation remains valid without the expectation operator, that is, for the random variables themselves.

Equation (11) is valid for arbitrary random variables, but we are interested in the case where X_i is phase-type distributed, $i = 1, \ldots, n$. Hence, we assume the existence of representations $(\underline{\alpha}_i, \mathbf{T}_i)$ for X_i, $i = 1, \ldots, n$. In this case, the interesting result is that the random variable $Y_i^{(i)}$ is phase-type distributed as well, with representation $(\underline{\alpha}_i', \mathbf{T}_i)$. The only difference between the distribution of X_i and $Y_i^{(i)}$ lies in the starting distributions; the transition structure within the phase-type distribution does not change! We will show this formally in Section 5, but give an intuitive explanation first.

Example: Frog Trapping. As an example we consider three frogs on a lily pond[3]. The pond is covered with n lily pads, and each frog starts from pad i with probability $\pi_0(i)$. Each frog hops from pad to pad, where on each pad i it thinks for a negative exponentially distributed time about which pad to hop next. Frogs are not good at thinking, therefore, the next pad is actually chosen at random. The thinking time and the hopping probabilities depend only on the pad the frog sits on. There is one pad that is actually a frog trap, probably installed by a representative of the French cuisine. Therefore, once a frog lands on the trap pad, it gets absorbed. We assume that the individual times of the frogs to reach the trap are random variables X_1, X_2 and X_3. All three random variables are of phase-type: the lily pond with trap represents an absorbing Markov chain; this Markov chain is described by its generator matrix $\mathbf{Q}$. The states of the absorbing Markov chain correspond to the lily pads. The probability vector $\underline{\pi}_0$ represent the starting distributions[4]. We assume that the frog hunt starts at time 0. We

[3] This example is adapted from [10].

[4] The only reason to consider all random variables as identically distributed is to keep the example simple. We could easily assume different lily ponds and different starting distributions.

start by considering frogs 2 and 3 at the time instant X_1, *i.e.*, at the time instant when frog 1 gets caught. Now we want to know how long the frog trapper still has to wait until frogs 2 and 3 also get caught. A good guess would be $X_i - X_1$, for $i = 2, 3$. However, these differences might be negative, if frog 1 was not the first to reach the trap. If frog 1 was even the last one the trapper does not have to wait at all. Therefore, the residual life times of the other frogs should be defined as $Y_i^{(2)} = 0 \boxplus \frac{X_i}{X_1}$, for $i = 2, 3$. The remaining time until all other frogs are caught can then be expressed as $Y_2^{(2)} \boxplus Y_3^{(2)}$.

We now might wonder what the distributions of $Y_2^{(2)}$ and $Y_3^{(2)}$ are? Since we assume that the lily pond does not change its structure, we can assume that $Y_2^{(2)}$ and $Y_3^{(2)}$ can also be described by means of $\mathbf{Q}$. However, frogs 2 and 3 have probably somehow changed their positions in the time interval $[0, X_1)$. This indicates that we have new starting probability distributions over the pads of the lily pond, to indicate the changed probability for the frogs to be on a particular pad. This distribution depends on the time that has passed, *i.e.*, on X_1, and it must be used as a new starting distribution for the frogs to derive their time to absorption.

We now can repeat the same reasoning, now only considering frogs 2 and 3 with their new starting distributions. So, how long does frog 3 still have to live if frog 1 *and* frog 2 have been trapped (without knowing the order of arrival)? The time from the start is $X_1 \odot Y_2^{(2)} = X_1 \boxplus X_2$. Hence, the remaining life time for frog 3 is

$$Y_3^{(3)} = 0 \boxplus \frac{X_3}{X_1 \boxplus X_2} = 0 \boxplus \frac{X_3}{X_1 \odot Y_2^{(2)}}.$$

Again, the distribution of $Y_3^{(3)}$ is represented by $\mathbf{Q}$, the lily pond, but the probability for frog 3 to be found on pad i has changed, since now the time that has passed is $X_1 \odot Y_2^{(2)}$. To conclude, the mean time until all three frogs are caught is given by

$$\mathsf{E}[X_1 \boxplus X_2 \boxplus X_3] = \mathsf{E}[X_1] + \mathsf{E}\left[Y_2^{(2)}\right] + \mathsf{E}\left[Y_3^{(3)}\right]$$

(End of Example)

The implications of the above derivations are straightforward. To compute $\mathsf{E}[Z]$, all we have to do is to compute the mean values of the variables $Y_i^{(i)}$. In the next section, we will derive the new starting distributions for the random variables $Y_i^{(i)}$. To conclude this section, we assume that we have a function $\mathcal{N}$ which takes two random variables X_1 and X_2 with representations $R_1 = (\underline{\alpha}_1, \mathbf{T}_1)$ and $R_2 = (\underline{\alpha}_2, \mathbf{T}_2)$ respectively and computes the new starting distribution $\mathcal{N}(R_1, R_2)$ of the random variable $Y = 0 \boxplus \frac{X_2}{X_1}$. Then the algorithm sketched in Figure 1 computes the mean value of the maximum of a set of phase-type distributed random variables.

· INPUT: set of representations $\{(\underline{\alpha}_i, \mathbf{T}_i) \mid i = 1, \ldots, n\}$ of *rv* $\{X_1, \ldots, X_n\}$
· OUTPUT: mean value $\mu = \mathsf{E}[X_1 \boxplus \cdots \boxplus X_n]$
1. $\mu \in \mathbb{R}; i, j \in \mathbb{N}$;
2. $\mu = 0$;
3. **for** $j = 1$ **to** $n - 1$ **do**
4.   **for** $i = j + 1$ **to** n **do** $\underline{\alpha}_i = \mathcal{N}((\underline{\alpha}_j, \mathbf{T}_j), (\underline{\alpha}_i, \mathbf{T}_i))$;
5. $\mu = \mu - \underline{\alpha}_j \mathbf{T}_j^{-1} \underline{1}$;
6. **od**;
7. $\mu = \mu - \underline{\alpha}_n \mathbf{T}_n^{-1} \underline{1}$;
8. **return** (μ);

Fig. 1. MEANMAX Algorithm

The algorithm works as follows:

1.–2. μ, i, j are three auxiliary variables. μ is initially set to zero.
3.–4. We have two loops: the outer loop ranges over j, the inner over i. j defines the random variable that defines the time for which the new starting distributions are computed. i defines the random variable for which a new starting distributions is computed by means of $\mathcal{N}$.
5. Here the contribution of representation j is added to μ. In terms of Equation (11), $-\underline{\alpha}_j \mathbf{T}_j^{-1} \underline{1} = \mathsf{E}\left[Y_j^{(j)}\right]$.
7.–8. The contribution of representation n is added to μ ($\mathsf{E}\left[Y_n^{(n)}\right]$); μ is returned.

5 Computing Starting Probability Distributions

In this section we present a method to compute the new starting probability distributions of an absorbing Markov chain under the assumption that a random, phase-type distributed amount of time has passed. As the algorithm in Figure 1 shows, it is not necessary to consider more than two random variables at a time.

5.1 Introduction

We assume two phase-type distributed, independent random variables A, B with representations $(\underline{\alpha}, \mathbf{A})$ for A and $(\underline{\beta}, \mathbf{B})$ for B. The associated generator matrices are denoted by $\dot{\mathbf{A}} = (a_{ij})_{k_a, k_a}$ and $\dot{\mathbf{B}} = (b_{ij})_{k_b, k_b}$, their starting distributions by $\underline{\dot{\alpha}}$ and $\underline{\dot{\beta}}$, respectively. The probability distributions at time t are denoted as

$\underline{\dot{\alpha}}(t)$ and $\underline{\dot{\beta}}(t)$, respectively. The column vectors $\underline{\mathbf{A}}_0$ and $\underline{\mathbf{B}}_0$ can be expressed as $\underline{\mathbf{A}}_0 = -\mathbf{A}\underline{1}$ and $\underline{\mathbf{B}}_0 = -\mathbf{B}\underline{1}$ (*cf.* Section 2.1). The distribution functions of A and B are denoted by F_A and F_B, respectively, and are defined as described in Section 2.1.

First recall that the transient behaviour of a CTMC is described by a matrix exponential:

$$\underline{p}(t) = \underline{p}e^{\mathbf{Q}t},$$

where $\mathbf{Q}$ is the generator matrix of the CTMC, and $\underline{p}(0) = \underline{p}$ the starting distribution.

We want to compute the distribution of Y, where $Y = 0 \boxplus \frac{B}{A}$. Y is the *residual time* of B given A. We can imagine that the probability mass which has been distributed over the states of the CTMC with generator $\dot{\mathbf{B}}$ according to $\underline{\dot{\beta}}$ at time 0 has shifted somehow towards the absorbing state in the time period $[0, A)$, as per the above matrix exponential. Hence, the new starting distribution of B given A, can be expressed as $\underline{\dot{\beta}}(A) = \underline{\dot{\beta}}e^{\dot{\mathbf{B}}A}$. Note that $\underline{\dot{\beta}}(A)$ is a random variable itself, depending on A. To take all different outcomes of A into account, we have to decondition $\underline{\dot{\beta}}(A)$, *i.e.*, we have to compute $\mathsf{E}\left[\underline{\dot{\beta}}(A)\right]$ as follows:

$$\underline{\dot{\beta}}' = \int_0^\infty \underline{\dot{\beta}}e^{\dot{\mathbf{B}}t}dF_A(t). \tag{12}$$

$\underline{\dot{\beta}}'$ is indeed the new starting distribution, so that $(\beta', \mathbf{B})$ is the representation of the random variable Y.

$\underline{\dot{\beta}}'$ must be computed effectively, and doing this by means of the integral (12) is not recommended (for numerical reasons). In the next section, we provide the means to compute $\underline{\dot{\beta}}'$ efficiently.

5.2 Computing the New Starting Distribution

The following theorem states that we can express the integral of Equation (12) by an infinite matrix sum.

Theorem 1. *Let $A, B, \mathbf{A}, \mathbf{B}, \underline{\alpha}, \underline{\beta}$, and $\underline{\dot{\beta}}'$ be defined as in Section 5.1. Let $\dot{\mathbf{P}}_B$ be defined such that $\dot{\mathbf{B}} = b(\dot{\mathbf{P}}_B - \mathbf{I})$ for $b = \max_{i=1,\ldots,k_b}\{|b_{ii}|\}$. Then*

$$\underline{\dot{\beta}}' = \frac{1}{b} \cdot \underline{\dot{\beta}} \sum_{n=0}^{\infty} (-1)^{n+1} \dot{\mathbf{P}}_B^n \underline{\alpha} \left(\frac{1}{b}\mathbf{A} - \mathbf{I}\right)^{-(n+1)} \underline{\mathbf{A}}_0 \tag{13}$$

The rest of this section is devoted to the proof of Theorem 1.

Proof. The proof of Theorem 1 comprises three steps.

Step 1. In the first step, we make use of the matrix $\dot{\mathbf{P}}_B$, which is defined as

$$\dot{\mathbf{P}}_B = \frac{\dot{\mathbf{B}}}{b} + \mathbf{I},$$

where $b = \max\{|b_{ii}|\}$. $\dot{\mathbf{P}}_B$ is the uniformised matrix of $\dot{\mathbf{B}}$. Substituting $\dot{\mathbf{B}} = b(\dot{\mathbf{P}}_B - \mathbf{I})$ in Equation (12) yields

$$\begin{aligned}
\dot{\beta}' &= \int_0^\infty \underline{\dot{\beta}} e^{b(\dot{\mathbf{P}}_B - \mathbf{I})t} dF_A(t) &&= \underline{\dot{\beta}} \int_0^\infty e^{-bt} e^{\dot{\mathbf{P}}_B bt} dF_A(t) \\
&= \underline{\dot{\beta}} \int_0^\infty e^{-bt} \sum_{n=0}^{\infty} \frac{(bt)^n}{n!} \dot{\mathbf{P}}_B^n dF_A(t) &&= \underline{\dot{\beta}} \sum_{n=0}^{\infty} \frac{b^n}{n!} \dot{\mathbf{P}}_B^n \int_0^\infty e^{-bt} t^n dF_A(t) \\
&= \underline{\dot{\beta}} \sum_{n=0}^{\infty} \frac{b^n}{n!} \dot{\mathbf{P}}_B^n \mathsf{E}[e^{-bA} A^n] && \qquad (14)
\end{aligned}$$

Step 2. $\mathsf{E}[e^{-bA}A^n]$ is the n-th derivative of F_A's moment generating function at point $-b$ (*cf.* [7]). The moment generating function $\mathsf{F}_A = \mathsf{E}[e^{sA}]$ of F_A can be derived as $\mathsf{F}_A(s) = -\underline{\alpha}(s\mathbf{I}+\mathbf{A})^{-1}\underline{\mathbf{A}}_0$ for $s < 0$. The n-th derivative can therefore be expressed as $\mathsf{F}_A^{(n)}(s) = (-1)^{n+1} \cdot n! \cdot \underline{\alpha}\,(s\mathbf{I} + \mathbf{A})^{-(n+1)}\,\underline{\mathbf{A}}_0$. Then,

$$\begin{aligned}
\mathsf{E}[e^{-bA}A^n] &= \mathsf{F}_A^{(n)}(-b) = (-1)^{n+1} n! \cdot \underline{\alpha}\,(\mathbf{A} - b\mathbf{I})^{-(n+1)}\,\underline{\mathbf{A}}_0 \\
&= (-1)^{n+1} n! \cdot \underline{\alpha} \left(b \left(\frac{1}{b}\mathbf{A} - \mathbf{I} \right) \right)^{-(n+1)} \underline{\mathbf{A}}_0 \\
&= \frac{n!}{(-b)^{n+1}} \cdot \underline{\alpha} \left(\frac{1}{b}\mathbf{A} - \mathbf{I} \right)^{-(n+1)} \underline{\mathbf{A}}_0 \qquad (15)
\end{aligned}$$

Step 3. Combining Equations (14) and (15) yields

$$\dot{\beta}' = \frac{1}{b} \cdot \underline{\dot{\beta}} \sum_{n=0}^{\infty} (-1)^{n+1} \dot{\mathbf{P}}_B^n \underline{\alpha} \left(\frac{1}{b}\mathbf{A} - \mathbf{I} \right)^{-(n+1)} \underline{\mathbf{A}}_0,$$

which concludes the proof of Theorem 1.

6 Numerical and Complexity Issues

Equation (13) can be seen as a guideline for the implementation of an algorithm that computes $\mathcal{N}$, the function that was needed in the MeanMax-Algorithm in Figure 1. However, careful considerations have to be made to ensure numerical stability and convergence. This is the topic of this section.

6.1 Uniformisation

Equation (13) makes use of the matrix $\dot{\mathbf{P}}_B$, defined as the matrix that satisfies the equation $\dot{\mathbf{B}} = b(\dot{\mathbf{P}}_B - \mathbf{I})$ for $b = \max_{i=1,\ldots,k_b}\{|b_{ii}|\}$. $\dot{\mathbf{P}}_B$ is the uniformised matrix of $\dot{\mathbf{B}}$, and is *probabilistic*: all entries are non-negative and less then or equal to 1. The reason to transform $\dot{\mathbf{B}}$ to $\dot{\mathbf{P}}_B$ is that computations on positive numbers reduce the occurrence of cancellation errors, a frequent source of numerical instability. Uniformisation is successfully used for the computation of transient measures of CTMCs [11].

6.2 Explicit Inversion

The computation of $\underline{\dot{\beta}}'$ by means of Equation (13) involves a matrix inversion. If $\mathbf{M}$ is a regular $k \times k$ matrix then $\mathbf{M}^{-1}$ is usually computed by solving k systems of linear equations: since $\mathbf{M}\mathbf{M}^{-1} = \mathbf{I}_k$, one has to solve $\mathbf{M}\underline{m}_i = \underline{e}_i$ for $i = 1, \ldots, k$ to obtain the k column vectors $\underline{m}_i$ of $\mathbf{M}^{-1}$. In our case,

$$\mathbf{M} = \left(\frac{\mathbf{A}}{b} - \mathbf{I}\right). \tag{16}$$

Although in (13) a power of $\mathbf{M}^{-1}$ occurs, it is never necessary to actually carry out the computations by means of matrix-matrix products. Instead, for each n of the sum, a matrix-*vector* multiplication has to be carried out. We can define two recursive functions which facilitate the computation of $\underline{\dot{\beta}}'$:

$$E(n) := \begin{cases} \frac{1}{b}\underline{\dot{\beta}}, & n = 0, \\ E(n-1)\dot{\mathbf{P}}_B, & \textit{otherwise}, \end{cases}$$

and

$$F(n) := \begin{cases} -\underline{\alpha}\mathbf{M}^{-1}, & n = 0, \\ -F(n-1)\mathbf{M}^{-1}, & \textit{otherwise}. \end{cases}$$

Then

$$\underline{\dot{\beta}}' = \sum_{n=0}^{\infty} E(n)(F(n)\underline{\mathbf{A}}_0).$$

Note that the vector $\underline{\mathbf{A}}_0$ has to be multiplied by $F(n)$ for every iteration, so we have for each n also a vector-vector multiplication. $F(n)$ can not be combined with $E(n)$.

$\mathbf{M}$ inherits some properties of $\mathbf{A}$. $\mathbf{A}$ is a representation matrix for a phase-type distribution. A consequence of the Theorem of Gerschgorin (*cf.* [6, Chapter 14.5]) is that a generator matrix, and therefore, also a representation matrix, has only eigenvalues on the negative complex half-plane. The eigenstructure of $\mathbf{A}/b$ is similar to that of $\mathbf{A}$, only that every original eigenvalue of $\mathbf{A}$ is divided by b. The matrix $\mathbf{M} = \mathbf{A}/b - \mathbf{I}$ has a very similar eigenstructure to that of $\mathbf{A}/b$, only that now every eigenvalue of $\mathbf{A}/b$ is shifted to the left by 1 on the complex plane. More concisely, if λ is an eigenvalue of $\mathbf{A}$, then $\lambda' = \lambda/b - 1$ is an eigenvalue of $\mathbf{M}$. The important conclusion is the following:

Lemma 1. *For all eigenvalues* μ *of* $\mathbf{M}$: $|\mu| > 1$.

If μ is an eigenvalue of a $\mathbf{M}$, than $1/\mu$ is an eigenvalue of $\mathbf{M}^{-1}$. This leads to the following lemma.

Lemma 2. *For all eigenvalues* μ *of* $\mathbf{M}^{-1}$: $|\mu| < 1$

Note that the relations in these two lemmas are strict, since $\mathbf{A}$ is invertible.

If μ is an eigenvalue for $\mathbf{M}^{-1}$, then μ^n is an eigenvalue of $\mathbf{M}^{-n}$ for $n > 1$. Therefore, we can see at once that $\mathbf{M}^{-n} \xrightarrow{n\to\infty} \mathbf{0}$,s ince all eigenvalues of $\mathbf{M}^{-n}$ tend to zero, as n goes to infinity.

Above, we have defined the function $F(n)$ with $\underline{F}(0) = -\underline{\alpha}\mathbf{M}^{-1}$. Apparently, $F(n)$ also tends to zero as n goes to infinity. When we write the vector $\underline{\alpha}$ as a linear combination of the eigenvectors $\underline{v}_1, \ldots, \underline{v}_k$ of $\mathbf{M}^{-1}$, *i.e.*,

$$\underline{\alpha} = \sum_{l=1}^{k} \nu_l \underline{v}_l$$

for $\nu_l \in \mathbb{C}$, for $l = 1, \ldots, k$, then we can write

$$F(n) = (-1)^n \cdot F(0)\mathbf{M}^{-n} = (-1)^{n+1} \sum_{l=1}^{k} \nu_l \lambda_l^{n+1} \underline{v}_l,$$

where λ_l is the eigenvalue belonging to eigenvector $\underline{v}_l$. If we assume that, without loss of generality, λ_1 is the dominant eigenvalue of $\mathbf{M}^{-1}$, then we see that the convergence of $F(n)$ mainly depends on $|\lambda_1|$, *i.e.*,, for some positive constant c, $c|\lambda_1^n|$ is an asymptote of $|F(n)|$.

As a consequence, λ_1 should be small. The other way round, since $1/\lambda_1$ is the smallest eigenvalue of $\mathbf{M}$, b should be chosen as small as possible: this prevents the eigenvalues of $\mathbf{A}/b$ becoming smaller than necessary and therefore also that of $\mathbf{M}$. However, in Theorem 1, b has already been chosen as small as possible, so we have no possibility to improve the convergence of F by another choice of b.

Convergence of the Sum. The sum $\sum_{n=0}^{\infty} E(n)(F(n)\underline{\mathbf{A}}_0)$, involves an infinite number of addends. To actually compute something, we can only do a finite summation. We therefore address the partial sum, $\sum_{n=0}^{N} E(n)(F(n)\underline{\mathbf{A}}_0)$, where N is chosen such that the sum is a good enough approximation of $\underline{\dot{\beta}}'$. We now consider the "rest" of the sum, *i.e.*, $\sum_{n=N+1}^{\infty} E(n)(F(n)\underline{\mathbf{A}}_0)$ and consider its absolute value:

$$\begin{aligned}\left|\sum_{n=N+1}^{\infty} E(n)(F(n)\underline{\mathbf{A}}_0)\right| &\leq \sum_{n=N+1}^{\infty} |E(n)(F(n)\underline{\mathbf{A}}_0)| \\ &\leq \sum_{n=N+1}^{\infty} |E(n)c^{n+1}|\end{aligned}$$

for some $c \in (0,1)$. We estimate $|F(n)\underline{\mathbf{A}}_0|$ as c^{n+1} since we can assume that N is large enough such that $|F(n)\underline{\mathbf{A}}_0| < 1$. Therefore we can always find a $c \in (0,1)$ such that $c^{n+1} > |F(n)\underline{\mathbf{A}}_0|$.

Since P_B is a probability matrix we can assume that $|E(n)| \leq 1$ (interpreting $|\cdot|$ as $\|\cdot\|_\infty$ for matrices), and therefore,

$$\begin{aligned} \sum_{n=N+1}^{\infty} |E(n)c^n| &\leq \sum_{n=N+1}^{\infty} c^n \\ &= \frac{c^{N+1}}{1-c} =: \rho(N) \end{aligned}$$

Now we want $\rho(N) \leq \varepsilon$, where ε is our desired accuracy of the finite sum. Transforming this, we find that

$$N > \frac{\ln(1-c) + \ln(\varepsilon)}{\ln c} \tag{17}$$

to achieve the required accuracy. We see that the precision ε has only a small influence on the number of iterations, if we assume that we will never want trying to achieve higher precision than machine precision. However, if $c \to 1$, then $\frac{\ln(1-c)+\ln(\varepsilon)}{\ln c} \to \infty$. If we set $c > |\lambda_1|$, the dominant eigenvalue of $\mathbf{M}^{-1}$, then we see that the number of iterations N grows exponentially, the nearer $|\lambda_1|$ comes to 1. If we assume a matrix for which the deviation of the dominant eigenvalue from 1 is barely within machine precision (which we can roughly assume to be of order 10^{-15}), we can estimate that $N > 7.13 \cdot 10^{18}$. We see that there are situation imaginable where our approach effectively does not converge. Exploring the circumstances under which this behaviour can be observed goes beyond the scope of this paper. However, one example can be easily constructed, where the convergence is bad.

Remember that the generator matrices we are dealing with have the form as the matrix $\mathbf{Q}$, depicted in (1). It is easy to see that $\mathbf{T} + \text{diag}(\underline{\mathbf{T}}_0)$ is a generator matrix that describes a CTMC, and let us assume that it is an irreducible one. Then the matrix $\mathbf{T} + \text{diag}(\underline{\mathbf{T}}_0)$ would have a zero eigenvalue. If we now assume that the rates in $\underline{\mathbf{T}}_0$ are very small compared to that in $\mathbf{T}$ (several orders of magnitude), then $\mathbf{T}$ would be very close to be a generator matrix of an irreducible CTMC, and therefore, $\mathbf{T}$ would have an eigenvalue very near to zero[5]. If we uniformise $\mathbf{T}$, *i.e.*, consider the matrix $\mathbf{P}_T = \mathbf{T}/q + \mathbf{I}$ for some properly chosen q, then $\mathbf{P}_T$ would have an eigenvalue very close to one. $\mathbf{P}_T$ is therefore an example for a probability matrix which would have bad convergence behaviour.

The above estimations are very rough, so usually the convergence will be better. We can not give a reasonable value for N in advance, since we do not know the dominant eigenvalue, and computing it in order to estimate N is much to costly for this purpose alone. However, since $F(n)\underline{\mathbf{A}}_0$ is positive for all n and does converge exponentially to zero, we can recognise the proper time when to break the summation quite easily.

[5] In the literature, such matrices are usually said to be *stiff*.

Complexity. This approach requires an explicit matrix inversion. The complexity of the solution of a system of linear equations is $\mathcal{O}(k^3)$, hence, a matrix inversion requires $\mathcal{O}(k^4)$ steps. On the other hand, the explicit computation of $\dot{\beta}'$ requires only two matrix-vector-multiplications per iteration step for the computation of $E(n)$ and $F(n)$. Therefore, the computation of $\dot{\beta}'$ is of order $\mathcal{O}(Nk^2 + k^4)$. Since N is unknown and depends on the eigenstructure of the considered matrix, we can **not** simplify this to $\mathcal{O}(k^4)$.

6.3 Implicit Inversion

Another way to compute $\underline{\dot{\beta}}'$ by means of Equation (13) is by successive solution of systems of linear equations: for each iteration step n, the vector $\underline{\alpha}$ is multiplied with $\mathbf{M}^{-(n+1)}$, where $\mathbf{M}$ is defined as in (16). For $n = 0$, we have to compute $\underline{x}^{(0)} = \underline{\alpha}\mathbf{M}^{-1}$, which is the solution of the system of linear equations

$$\underline{x}^{(0)}\mathbf{M} = \underline{\alpha}.$$

Then, for $n \geq 1$ we must compute $\underline{x}^{(n)} = \underline{\alpha}\mathbf{M}^{-(n+1)} = \underline{x}^{(n-1)}\mathbf{M}^{-1}$, which is the solution of the system of linear equations

$$\underline{x}^{(n)}\mathbf{M} = \underline{x}^{(n-1)}.$$

Hence, each step in the summation of (13) requires the solution of one system of linear equations, and

$$\underline{\dot{\beta}}' = \sum_{n=0}^{\infty} E(n)(\underline{x}^{(n+1)}\underline{\mathbf{A}}_0).$$

The complexity of this approach is then of order $\mathcal{O}(Nk^3 + Nk^2) = \mathcal{O}(Nk^3)$.

6.4 Conclusions

All the approaches we have described have the same convergence properties. However, the complexity of both approaches depends on the unknown N. This makes it difficult to make an a-priori choice. The question which approach is the most efficient must be determined empirically.

7 Comparing the Complexity

In this section we compare the worst-case complexity of our approach to compute $\mathsf{E}[X_1 \boxplus \cdots \boxplus X_n]$ with the worst-case complexity of more traditional approaches.

All traditional approaches, in which the overall state space is either explicitly or implicitly used, are based on the expression $\underline{\alpha}\mathbf{T}^{-1}\underline{1}$, where $(\alpha, \mathbf{T})$ is the representation of the phase-type distribution of $X_1 \boxplus \cdots \boxplus X_n$.

We assume n different absorbing CTMCs with representations $(\underline{\alpha}_i, \mathbf{A}_i)$, which describe the distributions of X_i, $i = 1, \ldots, n$. We assume that the generators have dimensions $k_i \times k_i$ and define $k = \max\{k_i\}$.

7.1 Kronecker Approach

We can assume a constant e which is the mean number of non-zero entries in the matrices $\dot{\mathbf{A}}_i$ *per row*. Hence, matrix $\dot{\mathbf{A}}_i$ has about $e \cdot k_i$ non-zero entries. We assume that the storage requirements of the matrices $\dot{\mathbf{A}}_i$ can be neglected.

Let $\dot{\mathbf{A}} = \bigoplus_{i=1}^n \dot{\mathbf{A}}_i$ and $\underline{\dot{\alpha}} = \bigotimes_{i=1}^n \underline{\dot{\alpha}}_i$. Then $\dot{\mathbf{A}}$ has about $e \cdot n$ non-zero entries per row and hence the number of non-zeros of $\dot{\mathbf{A}}$ is of order $\mathcal{O}(k^n)$.

The mean value $m = \mathsf{E}[X_1 \boxplus \cdots \boxplus X_n]$ of a random variable distributed according to $\dot{\mathbf{A}}$ can be computed as $m = -\underline{\alpha}\mathbf{A}^{-1}\underline{1}$. First a solution of $\underline{x}\mathbf{A} = \underline{\alpha}$ should be computed. Then $m = \underline{x}\,\underline{1}$. The computation of the mean value then reduces to the solution of a system of linear equations. If this is done numerically, for example by means of a Jacobi iteration, then one iteration step costs a number of operations that grows with the number of non-zero elements of $\dot{\mathbf{A}}$, *i.e.*, with $\mathcal{O}(k^n)$. Usually the number of iterations is negligible compared to the number of states (in [1], Bell and Haverkort report a number of about 4000 iterations for a (sparse) system of of 750 million equations). Hence, although the solution process might require only $\mathcal{O}(nk)$ memory (by exploiting the Kronecker structure), the time it requires grows exponentially with the number of random variables.

7.2 The MeanMax Approach

The new approach takes several steps:

1. Computing a mean value for all $\mathbf{A}_i$, *i.e.*, n times. If we assume complexity $\mathcal{O}(k_i^3)$ for case i, the overall effort is, in worst case, $\mathcal{O}(nk^3)$.
2. Computing starting vectors according to (13), either with explicit or implicit inversion. This has to be done $\frac{n(n-1)}{2}$ times (*cf.* MeanMax-Algorithm in Figure 1), *i.e.*, $\mathcal{O}(n^2)$.
3. From the considerations in Section 6 we see that the computation of the vector $\underline{\dot{\beta}}'$ depends on k and N. If we assume that the involved matrices are good-natured, it is fair to say that $N << k^n$ (several orders of magnitude). Then, in the worst case, we have to consider $\mathcal{O}(Nk^4)$.

The overall complexity of our approach is consequently of order $\mathcal{O}(nk^3 + n^2Nk^4)$.

7.3 Conclusions w.r.t. Complexity

We have assumed that the considered matrices are good-natured, *i.e.*, that they are not stiff, for example. If we assume that the considered matrices are not good-natured, then N can become very large, perhaps even $N > k^n$. This would suggest that in such cases the Kronecker approach would be superior to MeanMax. However, if we consider the case of stiff Markov chains, as described in Section 6.2, then probably also the convergence properties of the Kronecker approach would become very bad. The apparatus we have developed in this paper is not sufficient to permit a reliable comparison of the two approaches for such extreme cases.

For the good-natured case, however, we can conclude that the number of operations to compute the mean value m by means of MEANMAX is polynomial, whereas the effort to compute *even one* iteration step by means of the Kronecker approach is exponential in the number of the random variables. The memory consumption can be neglected for both approaches since it is never necessary to explicitly construct the global state space.

8 Conclusions

In this paper we have presented the technique MEANMAX to compute the mean value of the maximum of phase-type distributed random variables. MEANMAXs basic approach is to express the mean value of the maximum of the random variable $X_1, \ldots, X_n$, *i.e.*, $\mathsf{E}[X_1 \boxplus \cdots \boxplus X_n]$, as a sum of mean values of random variables $Y_1, \ldots, Y_n$, *i.e.*,$\mathsf{E}[X_1 \boxplus \cdots \boxplus X_n] = \mathsf{E}[Y_1] + \cdots + \mathsf{E}[Y_n]$. We have seen that the random variables Y_i can be represented by the same absorbing Markov chain as X_i, for $i = 1, \ldots, n$. Only the starting distribution corresponding to Y_i is different to that of X_i.

MEANMAX provides the efficient means to derive the appropriate starting distributions for the random variables Y_i for $i = 1, \ldots, n$. The actual mean value of the random variables Y_i can then be computed by standard measures.

The complexity to derive the starting distributions of the random variables Y_i is polynomial in the number of components and the size of considered Markov chains. MEANMAX has been used to efficiently solve semi-Markov chains that are composed by means of the $\diamond$ operator.

Acknowledgements

The authors would like to thank Joost–Pieter Katoen and Holger Hermanns for valuable suggestions. Special thanks are due to one of the anonymous reviewers, who proposed an interesting variant of Equation (13).

References

1. Alexander Bell and Boudewijn R. Haverkort. Serial and parallel out-of-core solution of linear systems arising from GSPNs. In Adrian Tentner, editor, *High Performance Computing (HPC 2001) — Grand Challenges in Computer Simulation*, pages 242–247, San Diego, CA, USA, April 2001. The Society for Modeling and Simulation International.
2. Henrik Bohnenkamp. *Compositional Solution of Stochastic Process Algebra Models.* PhD thesis, Department of Computer Science, Rheinisch-Westfälische Technische Hochschule Aachen, Germany, 2001. Preliminary Version.
3. Henrik C. Bohnenkamp and Boudewijn R. Haverkort. Semi-numerical solution of stochastic process algebra models. In Joost-Pieter Katoen, editor, *Proceedings of the 5th International AMAST Workshop, ARTS'99*, volume 1601 of *Lecture Notes in Computer Science*, pages 228–243. Springer-Verlag, May 1999.

4. Hendrik Brinksma. *On the design of extended LOTOS.* PhD thesis, University of Twente, The Netherlands, 1988.
5. M. Davio. Kronecker products and shuffle algebra. *IEEE Transactions on Computers*, C-30(2):116–125, February 1981.
6. F.R. Gantmacher. *Matrizentheorie.* Springer Verlag (translated from Russian; originally published in 1966), 1986.
7. Peter G. Harrison and Naresh M. Patel. *Performance Modelling of Communication Networks and Computer Architectures.* Addison-Wesley, 1993.
8. Holger Hermanns. *Interactive Markov Chains.* PhD thesis, Universität Erlangen-Nürnberg, Germany, 1998.
9. Ronald A. Howard. *Dynamic Probabilistic Systems*, volume 2: Semimarkov and Decision Processes. John Wiley & Sons, 1971.
10. Ronald A. Howard. *Dynamic Probabilistic Systems*, volume 1: Markov Models. John Wiley & Sons, 1971.
11. A. Jensen. Markoff chains as an aid in the study of Markoff processes. *Skand. Aktuarietidskrift*, 36:87–91, 1953.
12. Marcel F. Neuts. *Matrix-Geometric Solutions in Stochastic Models – An Algorithmic Approach.* Series in the Mathematical Sciences. Johns Hopkins University Press, Baltimore, 1981.
13. Robin A. Sahner. *A Hybrid, Combinatorial-Markov Method of solving Performance and Reliability Models.* PhD thesis, Duke University, Computer Science Department, 1986.
14. Robin A. Sahner, Kishor S. Trivedi, and Antonio Puliafito. *Performance and Reliability Analysis of Computer Systems. An Example-Based Approach Using the SHARPE Software Package.* Kluwer Academic Publishers, 1996.

Reduction and Refinement Strategies for Probabilistic Analysis

Pedro R. D'Argenio[1], Bertrand Jeannet[2], and
Henrik E. Jensen[3] and Kim G. Larsen[3]

[1] FaMAF, Universidad Nacional de Córdoba
Ciudad Universitaria. 5000 - Córdoba. Argentina
dargenio@mate.uncor.edu
[2] IRISA – INRIA
Campus de Beaulieu. F-35042 Rennes Cedex, France
Bertrand.Jeannet@irisa.fr
[3] BRICS - Aalborg University
Frederik Bajers vej 7-E. DK-9220 Aalborg, Denmark
ejersbo@least.dk, kgl@cs.auc.dk

Abstract. We report on new strategies for model checking quantitative reachability properties of Markov decision processes by successive refinements. In our approach, properties are analyzed on abstractions rather than directly on the given model. Such abstractions are expected to be significantly smaller than the original model, and may safely refute or accept the required property. Otherwise, the abstraction is refined and the process repeated. As the numerical analysis involved in settling the validity of the property is more costly than the refinement process, the method profits from applying such numerical analysis on smaller state spaces. The method is significantly enhanced by a number of novel strategies: a strategy for reducing the size of the numerical problems to be analyzed by identification of so-called *essential states*, and heuristic strategies for guiding the refinement process.

1 Introduction

Fully automatic verification of a given (traditionally) finite-state system with respect to a given temporal logic property is known as *model checking*. For finite state systems, model checking has reached a clear level of maturity as witnessed by a number of successful industrial cases (e.g. [18,11]). Model checking of finite state systems allows settlement of qualitative properties such as "the system will never reach an erroneous situation". However, it is often vital that additional quantitative properties are established in order for the system to be considered correct. Such properties include real-time requirements such as "a desired state will be reached within 105 seconds" and probabilistic properties of the type "a desired state will be reached with probability at least 99%". While real-time model checking tools have been subject to significant research efforts leading to mature tools such as UPPAAL [24] and Kronos [8], it was not until recently that attention was drawn to efficient tool implementations for probabilistic model checking

H. Hermanns and R. Segala (Eds.): PAPM-PROBMIV 2002, LNCS 2399, pp. 57–76, 2002.

despite theoretical studies carried out during the last decade [14,2,7,19,4, etc.]. In this paper we report on a significant technical improvement on a recent tool of ours [9] to model check probabilistic properties.

In our context systems are described in terms of Markov decision processes [29], also called probabilistic transition systems (PTS) or probabilistic automata. This model allows to combine probabilistic and non-deterministic steps providing a natural extension to traditional non-deterministic models. The choice of this model is partly due to the fact that it is closed under parallel composition (which facilitates modeling and compositional reasoning), but primarily because PTSs are amenable to abstractions. This is a key factor for the techniques introduced in this paper.

We focus on a restricted class of reachability properties. These properties allow to specify that the probability of reaching a particular final condition ϕ_f from any reachable state satisfying a given initial condition ϕ_i is smaller (or greater) than a given probability p *regardless* of how non-deterministic choices of the model are resolved. Though apparently restrictive, the use of test-automata allow the range of properties that can be specified to be broadened substantially.

Our method [9] is based on automatic abstraction and refinement. The basic idea is to use abstractions in order to reduce the high cost of the numerical analysis involved in computing the minimum and maximum reachability probabilities for PTSs. The abstractions considered are obtained via successive refinements, starting from an initial coarse partitioning of the state space derived from the property under study. For a given refinement the property is checked on the induced abstract model, hopefully settling the property. However, the verdict may be inconclusive, when threshold probability p happens to be between the calculated minimum and the maximum abstract probabilities. In this case, the abstraction is further refined and the property checked again. This process is successively repeated until either the property is settled, or no further refinement is possible. To efficiently store the state space, perform abstractions and process the refinement steps, we use BDDs and MTBDDs (or ADDs) [9,12,3].

The performance of our method depends intimately on the efficiency of the numerical analysis performed on abstract models as well as the choice of refinements and initial partitioning. As main contributions of this paper we provide strategies for reducing the size of the numerical problems to be analyzed (by identification of so-called *essential* states) as well as strategies for guiding the refinement process. Finally, a number of comparative experimental results demonstrate the effectiveness of our method and suggested strategies.

The paper is organized as follows: Sections 2, 3, and 4 recall the theoretical foundation of the method and the implemented tool, which has been presented in [9]. Sections 5 and 6 present our reduction and refinement strategies. Section 7 provides details on implementation and experimental results. Section 8 concludes.

Related Work. Other quantitative model checkers have been developed. The tool ProbVerus [15] allows to check the validity of a PCTL formula [14] on a (discrete time) Markov chain. Prism [22,28] is a quantitative model checker for PCTL formulas on (discrete time) Markov decision processes, i.e., nondeterminism is inherent to the model, and on continuous-time Markov chains. Like PRISM, we also do model-checking on Markov decision processes, but we restrict to a particular kind of PCTL formula. Another quantitative model checker is E ⊢ MC^2 [17], which model checks probabilistic timed properties on continuous-time Markov chains.

2 Probabilistic Transition Systems

Probabilistic transition systems (PTS for short) generalize the well-known transition systems with probabilistic information. In a PTS, a transition does not lead to a single state but to a probability space whose sample space is a set of states. The model we define is widely used (see, e.g. [30,7,21]) and is also known as Markov decision processes [29].

Let Distr5Ω0 denote the set of all probability distributions over the sample space Ω.

Definition 1 (Probabilistic Transition Systems). *A* probabilistic transition system *(PTS for short) is a structure* $T = 5S, \rightarrow 0$ *where* S *is a set of states, and* $\rightarrow \subseteq S \times$ Distr5S0 *is the transition relation. We write* $s \rightarrow \pi$ *for* $5s, \pi 0 \in \rightarrow$. *A PTS is said to be a* fully probabilistic transition system *(FPTS for short) if* $5s \rightarrow \pi \wedge s \rightarrow \rho 0 \Rightarrow \pi = \rho$. *A* rooted PTS (resp. FPTS) $5T, s_0 0$ *is a PTS (resp. FPTS) equipped with an initial state* $s_0 \in S$. *A PTS may be equipped with a proposition assignment* $p - S \rightarrow \mathcal{P}59\,\mathrm{P}0$, *where* 9 P *is a finite set of atoms and* $\mathcal{P}59\,\mathrm{P}0$ *the set of propositional formula on* 9 P. *We define* $\models \subseteq S \times \mathcal{P}59\,\mathrm{P}0$ *by* $s \models g$ *iff* $p5s0 \Rightarrow g$ *is a tautology.*

We write $s \rightarrow$ whenever there is a π such that $s \rightarrow \pi$; otherwise, we write $s \not\rightarrow$. We let supp$5\pi 0 = \{s' \mid \pi 5s'0 > \mathrm{D}\}$. We call s a *sink state* if $s \not\rightarrow$.

Figure 1 shows (a symbolic representation) of an example PTS $T = 5S, \rightarrow 0$ where $S = \{\mathsf{i}, \mathsf{f}\} \times \{\mathrm{D}, \mathrm{L}, \mathrm{z}\}$. An example transition is $5\mathsf{i}, \mathrm{D}0 \rightarrow \{5\mathsf{i}, \mathrm{L}0 \mapsto \mathrm{DW}, 5\mathsf{f}, \mathrm{D}0 \mapsto \mathrm{DzI}\}$. T is nondeterministic as exhibited by the additional transition $5\mathsf{i}, \mathrm{D}0 \rightarrow \{5\mathsf{i}, \mathrm{L}0 \mapsto \mathrm{DzI}, 5\mathsf{f}, \mathrm{D}0 \mapsto \mathrm{DW}\}$ originating from state $5\mathsf{i}, \mathrm{D}0$. We can assume that T is equipped with a proposition assignment p such that $p5l, v0 = 5l \wedge x = v0$.

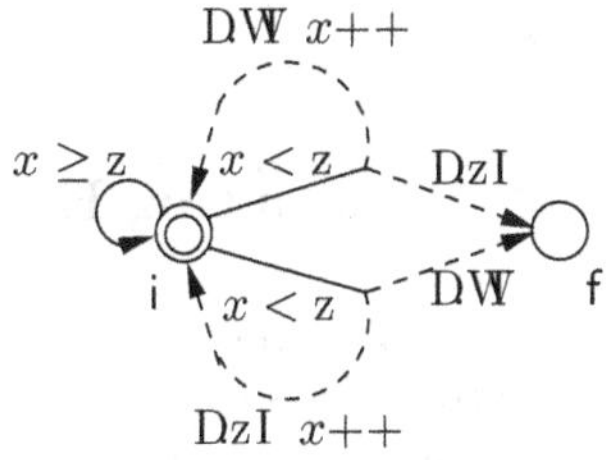

Fig. 1. A PTS

Let $T = 5S, \rightarrow 0$ be a PTS. A *simple path in* T is a finite sequence of states $\sigma = s_0 s_1 s_2 \ldots s_n$, where for each $\mathrm{D} \leq i < n$ there exists $\pi_i \in$ Distr5S0 such that $s_i \rightarrow \pi_i$ and $\pi_i 5 s_{i+1} 0 > \mathrm{D}$. Let $\sigma 5i0$ denote the state in the i-th position. Let $|\sigma|$ be the length of σ and let *first*$5\sigma 0 = \sigma 5\mathrm{L}0$ and *last*$5\sigma 0 = \sigma 5|\sigma|0$. A *simple path starting from* $s \in S$ is a simple path σ with $\sigma 5\mathrm{L}0 = s$. A state t is *reachable* from another state s in T if there is a simple path in T with $s =$ *first*$5\sigma 0$ and $t =$ *last*$5\sigma 0$.

A *full path in* T is a sequence of states σ being either a simple path with $last5\sigma0 \not\rightarrow$, or an infinite sequence. We denote by $s\text{-}paths5T0$ and $f\text{-}paths5T0$ the sets of simple paths and fullpaths in T, and by $s\text{-}paths5T, s0$ and $f\text{-}paths5T, s0$ the sets of simple paths and fullpaths in T starting from s. Let $\text{reach}5T, s0$ denote the set of all states reachable from s in T.

We now define a probability measure on the full paths of a FPTS F. For any simple path $\sigma \in s\text{-}paths5F0$, define $\sigma^{\uparrow} = \{\pi \in f\text{-}paths5F0 \mid \sigma \leq \pi\}$ where $\leq$ is the classical prefix order on sequences. Let $\mathcal{F}5F0$ be the smallest σ-field on $f\text{-}paths5F0$ which contains $\sigma^{\uparrow}$ for each $\sigma \in s\text{-}paths5F0$. Then for any state s of F, $\mathsf{P}_{F,s}$ is the uniquely defined probability measure on $\mathcal{F}5F0$ such that for any $\sigma = s_0 s_1 \dots s_n \in s\text{-}paths5F0$ such that $s_i \rightarrow_F \pi_i$ for all i, $\text{D}\leq i < n$:

$$\mathsf{P}_{F,s}5\sigma^{\uparrow}0 \quad \triangleq \quad \textbf{if } 5s = s_0 0 \textbf{ then } \pi_0 5s_1 0 \cdot \pi_1 5s_2 0 \cdot \ldots \cdot \pi_{n-1} 5s_n 0 \textbf{ else } \text{D}$$

We will write $\mathsf{P}_{F,s}5\sigma0$ to denote $\mathsf{P}_{F,s}5\sigma^{\uparrow}0$. Intuitively, $\mathsf{P}_{F,s}5\sigma0$ is the probability of σ in F starting from s.

Any given PTS T defines a set of *probabilistic executions*, each one obtained by iteratively scheduling one of the possible post-state distributions from each pre-state, starting from a given state $s_0 \in S$. Notice that the same state s of T may occur more than once during a probabilistic execution and each time a different distribution from s may be scheduled. In order to distinguish such occurrences we include in all states s of a probabilistic execution the past history of s which is the unique path leading from the start state to s. Thus, a probabilistic execution essentially defines a finite or infinite *tree*.

Definition 2 (Probabilistic Execution). *A* probabilistic execution *of a PTS* $T = 5S, \rightarrow_T 0$ *is a FPTS* $F = 5s\text{-}paths5T0, \rightarrow_F 0$ *such that* $5q \rightarrow_F \rho0 \Leftrightarrow 5\exists\pi - last5q0 \rightarrow_T \pi \wedge \forall s \in S - \rho5qs0 = \pi5s00$

We denote by $execs5T, s_0 0$ the set of all probabilistic executions of T rooted in s_0.

3 Computing Extremum Probabilities

For a given rooted PTS $5T, s_0 0$ we are interested in the extremum probabilities of reaching some final condition from a given initial condition. For any given formula $\phi \in \mathcal{P}59\ \text{P}0$ we define the set of all minimal simple paths of T that end in a state satisfying condition ϕ as:

$$\Sigma_{\phi}^{T} \triangleq \{\sigma \in s\text{-}paths5T0 \mid last5\sigma0 \models_T \phi \wedge \forall i, \text{D}< i < |\sigma| - \sigma5i0 \models_T \neg\phi\}$$

By recording history information in states, the above set characterizes uniquely a set of simple paths of probabilistic executions of T. We also use Σ_{ϕ}^{T} to denote this alternative characterization. It should be clear from the context which alternative is used. We omit T in the notation whenever clear from context.

Definition 3 (Extremum Probabilities). *The* minimum *and* maximum probabilities *of reaching a final condition ϕ_f from an initial condition ϕ_i in a rooted PTS $5T, s_0 0$ equipped with a proposition assignment are defined respectively by*

$$\mathsf{P}^{\inf}_{T,s_0} 5\phi_i, \phi_f 0 \triangleq \inf \left\{ \mathsf{P}_{F,s} 5\Sigma_{\phi_f} 0 \mid s \in \text{reach}5T, s_0 0 \wedge \right. \\ \left. s \models \phi_i \wedge 5F, s0 \in execs5T, s0 \right\} \tag{1}$$

$$\mathsf{P}^{\sup}_{T,s_0} 5\phi_i, \phi_f 0 \triangleq \sup \left\{ \mathsf{P}_{F,s} 5\Sigma_{\phi_f} 0 \mid s \in \text{reach}5T, s_0 0 \wedge \right. \\ \left. s \models \phi_i \wedge 5F, s0 \in execs5T, s0 \right\} \tag{2}$$

We talk of an extremum probability *to refer to either the infimum or supremum probability.*

We use the shorthand $\mathsf{P}^{\inf} 5s0$ and $\mathsf{P}^{\sup} 5s0$ for $\mathsf{P}^{\inf}_{T,s_0} 5s = s_0, \phi_f 0$ and $\mathsf{P}^{\sup}_{T,s_0} 5s = s_0, \phi_f 0$, respectively. We denote by I and F the sets of states satisfying ϕ_i and ϕ_f, respectively. Our aim is to efficiently compute $\mathsf{P}^{\inf} 5I0 \triangleq \inf_{s \in I} \mathsf{P}^{\inf} 5s0$ and $\mathsf{P}^{\sup} 5F0 \triangleq \sup_{s \in I} \mathsf{P}^{\sup} 5s0$.

Consider again the PTS T of Figure 1. Take the initial condition $\phi_i = 5\mathsf{i} \wedge x = \mathrm{D}0$ (which correspond to the initial state) and the final condition $\phi_f = \mathsf{f}$. It is easy to see that $\mathsf{P}^{\inf} 5\phi_i, \phi_f 0$ is obtained by always resolving the nondeterminism from state i in favor of the upper transition. Thus, $\mathsf{P}^{\inf} 5\phi_i, \phi_f 0 = \mathrm{DzI} + 5\mathrm{DW} \cdot \mathrm{DzI}\,0 = \mathrm{D43W}$. Analogously, $\mathsf{P}^{\sup} 5\phi_i, \phi_f 0$ is obtained by always resolving in favor of the lower transition and thus $\mathsf{P}^{\sup} 5\phi_i, \phi_f 0 = \mathrm{DW} + 5\mathrm{DzI} \cdot \mathrm{DW}0 = \mathrm{DJ3W}$.

The equations 1 and 2 of definition 3 define extremum probabilities, but do not provide an effective way of computing them. However, it is well known [7,5] that $\mathsf{P}^{\inf}$ and $\mathsf{P}^{\sup}$ can be characterized as the least fixpoints of operators $F^{\inf}, F^{\sup} - 5S \to [\mathrm{D}, \mathrm{L}]0 \to 5S \to [\mathrm{D}, \mathrm{L}]0$ defined as follows. If $s \in F$ then $F^{\inf} 5f05s0 = F^{\sup} 5f05s0 = \mathrm{L}$. If $s \notin F$ then

$$F^{\inf} 5f05s0 = \min_{s \to \pi} \sum_{s' \in S} \pi 5s'0 \cdot f5s'0 \quad \text{and} \quad F^{\sup} 5f05s0 = \max_{s \to \pi} \sum_{s' \in S} \pi 5s'0 \cdot f5s'0 \tag{3}$$

Based on the above equations, two methods have been explored to compute $\mathsf{P}^{\inf} 5s0$ and $\mathsf{P}^{\sup} 5s0$. One can either compute the least fixpoints by iterative methods, or the equations can be transformed into a linear optimization problem that can be solved using classical techniques of linear programming. In this work we choose the linear programming method.

We use a standard precomputation of certain sets of system states in order to simplify the system before applying linear programming techniques. These sets are: the set of all reachable states *Reach*, and for each $p \in \{\mathrm{D}, \mathrm{L}\}$ the set of states having infimum (resp. supremum) probability p of reaching ϕ_f. These latter sets of states are denoted $\mathsf{P}^{\inf}_{=0}$, $\mathsf{P}^{\inf}_{=1}$, $\mathsf{P}^{\sup}_{=0}$, and $\mathsf{P}^{\sup}_{=1}$, respectively. All of the above sets can be computed using discrete fixpoint analysis [10] on a boolean abstraction of the system.

Based on the above precomputations our linear programming problems for computing $\mathsf{P}^{\mathrm{inf}}$ and $\mathsf{P}^{\mathrm{sup}}$ become as follows:

$$\text{maximize } \mathsf{P}^{\mathrm{inf}} \text{ under the constraints}$$
$$\begin{cases} \mathsf{P}^{\mathrm{inf}} \leq \mathsf{P}^{\mathrm{inf}}5s0, & s \in I \\ \mathsf{P}^{\mathrm{inf}}5s0 = \mathrm{D} & s \in \mathsf{P}^{\mathrm{sup}}_{=0} \\ \mathsf{P}^{\mathrm{inf}}5s0 = \mathrm{L}, & s \in 5F \cup \mathsf{P}^{\mathrm{inf}}_{=1}0 \\ \mathsf{P}^{\mathrm{inf}}5s0 \leq \sum_{s' \in S} \pi 5s'0 \cdot \mathsf{P}^{\mathrm{inf}}5s'0, & s \to \pi, s \in S \setminus 5\mathsf{P}^{\mathrm{sup}}_{=0} \cup \mathsf{P}^{\mathrm{inf}}_{=0} \cup F0 \end{cases} \tag{4}$$

$$\text{minimize } \mathsf{P}^{\mathrm{sup}} \text{ under the constraints}$$
$$\begin{cases} \mathsf{P}^{\mathrm{sup}} \geq \mathsf{P}^{\mathrm{sup}}5s0, & s \in I \\ \mathsf{P}^{\mathrm{sup}}5s0 = \mathrm{D} & s \in \mathsf{P}^{\mathrm{sup}}_{=0} \\ \mathsf{P}^{\mathrm{sup}}5s0 = \mathrm{L}, & s \in F \cup \mathsf{P}^{\mathrm{sup}}_{=1} \\ \mathsf{P}^{\mathrm{sup}}5s0 \geq \sum_{s' \in S} \pi 5s'0 \cdot \mathsf{P}^{\mathrm{sup}}5s'0, & s \to \pi, s \in S \setminus 5\mathsf{P}^{\mathrm{sup}}_{=0} \cup \mathsf{P}^{\mathrm{sup}}_{=1} \cup F0 \end{cases} \tag{5}$$

4 Simulations and Partitioning

Probabilistic simulation [20,30] is central to state the correctness of the abstraction technique proposed in this paper. For any $\delta \in \mathsf{Distr}5S \times S0$, $s \in S$ and $X \subseteq S$, $\delta 5s, X0$ and $\delta 5X, s0$ will denote resp. $\sum_{x \in X} \delta 5s, x0$ and $\sum_{x \in X} \delta 5x, s0$.

Definition 4 (Simulation). *Let $C \subseteq S \times S$ be a relation on states defining a* discrimination criterion. *A relation $R \subseteq S \times S$ is a C-*(probabilistic) simulation *if, whenever sRt,*

1. *$5s, t0 \in C$, and*
2. *if $s \to \pi$, there exist ρ such that $t \to \rho$ and $\pi \sqsubseteq_R \rho$.*

*where $\pi \sqsubseteq_R \rho$ if there is $\delta \in \mathsf{Distr}5S \times S0$ such that for all $s, t \in S$, (i) $\pi 5s0 = \delta 5s, S0$, (ii) $\rho 5t0 = \delta 5S, t0$, and (iii) $\delta 5s, t0 > \mathrm{D} \Rightarrow sRt$. s is C-*simulated *by t, notation $s \preceq_C t$, if there is a C-simulation R with sRt.*

Our interest is to check when a PTS reaches a goal ϕ_f starting from any state satisfying some initial condition ϕ_i ($\phi_i, \phi_f \in \mathcal{P}59\,\mathrm{P}0$). Let C_{ϕ_i,ϕ_f} be the discriminating criterion defined by

$$5s, t0 \in C_{\phi_i,\phi_f} \iff 5s \models \phi_f \Leftrightarrow t \models \phi_f 0 \wedge 5s \models \phi_i \Leftrightarrow t \models \phi_i 0$$

We write only C whenever ϕ_i and ϕ_f are clear from the context. Notice that C is equivalence relation. Simulation $\preceq_C$ provides a sufficient condition for preservation of extremum probabilities, as made precise by the following theorem.

Theorem 1. *Let $5T_1, s_0^1 0$ and $5T_2, s_0^2 0$ be two rooted PTSs such that none of them contains a sink state, and let $C = C_{\phi_i,\phi_f}$. Then $5T_1, s_0^1 0 \preceq_C 5T_2, s_0^2 0$ implies $\mathsf{P}^{\mathrm{sup}}_{T_1,s_0^1} 5\phi_i, \phi_f 0 \leq \mathsf{P}^{\mathrm{sup}}_{T_2,s_0^2} 5\phi_i, \phi_f 0$ and $\mathsf{P}^{\mathrm{inf}}_{T_1,s_0^1} 5\phi_i, \phi_f 0 \geq \mathsf{P}^{\mathrm{inf}}_{T_2,s_0^2} 5\phi_i, \phi_f 0$.*

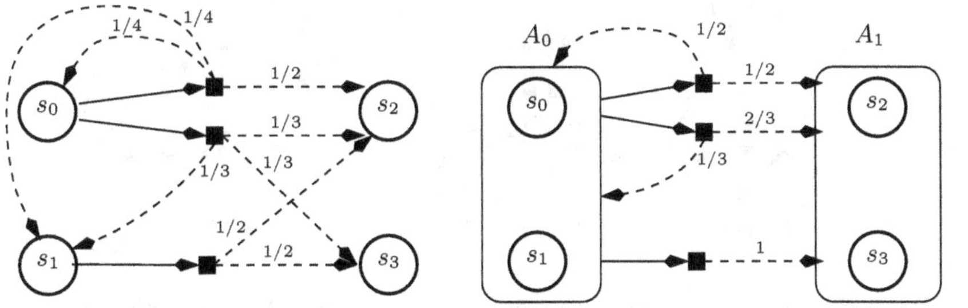

(a) (b)

Fig. 2. A PTS and its quotient by the partition $\mathcal{A} = \{\{s_0, s_1\}, \{s_2, s_3\}\}$

The requirement that every state has a transition is not really harmful as each sink state can always be completed with a self-looping transition without affecting the properties of our interest on the original PTS.

We can abstract a PTS by partitioning its state space, and any such partitioning will induce an abstract PTS which will simulate the original (concrete) one. As a result extremum properties will be preserved by the abstract system.

Definition 5 (Quotient PTS). *Let $T = 5S, \to_T 0$ a PTS equipped with the proposition assignment p. Let $\mathcal{A} = 5A_k0_{k\in K}$ be a partition of S. The* quotient PTS *according to $\mathcal{A}$ is the PTS $T/\mathcal{A} = 5\mathcal{A}, \to_{\mathcal{A}}, p/\mathcal{A}0$, where*

1. $A \to_{\mathcal{A}} \pi/\mathcal{A} \Leftrightarrow \exists s \in A - s \to \pi \wedge \forall A' \in \mathcal{A} - 5\pi/\mathcal{A}05A'0 \triangleq \sum_{s' \in A} \pi 5s'0$, *and*
2. $p/\mathcal{A}5A0 \triangleq \bigvee_{s \in A} p5s0$.

For a rooted PTS $5T, s_00$, its quotient is given by $5T, s_00/\mathcal{A} \triangleq 5T/\mathcal{A}, A0$ provided $s_0 \in A \in \mathcal{A}$.

Figure 2 gives an example of a quotient PTS. In the following we state the formal relationships between abstraction by partitioning and simulation. For any two partitions $\mathcal{A}$ and $\mathcal{B}$ of the same set, define $\mathcal{A} \leq \mathcal{B} \Leftrightarrow \forall A \in \mathcal{A}.\ \exists B \in \mathcal{B}.\ A \subseteq B$.

Theorem 2. *Let $5T, s_00$ be a rooted PTS with a set of states S and let C be an equivalence relation defining a partition $\mathcal{A}$ of S. Then for any partition $\mathcal{B}$ of S such that $\mathcal{B} \leq \mathcal{A}$, $5T, s_00/\mathcal{B} \preceq_C 5T, s_00/\mathcal{A}$*

Notice the special case of the theorem where $\mathcal{B}$ partitions S into singleton sets. In this case $5T, s_00/\mathcal{B}$ is isomorphic to $5T, s_00$. The following corollary states the relationship of abstraction by partitioning and preservation of extremum probabilities.

Corollary 1. *Let $5T, s_00$ be a rooted PTS equipped with a proposition assignment. Let ϕ_i and ϕ_f be the initial and final conditions. Let C be the equivalence relation $C = C_{\phi_i,\phi_f}$ defining a partition $\mathcal{C}$ of S. Then for any two partitions $\mathcal{A}$ and $\mathcal{B}$ such that $\mathcal{B} \leq \mathcal{A} \leq \mathcal{C}$,*

$$\mathsf{P}^{\sup}_{(T,s_0)/\mathcal{B}} 5\phi_i, \phi_f 0 \leq \mathsf{P}^{\sup}_{(T,s_0)/\mathcal{A}} 5\phi_i, \phi_f 0 \quad \textit{and} \quad \mathsf{P}^{\inf}_{(T,s_0)/\mathcal{B}} 5\phi_i, \phi_f 0 \geq \mathsf{P}^{\inf}_{(T,s_0)/\mathcal{A}} 5\phi_i, \phi_f 0$$

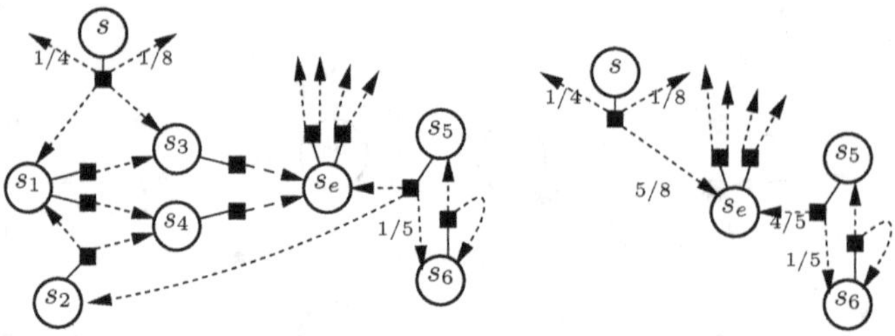

(a) Initial situation (b) After reduction

Fig. 3. Example of an essential state

5 Reduction Strategies

In this section we present new reduction strategies, that allow for simplifications of the linear programming problem characterizing the extremum probabilities of a PTS. First, we describe some optimizations which can be applied to the obtained linear programming problem. Then, we describe a new reduction technique which can be applied to a PTS before generating the corresponding linear programming problem. This reduction removes all but a particular type of essential states.

Our reduction techniques are orthogonal to our abstraction and refinement techniques and are therefore generally applicable as preprocessing steps for computing extremum probabilities of any PTS.

Optimizing Linear Programming Problems. Some optimizations are possible before solving an obtained linear programming problem. First, it is worth substituting the constant values Dor L associated to special states. Notice that this a very simple case of Gaussian elimination and the substitution can make some inequations redundant; such inequations can be removed. Last and most important, when we obtain a single inequation of the form $\mathsf{P}^{\mathrm{inf}}5s0 \geq \ldots$ or $\mathsf{P}^{\mathrm{sup}}5s0 \leq \ldots$, we can replace the inequality by an equality, then perform Gaussian elimination, and finally remove new redundant inequations. Such transformations reduce the number of both variables and constraints.

Abstracting to Essential States. Suppose that a fragment of a PTS looks like the one depicted in Fig. 3(a). All probabilistic paths starting from s_1, s_2, s_3, s_4 are leading to the state s_e with probability 1 in a finite number of steps and therefore $\mathsf{P}^{\mathrm{inf}}5s_i0 = \mathsf{P}^{\mathrm{inf}}5s_e0$ and $\mathsf{P}^{\mathrm{sup}}5s_i0 = \mathsf{P}^{\mathrm{sup}}5s_e0$, for $i = \mathrm{L}, \mathrm{z}, 3, 4$. Thus, we could reduce the system by representing all of the above states via the single state s_e, and in addition merge (add) the probabilities for any distribution to enter the states represented by s_e. This analysis may at first seem quite identical to performing Gaussian elimination on the induced linear programming problem. However, this is not completely true. Besides reducing the size of the PTS, our analysis can also remove some non-determinism in the system, thus allowing to replace inequations by equations in the corresponding linear programming problem. In the figure on the right, we have $\mathsf{P}^{\mathrm{inf}}5s_1 0 \leq \mathsf{P}^{\mathrm{inf}}5s_3 0 = \mathsf{P}^{\mathrm{inf}}5s_e 0$ and $\mathsf{P}^{\mathrm{inf}}5s_2 0 \leq \mathsf{P}^{\mathrm{inf}}5s_4 0 = \mathsf{P}^{\mathrm{inf}}5s_e 0$. Gaussian elimination would remove $\mathsf{P}^{\mathrm{inf}}5s_3 0$ and $\mathsf{P}^{\mathrm{inf}}5s_4 0$, but would not replace the inequations on $\mathsf{P}^{\mathrm{inf}}5s_1 0$ and $\mathsf{P}^{\mathrm{inf}}5s_4 0$ by equations, thus limiting further eliminations.

Consider states s_5 and s_6 on Fig. 3(a). They also agree with state s_e on their infimum and supremum probabilities. However, due to the loop between s_5 and s_6, these properties appear only when taking into account infinite paths. The corresponding analysis is thus more expensive than the one in which only finite paths needs consideration. We consider only the analysis based on finite paths.

We now formalize the intuitive ideas of the previous paragraph. Let $T = 5S, \to 0$ be a PTS equipped with a set of final states F that are supposed to be sink. We define a *domination relation* $\preceq \subseteq S \times S$ with the intuition that $s_1 \preceq s_2$ whenever all probabilistic paths starting from s_1 will pass through s_2 with probability L. This intuitive semantic definition can be characterized as a least fixpoint as follows.

Definition 6 (Domination Relation, Essential States, and Domination Equivalence). *Let $\preceq$ be the smallest relation satisfying the following. For all $s \in S$, $s \preceq s$, and for all $s, t \in S$ s.t. $s \neq t$,*

$$s \preceq t \;\Leftarrow\; 5s \to 0 \wedge \forall \pi - 5s \to \pi \Rightarrow 5\forall s' \in \mathrm{supp} 5\pi 0 - s' \preceq t 00 \tag{6}$$

The relation $\preceq$ can be shown to be a partial order under the condition all states in $\mathsf{P}^{\mathrm{sup}}_{=0}$ have been removed. States that are maximal with respect to $\preceq$ are called essential states. *We let $\sim$ denote the relation induced from $\preceq$ as follows:*

$$s_1 \sim s_2 \;\Leftrightarrow\; \exists t - s_1 \preceq t \wedge s_2 \preceq t \tag{7}$$

In other words, states related by $\sim$ are dominated by the same state. Relation $\sim$ is an equivalence relation.

We will reduce a concrete PTS to an abstract one containing only essential states, and such that the abstract PTS preserves the exact extremum probabilities of the concrete PTS. We call the abstract PTS an *essential state abstraction*. To compute the essential state abstraction, we could try to use the quotient construction of Definition 5 with respect to the partition corresponding to the equivalence relation $\sim$. However, this will not guarantee an exact abstraction. Intuitively, the transitions linking states inside an equivalence class will generate loops on the abstract graph, and thus $\mathsf{P}^{\mathrm{inf}}$ may be lower than on the initial system. For states inside a class with a looping transition $\mathsf{P}^{\mathrm{inf}}$ will be D

However, by definition of $\sim$, we know that each equivalence class has a unique essential state. These essential states can be taken as abstract states representing all the equivalence classes. Also, any concrete distribution can be abstracted by merging (adding) all probabilities for states in the same equivalence class onto the essential state representing this class. The following definition formalizes this construction.

Definition 7 (Essential State Abstraction of a PTS). *Let $T = 5S, \to 0$ be a PTS equipped with a set of final states F supposed to be sink. Let $\mathcal{E}$ be the partition of S associated to $\sim$, and let $E \subseteq S$ be the set of essential states. For any $e \in E$ let $[e]$ denote the equivalence class of e in $\mathcal{E}$. The* essential PTS of T *is the PTS $T_\epsilon = 5E, \to_\epsilon 0$ with final condition $F \subseteq E$, and where $\to_\epsilon$ is defined by: $e \to_\epsilon \pi/\mathcal{E} \Leftrightarrow e \to \pi$ where $\pi/\mathcal{E} \in \mathsf{Distr}5E0$ is the distribution st. $\pi/\mathcal{E}5e0 = \sum_{s\in[e]} \pi 5s0$.*

Proposition 1 (Preservation of Extremum Probabilities). *Let $T = 5S, \to 0$ be a PTS equipped with a set of final states F that are sink, and let $T_\epsilon = 5E, \to_\epsilon 0$ be its essential abstraction. For $s \in S$ dominated by $e \in E$, we have:*

$$\mathsf{P}_T^{\mathrm{inf}} 5s, F0 = \mathsf{P}_{T_\epsilon}^{\mathrm{inf}} 5e, F0 \quad \textit{and} \quad \mathsf{P}_T^{\mathrm{sup}} 5s, F0 = \mathsf{P}_{T_\epsilon}^{\mathrm{sup}} 5e, F0$$

Algorithm and Complexity. Given a PTS $T = 5S, \to 0$, we want to compute the essential elements of T, as well as their associated set of dominated elements. This is done by using the fixpoint definition of $\preceq$, and using the UNION-FIND data structure, with path compression and weighted union techniques [27]. The basic UNION-FIND data structure is an inverted tree (Fig. 4) where sons are pointing to their father and where the root represents both itself and the set of the nodes pointing to it. Figure 4 depicts the situation of the algorithm applied to the PTS of Fig. 3(a) after two steps. At that point we know that s_1, s_3, s_4 are dominated by s_e, and in the next step, s_2 will also be incorporated to the elements dominated by s_e, because all paths starting from s_2 lead to states that are dominated by s_e.

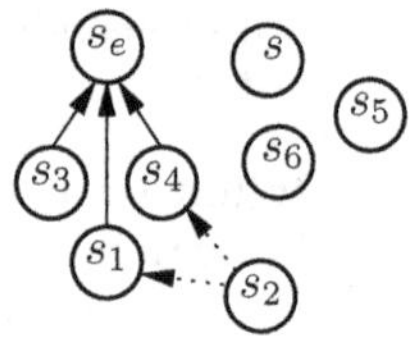

Fig. 4.

The only modification we have to bring to this algorithm is that we need to maintain, together with each tree, the essential state dominating all elements of the tree. Indeed, the root of a tree is not necessarily the essential state dominating all the elements of the tree, when weighted union is performed. According to [27], if m FIND operations and n UNION operations are performed, the global time to carry out these operations is $\mathcal{O}5n + m\alpha 5m, n00$, where α is the inverse of the Ackerman function. Let us evaluate n and m in the algorithm sketched above. We have obviously $n \leq |S|$. For m, at each step, we do in the worst case $N \cdot D \cdot d$ FIND operations, where N is the number of equivalent classes, D is the maximum number of distributions outgoing from a state, and d the maximum size of their support. The maximum number of steps is $|S|$, and at each step we have $N \leq |S|$, so $m \leq |S|^2 \cdot D \cdot d$. As a result, an upper bound of the complexity of the algorithm is $\mathcal{O}5|S| + |S|^2 \cdot D \cdot d \cdot \alpha 5|S|^2 \cdot D \cdot d, |S|00$. Notice that in practice the number of steps is likely to be much smaller than $|S|$.

6 Refinement Strategies

Our method of verification via abstraction follows the classical partition refinement scheme and it has been presented earlier in [9]. It starts with a coarse abstraction of the concrete system under investigation. If the analysis of the abstract system allows to conclude on the properties under investigation then the verification process is finished. Otherwise, a partition refinement step is performed in order to obtain more precise information. This process is iterated up to success or until all classes of the partition are stable.

This section reports on new results of ours to enhance the method of [9] with new strategies for constructing an initial partition, for refining individual classes, and for choosing intelligently the classes to refine.

Initial Partition. Before constructing the initial partition, we use the precomputations described in section 3 to simplify the system as follows: (1) We first restrict to the reachable (from the root) state space, since the definition of $\mathsf{P}^{\mathrm{inf}}$ and $\mathsf{P}^{\mathrm{sup}}$ involves only reachable states. (2) We then augment the set of final states F with $\mathsf{P}^{\mathrm{inf}}_{=1}$, we compute $\mathsf{P}^{\mathrm{sup}}_{=0}$ and we make all these states sink. (3) Finally, we restrict the state space to states reachable from the initial ones. Each of these transformations preserves extremum probabilities and can give large reductions in the state space.

Let $T = 5S, \to 0$ be a PTS on which the above preprocessing has been performed. Let I and F be the sets of states satisfying ϕ_i and ϕ_f. Since we want to compute safe approximations of $\mathsf{P}^{\mathrm{inf}}_{T,s_0} 5\phi_i, \phi_f 0$ and $\mathsf{P}^{\mathrm{sup}}_{T,s_0} 5\phi_i, \phi_f 0$, our initial partition should be at least a refinement of the partition $\{I,\ F,\ S \setminus 5I \cup F0\}$. In addition, we distinguish the sets of states $\mathsf{P}^{\mathrm{sup}}_{=0}$, $\mathsf{P}^{\mathrm{inf}}_{=0}$ and $\mathsf{P}^{\mathrm{sup}}_{=1}$ (the states $\mathsf{P}^{\mathrm{inf}}_{=1}$ have already been merged with the final states). This is wise since we already have these sets of states from the precomputation, and since joining these states with other states will inevitably lead to a loss of precision in the extremum probabilities of the latter ones. In the case that the above sets are not disjoint, we make a proper partition.

We further refine the obtained partition according to the explicit control structure of our PTS, such that only the values of variables are abstracted. In our tool, this default behavior can be user modified by specifying the variables and the processes which should not be abstracted.

Refining Individual Classes. Our refinement method tries to stabilize classes in the standard way by splitting classes based on different futures for their contained sets of states. However, we allow for a more strategic splitting of a class than simply splitting it with respect to all its outgoing transitions.

Let T be the concrete PTS and let $T/\mathcal{A} = 5\mathcal{A}, \to_{\mathcal{A}} 0$ be its quotient in the current refinement step. For any abstract transition $t - A \to_{\mathcal{A}} \Pi$ where $A \in \mathcal{A}$ and $\Pi \in \mathsf{Distr}5\mathcal{A}0$, we define the *guard* of t as: $g5t0 = \{s \in A \mid \exists s \to_T \pi.\ \forall i.\ \pi 5A_i 0 = \Pi 5A_i 0\}$ where $\pi 5A0 = \sum_{s \in A} \pi 5s0$. Now, if $g5t0 \neq \emptyset$ and $g5t0 \neq A$, then the class A can be split into two new classes: $g5t0$ and $A \backslash g5t0$.

For a given class A our method allows different strategies for choosing the transitions $A \to_{\mathcal{A}} \Pi$ to serve as basis for splitting A.

1. *Binary splitting:* This means that we do not split A with respect to all outgoing transitions, but we choose one particular transition. Candidates are transitions $A \to_{\mathcal{A}} \Pi$ where

(a) $\Pi 5A0 =$ L: Splitting with respect to these transitions is an attempt to raise the abstract infimum probabilities. If $\Pi 5A0 =$ L then at least class A will have infimum probability D Transitions of this type are denoted L-transitions (L for looping).
(b) $\exists B \neq A$. $\Pi 5B0 =$ L: Splitting with respect to these transitions is an attempt to lower the abstract supremum probabilities. If $\Pi 5B0 =$ L and the supremum probability of B is L then so is the supremum probability of A. Transitions of this type are denoted S-transitions (S for single).
(c) neither of the above hold. Transitions of this type are denoted O-transitions (O for other).

2. *N-ary splitting:* This means that we split A with respect to all its outgoing transitions, or to all the transitions of exactly one of the above types.

These types of splitting will be discussed in section 7.

Choosing Classes to Refine. We discuss here how we choose the classes to be refined. Our standard strategy tries to split once every class for which there exist a splitting guard, using the strategies from above to decide which transition to use as the basis for the split. However, we can do better if we take advantage of the dominance relation (section 5). Indeed, if $A \preceq A_e$ then any finite probabilistic path starting from A leads with probability L to A_e. As a consequence, refining class A without refining class A_e will not change the computed values of $\mathsf{P}^{\text{inf}}5A0$ and $\mathsf{P}^{\text{sup}}5A0$. For that reason, we allow the possibility to refine only essential classes in an abstract PTS.

Tradeoff between Refinement and Analysis. Stopping the refinement process as soon as possible, i.e. with the fewest number of classes enabling the proof of the property or its negation, requires to split only one class at a time, and then to perform an analysis to check the property. However, as probabilistic analysis is rather expensive, even on an abstract system, it should not be applied too often.

Our technique is to refine a partition $\mathcal{A}$ into a partition $\mathcal{A}'$ such that $\frac{|\mathcal{A}'|}{|\mathcal{A}|} \geq r$, with r specified by the user. A small value of r (close to L) produces frequent analysis, whereas a large value favorizes refinement over analysis. If the goal is to produce the minimal stable partition of the concrete PTS, a infinite ratio allows to obtain it without any intermediate probabilistic analysis.

7 Implementation and Experiments

A tool called RAPTURE and based on the principles presented in this paper has been implemented. Its architecture (see Fig. 5) is the following: (1) the front-end parses the input language, that specifies both the system to be analyzed, the property and possibly the components (processes and variables) not abstracted in the initial abstraction. The output is a symbolic representation of the system (i.e., the probabilistic transition function and sets of root, initial and final states). (2) Boolean analysis is then performed. If it allows to prove or disprove the

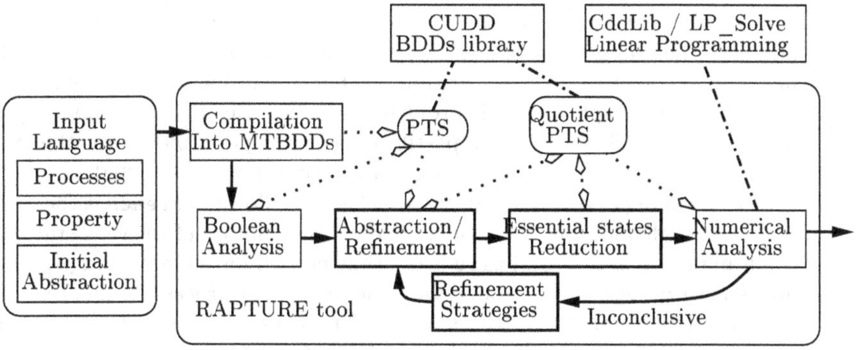

Fig. 5. Architecture of the RAPTURE tool

property, the verdict is emitted. (3) Otherwise, the initial abstraction is build, and the verification process alternating numerical analysis and refinement steps starts. The thick boxes on figure 5 indicate the new or updated modules since the prototype reported in [9].

As stated before, we use linear programming to compute extremum probabilities. Because of precision problems arising with some case studies, we offer the following possibilities in our tool :

- Use of the sparse matrix based solver LP_SOLVE [6], with coefficients being ordinary real numbers (1);
- Use of the dense matrix based solver CDDLIB [13], with coefficients being either exact rational numbers (2), multi-precision floating point numbers (3), or ordinary real numbers (4).

The possibilities that give the best results are (1) and (2). (1) is better whenever there is no precision problems, because it uses sparse matrices and our LP problems are sparse, whereas (2) is very useful when precision problems arise and/or when *exact results* are wanted.

Experiments. We have conducted several experiments in order to evaluate our reduction and refinement strategies as well as our implementation. The aim is to analyze the practical usefulness of our reduction strategies and the efficiency of our refinement strategies.

Our first case study is the Bounded Retransmission Protocol (BRP) [16]. The BRP is based on the well-known alternating bit protocol but allows for a bounded number of retransmissions of a chunk, i.e., part of a file, only. So, eventual delivery is not guaranteed and the protocol may abort the file transfer. We use the version presented in [9], where probabilities model the possible failures of the two channels used for sending chunks and acknowledgments, respectively. In Table 1 we check the maximum probabilities that the sender does not report a successful transmission. We consider a file composed of either 16 or 64 chunks, and N is the number of allowed retransmissions. We have to use here the dense matrix based solver with exact arithmetic, because probabilities of very different magnitude order appear in LP problems, which makes the usual floating point arithmetic unusable. The initial partitioning is here performed w.r.t. the explicit control structure of the specification: only variables are abstracted.

The meaning of row labels in the table is the following: `#reach` is the number of states reachable from root states, `#rel` is the number of relevant states after Boolean preprocessing, and `time` is the time needed for building and preprocessing. The three next sets of rows details the refinement process for different upper bounds for $\mathsf{P}^{\mathsf{sup}}$. `#refin` is the number of refinement steps, `#abst` the number of states of the most refined abstract PTS, `#ess` the number of its essential states, `psup` the computed probability, `verd` is the verdict (true or false), and `time(a+r)` gives the time spent in numerical analysis and in refinement process. When the verdict is false, the refinement has gone to the stable partitioning of the PTS and gives the actual $\mathsf{P}^{\mathsf{sup}}$ of the concrete PTS.

The first observation is that Boolean preprocessing is here very efficient to reduce the state space: the reduction is one third in average. Second, the essential state reduction allows again a reduction of one third. The third observation is that it is nearly as easy to prove $\mathsf{P}^{\mathsf{sup}} \leq \mathrm{1D}^{-3}$ for big instances of BRP than for small ones: that means that the refinement strategy works well and will not perform too many useless splits. It can also be observed that checking smaller upper bounds can still be performed on very small abstract PTSs, compared to the concrete one, even reduced by preprocessing, and also compared to the stable partitioning (row $\mathsf{P}^{\mathsf{sup}} \leq \mathrm{1D}^{-90}$). Last, as we try to check bigger instances, time spent in refinement is much smaller than time spent in analysis, if we refine up to stabilization. Again, this shows the relevance of our method. We do not illustrate here the effect of the different options for refinement, because they give equivalent results on this example.

Our second example is the Probabilistic Dining Philosopher from [25], studied by [26] and analyzed by the the PRISM team [28]. In this example N philosophers are trying to eat and we want to prove a lower bound on the probability for some process to eat after a number of time units specified by value of deadline. As the philosophers perform asynchronous moves, we add the following *bounded fairness constraint*: a philosopher cannot stay idle for more than K steps[1]. Table 2 shows results for $N = 3$ and different values of K. The chosen deadline corresponds to the smallest one for which the property holds with a probability more than 0. We try to prove successively $\mathsf{P}^{\mathsf{inf}} \geq \frac{1}{16}$ and $\mathsf{P}^{\mathsf{inf}} \geq \frac{1}{8}$ (this last bound is the real one, according to [28]). Compared to the previous case study, the number of states is much bigger, as well as the BDDs and MTBDDs representing respectively sets of states and the transition relation. We give in the table not only the number of abstract and essential states, but also in each case the number of abstract distributions. We use here the sparse matrix based solver with ordinary

[1] Without the fairness constraint, the lower bound is zero and can be checked by the discrete fixpoint computations mentioned in section 3.

Table 1. Results in BRP

		file length 16				file length 64
	MAX	2	4	8	15	15
	#reach.	3908	6060	10364	17896	58024
	#relev.	1014	1790	3342	6058	26362
	time	0.94	1.10	1.59	1.75	5.20
$\leq 10^{-3}$	#refin.	4	5	6	6	6
	#abst.	52	89	161	161	161
	#ess.	24	42	85	85	85
	psup	3.09e-04	4.27e-06	7.89e-06	7.89e-06	7.89e-06
	verd.	T	T	T	T	T
	time(a+r)	0.07+0.88	0.72+0.83	2.96+1.68	2.95+1.72	2.97+2.62
$\leq 10^{-10}$	#refin.	9	10	7	7	7
	#abst.	375	675	242	247	247
	#ess.	152	272	108	115	115
	psup	2.65e-05	2.35e-08	7.01e-12	7.01e-12	7.01e-12
	verd.	F	F	T	T	T
	time(a+r)	0.58+2.56	3.30+5.42	3.15+2.07	3.54+2.33	3.51+3.55
$\leq 10^{-90}$	#refin.	9	10	11	12	16
	#abst.	375	675	1275	2325	9765
	#ess.	152	272	512	932	3908
	psup	2.65e-05	2.35e-08	1.85e-14	3.87e-25	3.87e-25
	verd.	F	F	F	F	F
	time(a+r)	0.58+2.56	3.30+5.42	15.87+11.00	186.58+22.06	1209.72+165.3

floating point arithmetic. We choose as the initial partition the one obtained by abstracting everything but the counter used for the deadline, as it is clear the value of the deadline is of fundamental importance for the studied property. Most of the encouraging observations made for the BRP are still true. The only exception is that essential state reduction is not very useful here. Execution times are much higher, because the abstract PTSs are much more complex, and the corresponding LP problems are very big. Still, refinement remains much cheaper than analysis, and state space reduction between the concrete PTS and the abstract one allowing to prove the property is impressive.

Table 3 shows verification of this case study for $K = 3$, with various refinement options and initial control structures. The first column corresponds to the options that work best and that were used in the previous table: the initial partition detail only the counter for the deadline, and we use n-ary division, giving priority to respectively O S and L types of probabilistic transitions, as described in section 6. Using binary divisions gives similar results (second column). Column 3 shows that inverting the priority of the different types of split in column 1 gives very bad results: a much more refined system is needed to prove the property. This is quite counter-intuitive (cf. our remark in section 6) but has been observed on nearly all case studies we performed. We conjecture that splitting a abstract state according to its looping transition most often leads to an unbal-

Table 2. Results in Dining Philosophers with $N = 3$

	K	4	5	6
	deadline	23	27	31
	#reach.	1.00e06	1.97e06	3.40e06
	#relev.	121041	271287	488859
	time	14.4	23.6	34
$\leq \frac{1}{16}$	#refin.	5	7	8
	#abst.	3064/11536	16903/52435	35780/111084
	#ess.	2778/11250	14442/49974	30361/105665
	pinf	0.0625	0.0625	0.0625
	verd.	T	T	T
	time(a+r)	49.6+79.5	2120+590	10353+1462
$\leq \frac{1}{8}$	#refin.	7	8	9
	#abst.	8512/22757	21011/59866	37542/114703
	#ess.	6668/20913	16996/55851	31656/108817
	pinf	0.125	0.125	0.125
	verd.	T	T	T
	time(a+r)	290+220	3683+712	20335+1575

anced division, that separates a few concrete states from a huge set of concrete states. Column 4 corresponds to the strategy where we fully stabilize all abstract states wrt. the current partition. Last, column 5 illustrates the importance of a good initial partition. Here, we generated it according to the explicit control structure of the philosopher, and it produces very bad result.

We also tried the IEEE FireWire root contention protocol [32,31], using the model developed by [28]. The property we want to prove is that a leader (root) is chosen before the time bound deadline is reached with some probability. Results are depicted in table 4, where we refine until we reach the probability given in row `pinf`. Here Boolean analysis is very expensive, because counters involved in the model make the BDDs huge ($\sim 4 \cdot 10^5$ nodes) and the number of iterations is also big ($>$ 200, 300, 400). For deadline=200, Boolean analysis allows to show that $\mathsf{P}^{\mathrm{inf}} = 0$. The table shows also that the number of relevant states is here really smaller than the number of reachable states. That comes from the fact that at least all the states corresponding to deadline less than 200 satisfy the property with probability 0 and are removed, as shown by column 1.

8 Conclusion

In this paper we have introduced new efficient strategies for model checking quantitative reachability properties of Markov decision processes. The fundamental method of our approach is based on automatic abstraction and refinement [9], where properties are analyzed on abstractions rather than on the original system. The abstractions are safe with respect to the property under consideration and moreover they are expected to have significantly smaller state spaces than

Table 3. Results in Dining Philosophers with $N = K = 3$ and deadline $= 19$

	control option	deadline nary+osl	deadline bin+osl	deadline nary+lso	deadline nary+a	ctrl. struct. nary+osl
	#reach.	408397				
	#relev.	30018				
	time	6.14				
$\geq \frac{1}{4}$	#refin.	2	2	4	2	4
	#abst.	51/87	35/122	882/2880	140/680	5861/12972
	#ess.	51/87	35/122	827/2825	140/680	4109/11196
	pinf	0.25	0.25	0.25	0.25	0.25
	verd.	T	T	T	T	T
	time(a+r)	0.03+3.85	0.02+3.48	5.16+16.79	0.19+5.09	53.6+44.8

Table 4. Results in FireWire

deadline	200	300	400	400
#reach.	6.8e6	2.3e07	4.4e07	4.4e07
#relev.	-	21	129661	129661
time	512	3240	8018	8018
#refin.	-	2	2	11
#abst.	-	6/4	11/15	530/1295
#ess.	-	5/3	5/7	71/178
pinf	0.0	0.5	0.5	0.625
time(a+r)	-	0.01 + 0.05	0.01+39	0.26+104

their corresponding concrete system. If an abstraction cannot prove or disprove a considered property, the abstraction is refined and the analysis is repeated on the refined system.

The overall performance of our method depends crucially on the efficiency of the numerical analysis performed on abstract systems as well as on the choice of refinement method. A main contribution of this paper has been the development of strategies for reducing the size of the numerical problems to be analyzed as well as strategies for guiding the refinement process.

Our experiments have shown that our reduction and refinement strategies are relevant for several case studies of various type, and the method of abstraction and refinement allows considerable simplifications. This is especially true when the property of interest does not require the computation of exact probabilities, but also holds in general.

It is worth to mention that the techniques reported in this article considerably improved the performance of our tool with respect to the prototype reported in [9]. We highlight that the processing speed has increased around 3 orders of magnitudes: whereas checking the BRP (MAX=4, length=16) for unsuccessful transmission with probability $\leq 10^{-5}$ took about half an hour, now it only takes around 2 seconds. There are two fundamental reasons. The first one is due to the fact that the new implementation is smarter on doing the refinements and therefore less number of steps are required to reach the result. Second, the LP problem in [9] was constructed with one variable per abstract state and one inequality per transition on this abstract representation. Now, instead, considering a variable per essential state reduces the number of variable to the 50% (in the BRP, see Table 1). Of course this percentage is sensible to the problem under study. Compare the 90% in the Dining Philosophers to the 13% in the FireWire (Tables 2 and 4, respectively). Last but not least, we tuned more carefully the abstraction and refinement algorithms described in [9] (data-structures, custom operations and variable ordering in MTBDDs). These improvements were however too technical to be described here.

Our near future goal is to apply the results of this paper to model check probabilistic timed automata. Decidability of the model checking of properties on such automata has been proven, by resorting to their region graphs [23]. However, region graphs are known to be practically unusable, and our technique would allow to generate progressively a minimal *probabilistic* model, in the spirit of [1]. The different structure of the state space naturally requires the design of new abstraction algorithms, compared to the ones currently implemented in our tool.

References

1. R. Alur, C. Courcoubetis, D. Dill, N. Halbwachs, and H. Wong-Toi. Minimization of timed transition systems. In *Proceedings of the Third Conference on Concurrency Theory CONCUR '92*, volume 630 of *LNCS*, pages 340–354. Springer-Verlag, 1992.
2. A. Aziz, V. Singhal, F. Balarin, R.K. Bryton, and A.L. Sangiovanni-Vincentelli. It usually works:the temporal logics of stochastic systems. In *Computer Aided Verification, CAV'95*, volume 939 of *LNCS*, July 1995.
3. R. Bahar, E. Frohm, C. Gaona, G. Hachtel, E. Macii, A. Pardo, and F. Somenzi. Algebraic decision diagrams and their applications. *Formal Methods in System Design*, 10(2/3):171–206, April 1997.
4. C. Baier, J.-P. Katoen, and H. Hermanns. Approximate symbolic model checking of continuous-time Markov chains. In *CONCUR'99*, volume 1664 of *LNCS*, 1999.
5. Christel Baier. *On Algorithmic Verification Methods for Probabilistic Systems.* Habilitation thesis, Faculty for Mathematics and Informatics, University of Mannheim, 1998.
6. M. Berkelaar. LP_SOLVE: Mixed integer linear program solver. `ftp://ftp.ics.ele.tue.nl/pub/lp_solve`.
7. A. Bianco and L. de Alfaro. Model checking of probabilistic and nondeterministic systems. In *Foundations of Software Technology and Theoretical Computer Science, FSTTCS'95*, volume 1026 of *LNCS*, 1995.
8. M. Bozga, C. Daws, O. Maler, A. Olivero, S. Tripakis, and S. Yovine. KRONOS: A model-checking tool for real-time systems. In *Computer Aided Verification, CAV'98*, volume 1427 of *LNCS*, July 1998.

9. P.R. D'Argenio, B. Jeannet, H.E. Jensen, and K.G. Larsen. Reachability analysis of probabilistic systems by successive refinements. In *Process Algebra and Probabilistic Methods - Performance Modelling and Verification, PAPM-PROBMIV 2001*, volume 2165 of *LNCS*, pages 39–56, 2001.
10. Luca de Alfaro. *Formal Verification of Probabilistic Systems*. PhD thesis, Stanford University, 1997.
11. J.-C. Fernandez, H. Garavel, A. Kerbrat, R. Mateescu, L. Mounier, and M. Sighireanu. CADP (Cæsar / Aldébaran Development Package) : A protocol validation and verification toolbox. In *Computer Aided Verification, CAV'96*, volume 1102 of *LNCS*, July 1996.
12. M. Fujita, P.C. McGeer, and J.C.-Y. Yang. Multi-terminal binary decision diagrams: An efficient data structure for matrix representation. *Formal Methods in System Design*, 10(2/3):149–169, April 1997.
13. K. Fukuda. CDDLIB. `ftp://ftp.ifor.math.ethz.ch/pub/fukuda/cdd`.
14. H.A. Hansson and B. Jonsson. A logic for reasoning about time and reliability. *Formal Aspects of Computing*, 6:512–535, 1994.
15. V. Hartonas-Garmhausen and S. Campos. ProbVerus: Probabilistic symbolic model checking. In *AMAST Workshop on Real-Time and Probabilistic Systems, AMAST'99*, volume 1601 of *LNCS*, 1999.
16. L. Helmink, M. Sellink, and F. Vaandrager. Proof-checking a data link protocol. In *Proc. International Workshop TYPES'93*, volume 806 of *LNCS*, 1994.
17. H. Hermanns, J.-P. Katoen, J. Meyer-Kayser, and M. Siegle. A Markov chain model checker. In *Tools and Algorithms for the Construction and Analysis of Systems, TACAS 2000*, volume 1785 of *LNCS*, 2000.
18. G. Holzmann. The model cheker SPIN. *IEEE Transactions on Software Engineering*, 23(5):279–295, 1997.
19. M. Huth and M. Kwiatkowska. Quantitative analysis and model checking. In *Logic in Computer Science, LICS'97*. IEEE Computer Society Press, 1997.
20. B. Jonsson and K.G. Larsen. Specification and refinement of probabilistic processes. In *Procs. 6th Annual Symposium on Logic in Computer Science*, pages 266–277. IEEE Press, 1991.
21. B. Jonsson, K.G. Larsen, and W. Yi. Probabilistic extensions in process algebras. In J.A. Bergstra, A. Ponse, and S. Smolka, editors, *Handbook of Process Algebras*. Elsevier, 2001.
22. M. Kwiatkowska, G. Norman, and D. Parker. PRISM: Probabilistic symbolic model checker. In *TOOLS'2002*, volume 2324 of *LNCS*, April 2002.
23. M. Kwiatkowska, G. Norman, R. Segala, and J. Sproston. Automatic verification of real-time systems with probability distributions. In J.P. Katoen, editor, *Procs. of the 5th ARTS*, volume 1601 of *LNCS*, pages 75–95. Springer, 1999.
24. Kim G. Larsen, Paul Pettersson, and Wang Yi. UPPAAL in a nutshell. *Springer International Journal of Software Tools for Technology Transfer*, 1(1/2), 1997.
25. D. Lehmann and M. Rabin. On the advantages of free choice: A symmetric fully distributed solution to the dining philosophers problem. In *Proc. 8th Symposium on Principles of Programming Languages*, 1981.
26. N. Lynch, I. Saias, and R. Segala. Proving time bounds for randomized distributed algorithms. In *Proc. 13th ACM Symposium on Principles of Distributed Computing*, 1984.

27. K. Mehlhorn and A.K. Tsakalidis:. Data structures. In J. van Leeuwen, editor, *Handbook of Theoretical Computer Science*, volume A : Algorithms and Complexity, pages 301–342. Elsevier, 1990.
28. PRISM Web Page. `http://www.cs.bham.ac.uk/~dxp/prism/`.
29. M.L. Puterman. *Markov Decision Processes: Discrete Stochastic Dynamic Programming*. Wiley series in probability and mathematical statistics. John Wiley & Sons, 1994.
30. R. Segala. *Modeling and Verification of Randomized Distributed Real-Time Systems*. PhD thesis, Department of Mathematics, Massachusetts Institute of Technology, 1995.
31. D. Simons. and M. Stoelinga. Mechanical verification of the IEEE1394a root contention protocol using Uppaal2k. To appear in *International Journal on Software Tools for Technlogy Transfer*, 2001.
32. M. Stoelinga and F. Vaandrager. Root contention in IEEE 1394. In *Proc. 5th AMAST Workshop on Real-Time and Probabilistic Systems (ARTS'99)*, volume 1601 of *LNCS*, 1999.

Action Refinement for Probabilistic Processes with True Concurrency Models

Harald Fecher, Mila Majster-Cederbaum, and Jinzhao Wu

Universität Mannheim
Fakultät für Mathematik und Informatik D7, 27
68131 Mannheim, Germany
{hfecher,mcb,wu}@pi2.informatik.uni-mannheim.de

Abstract. In this paper, we develop techniques of action refinement for probabilistic processes within the context of a probabilistic process algebra. A semantic counterpart is carried out in a non-interleaving causality based setting, probabilistic bundle event structures. We show that our refinement notion has the following nice properties: the behaviour of the refined system can be inferred compositionally from the behaviour of the original system and from the behaviour of the systems substituted for actions; the probabilistic extensions of pomset trace equivalence and history preserving bisimulation equivalence are both congruences under the refinement; and with respect to a cpo-based denotational semantics the syntactic and semantic refinements coincide with each other up to the aforementioned equivalence relations when the internal actions are abstracted away.

1 Introduction

Process algebras are a frequently used tool for the specification and verification of concurrent systems. In process algebras, actions are used to denote the basic entities of systems. By an action we understand here any activity which is considered as a conceptual entity on a chosen level of abstraction. This allows the representation of systems in a hierarchical way, i.e. actions on a higher level are interpreted by more complicated processes on a lower level. Such a change in the level of abstraction is usually referred to as *action refinement*, which is a core operation in the methodology of top-down system design.

Action refinement for classical concurrent systems without probabilistic constraints has been thoroughly discussed in the literature [2,13]. The models of concurrency can roughly be distinguished in two kinds: interleaving based models, in which the independent execution of two processes is modelled by specifying the possible interleaving of their actions; and true concurrency models, in which the causal relations between actions are represented explicitly.

Without further restriction or relaxation on actions or refinement operations, most of the equivalence relations are not preserved under refinement in the interleaving approach [10]. These equivalence relations, however, are often used

H. Hermanns and R. Segala (Eds.): PAPM-PROBMIV 2002, LNCS 2399, pp. 77–94, 2002.

to establish the correctness of the implementation with respect to the specification of concurrent systems. This problem can be overcome by moving to true concurrency models. Also, in the system design phase the local causal dependencies between actions are important. Interleaving with actions of other parts of the system burdens the design. Partial order models are considered to be much more appropriate here. Moreover, a causality based model does not suffer from the state explosion problem. Parallelism leads to the sum of the components, rather than to their product as in the interleaving approach.

We study action refinement for truly concurrent systems with probabilistic constraints. Probabilistic phenomena are important in many areas of computing. Therefore, formal tools for describing and reasoning about such systems are needed. A number of probabilistic process calculi have been proposed [11,12]. Our aim is to achieve higher reliability with respect not only to the functional correctness but also to the performance issue. In our setting, probabilistic concurrent systems are described in a probabilistic LOTOS-like process algebra [3] proposed in [6,15]. It is actually a synthesis of CCS [23] and CSP [14]. We use probabilistic bundle event structures [15,18] as the semantic model. Note that bundle event structures have been shown to adequately deal with e.g. parallel composition, and the method to attach probabilistic information does not complicate the theory and it is simple and practical. We are unaware of any other work on action refinement in a partial-order setting associated with probabilistic information.

Current probabilistic process algebras usually use probabilistic extensions of labelled transition systems as an underlying semantic model. It is quite common to distinguish between probabilistic and nonprobabilistic transitions in these models. The main problem with this approach is the intertwining of these types of transitions. This is to say, it is not clear what the intended meaning of a probability attached to a transition is in the presence of a competitive nonprobabilistic transition. Such situations can happen if e.g. combinations of parallel composition and probabilistic choice are taken. Several solutions have been proposed for this problem, but most of them loose the property of backward compatibility with the nonprobabilistic semantics. A causality-based model does not have such problems [7,15].

There are essentially two interpretations of action refinement, called syntactic and semantic. Syntactic action refinement, where actions occurring in a process are replaced by more complicated ones, yields a more detailed process description [2,13]. Due to its definitional clarity, syntactic action refinement can be easily used without too much insight on the semantics. Semantic action refinement is carried out in the semantic domain. It avoids a confusion of the abstraction levels, which can happen in syntactic refinement. Such a confusion may result in undesirable situations [10,13].

We adopt the methodology that refinement is modelled as an operator. After defining syntactic and semantic action refinement in the probabilistic process algebra, we discuss three common issues of interest: *safety*, *congruence* and *coincidence* problems.

A refinement operation should be safe. That is, the behaviour of the refined system should be the refinement of the behaviour of the original system (in some sense), and vice versa. Equivalence relations are often used to capture when two systems are considered to exhibit the same behaviour. To examine the well-definedness of the refinement operator, we have to try to find such equivalence relations which are congruences under it. The result that syntactic and semantic refinements coincide can give a clear understanding of the concept of action refinement, and has important applications in system verifications [20,21]: the refined syntactic specification can be modelled by semantic action refinement, and the refined semantic model can be specified by syntactic action refinement.

The main contributions of this paper are:

- the notions of syntactic and semantic refinement operators in the probabilistic process algebra;
- the verification of safety of refinement;
- the congruence results about pomset trace and history preserving bisimulation equivalences; and
- the coincidence result of semantic refinement with syntactic refinement.

This paper is a further development of [10], where action refinement in event structures without probabilistic information is investigated.

2 A Probabilistic Process Algebra

In this section, we fix the process algebraic framework that is used to develop probabilistic concurrent systems, and we describe a cpo-based denotational semantics.

2.1 Syntax

Assume a given set Θ of observable actions and an invisible internal action τ ($\tau \notin \Theta$). Action $\surd$ ($\surd \notin \Theta \cup \{\tau\}$) indicates the successful termination of a process. Let $\Omega = \Theta \cup \{\tau, \surd\}$.

Definition 2.1.1 (*Probabilistic Formalism L*).

$$B ::= 0 \mid 1 \mid a.B \mid B; B \mid B \parallel_A B \mid B + B \mid B +_p B \mid B \backslash A \mid B[\lambda] \mid x \mid \mu x.B.$$

Here, $a \in \Theta \cup \{\tau\}$, $A \subseteq \Theta$, $\lambda\colon \Omega \to \Omega$ is a relabelling function such that $\lambda(\tau) = \tau$, $\lambda(\surd) = \surd$ and $\lambda(a) \neq \surd$ for $a \in \Theta$, $x \in V$, V is a set of process variables, and $p \in (0,1)$, i.e. $0 < p < 1$.

We assume $dom(\lambda) = \{a \in \Omega \mid \lambda(a) \neq a\}$, representing the set of actions that are really relabelled by λ. The operators have the following intuitive meaning:

0 denotes inaction. 1 represents the process that terminates successfully.

$a.B_1$ denotes the prefix of action a before process B_1. $B_1[\lambda]$ denotes the relabelling of B_1 according to λ. $B_1 \backslash A$ behaves as B_1, except that the actions in A are abstracted, i.e., turned into τ-actions.

$B_1; B_2$ denotes the sequential composition of processes B_1 and B_2, where the control is passed to B_2 by the successful termination of B_1. $B_1 \parallel_A B_2$ denotes the parallel composition of B_1 and B_2, where B_1 and B_2 must perform any actions in $A \cup \{\surd\}$ simultaneously, while the other actions are executed independently from each other.

We distinguish between a non-deterministic and a probabilistic choice. $B_1 + B_2$ indicates the non-deterministic choice between the behaviours described by processes B_1 and B_2. $B_1 +_p B_2$ behaves like B_1 with probability p or like B_2 with probability $1-p$. This distinction is important. From a design perspective it is necessary to express choices for which the probability of an alternative is left unspecified. Such quantitative knowledge may either be absent at the current stage of design or it may be deliberately left unspecified. Therefore, one should not be forced to associate such quantity with an alternative. When going from an abstract specification to a more concrete specification, it is useful to consider the replacement of some non-deterministic choices by probabilistic choices.

Recursive behaviour is described by $\mu x.B_1$. It can be understood through $x := B_1$, where x may occur in the body B_1.

The probabilistic choice between processes B_1 and B_2 should not be influenced by the environment. So we have to add some constraints on B_1 and B_2. These constraints guarantee that $B_1 +_p B_2$ induces an independent stochastic experiment.

We first introduce two subsidiary predicates *pc* and *ppc*.

Let $ppc : L \to \{true, false\}$ be defined as follows:

$$ppc(B_1 +_p B_2) = (ppc(B_1) \vee B_1 = \tau.B_1') \wedge (ppc(B_2) \vee B_2 = \tau.B_2'),$$
$$ppc(B_1; B_2) = ppc(B_1),$$
$$ppc(\circ B_1) = ppc(B_1) \text{ for } \circ \in \{\backslash A, [\lambda]\},$$
$$ppc(\mu x.B_1) = ppc(B_1),$$
ppc is $false$ for all other syntactical constructs.

Let $pc : L \to \{true, false\}$ be defined as follows:

$$pc(B_1 +_p B_2) = true,$$
$$pc(B_1; B_2) = pc(B_1),$$
$$pc(B_1 \parallel_A B_2) = pc(B_1) \vee pc(B_2),$$
$$pc(\circ B_1) = pc(B_1) \text{ for } \circ \in \{\backslash A, [\lambda]\},$$
$$pc(\mu x.B_1) = pc(B_1),$$
pc is $false$ for all other syntactical constructs.

In fact, $ppc(B)$ holds if a pure probabilistic choice has to happen next in B. $pc(B)$ holds if a probabilistic choice occurs in a proper component of B.

Definition 2.1.2 (*Probabilistic Process Algebra PPA*). $PPA = \{B \in L \mid ppa(B)\}$, where $ppa : L \to \{true, false\}$ is defined as:

$$ppa(0) = true,$$
$$ppa(1) = true,$$

$ppa(x) = true$ for $x \in V$,
$ppa(\circ B_1) = ppa(B_1)$ for $\circ \in \{a., \backslash A, [\lambda]\}$,
$ppa(B_1 \circ B_2) = ppa(B_1) \wedge ppa(B_2)$ for $\circ \in \{;, \|_A\}$,
$ppa(B_1 + B_2) = \neg pc(B_1) \wedge \neg pc(B_2) \wedge ppa(B_1) \wedge ppa(B_2)$,
$ppa(B_1 +_p B_2) = ppc(B_1 +_p B_2) \wedge ppa(B_1) \wedge ppa(B_2)$,
$ppa(\mu x.B_1) = ppa(B_1)$.

Hereafter, the elements of PPA, denoted as B and B_k, are called *(probabilistic) expressions.*The alphabet of expression B, namely the set of all observable actions occurring in B, is denoted by A_B.

Example 2.1.3. $B' = \tau.a.1 + (\tau.b.0 +_{0.5} \tau.c.1)$ is not an expression, whereas the following B and B_c are expressions:
$B = r.((\tau.a.s.0 +_{0.4} \tau.b.s.0) +_{0.7} \tau.c.s.0)$
$B_c = \tau.c_1.1 +_{0.8} \tau.c_2.1$

2.2 Semantic Model

We use probabilistic bundle event structures as the semantic model of the probabilistic process algebra.

Probabilistic Event Structures. A bundle event structure [18] consists of events labelled with actions and two relations between events for causality and for conflict. When an event occurs, the action labelled to it occurs. The symmetric conflict relation, denoted $\sharp$, is a binary relation between events, and the intended meaning of $e \sharp e'$ is that events e and e' cannot both occur in a single system run. Causality is represented by a binary bundle relation, denoted $\mapsto$. It relates a set of events, where all events have to be pairwise in conflict, and an event. $X \mapsto e$ means that if event e happens in a system run, exactly one event in event set X has happened before and caused e. X is called a *bundle-set.*

The basic idea of probabilistic event structures is to incorporate fixed probabilities in event structures by associating probabilities with some events [15]. Suppose we have an event e and we decorate this event with probability $p \in (0, 1)$. The intuitive interpretation is that e happens with likelihood p provided that it is enabled. Thus, p is a conditional probability. A partial mapping π is assumed that decorates an event with a probability in $(0, 1)$. Some events are grouped into clusters in order to model a stochastic experiment.

Definition 2.2.1 (*Probabilistic Event Structure, pes* for Short). A probabilistic event structure Π is a tuple $(E, \sharp, \mapsto, l, \pi)$ with

E, a set of events,
$\sharp \subseteq E \times E$, the irreflexive and symmetric conflict relation,
$\mapsto \subseteq \mathcal{P}(E) \times E$, the bundle relation,
$l : E \rightarrow \Omega$, the action labelling function,

$\pi : E \longrightarrow_P (0,1)$ the probability function,
such that for any bundle-set X and event $e \in dom(\pi)$
(1) $(X \times X) \setminus Id_E \subseteq \sharp$, and
(2) $\exists Q \subseteq dom(\pi) : (e \in Q) \wedge (Q \text{ is a cluster }) \wedge (\sum_{e' \in Q} \pi(e') = 1)$.

Here $\mathcal{P}(\cdot)$ denotes the power-set function, $Id_E = \{(e,e) \mid e \in E\}$, $\longrightarrow_P$ indicates a partial function, $dom(\pi)$ denotes the domain of π, and a cluster is defined as follows:

$Q \subseteq E$ is called a *cluster*, if

(i) $|Q| > 1$,
(ii) $\forall e \in Q : l(e) = \tau$,
(iii) $\forall e, e' \in Q : (e \neq e') \Rightarrow (e \sharp e')$,
(iv) $\forall e \in Q, e' \in E : (e \sharp e') \Rightarrow (e' \in Q)$,
(v) $\forall e, e' \in Q, X \subseteq E : (X \mapsto e) \Rightarrow (X \mapsto e')$.

$|Q|$ denotes the number of the elements of Q. Constraint (i) requires a cluster to consist of more than one event. This is convenient for technical reasons and poses no real practical constraint. In order to guarantee that stochastic experiments represented by clusters are indeed independent from their context, Constraint (ii) requires all events in a cluster to be internal, i.e. labelled with action τ. In this way, we are sure that such events are not subject to interaction anymore, which would make their probability dependent on the context in which they are embedded. According to Constraint (iii), events in a cluster mutually exclude each other such that only one event can happen. This is because the realization of a stochastic experiment usually has a single outcome. Constraint (iv) requires that events in a cluster are not in conflict with events outside the cluster. Allowing such conflicts will destroy the interpretation that an event probability represents the likelihood that this event happens once enabled. Finally, Constraint (v) requires all events in a cluster must be pointed to by the same bundle-sets. Together with the fourth constraint, this guarantee that if an event in a cluster is enabled all events in this cluster are enabled.

Constraint (1) in the definition of pes's enables us to uniquely define a causal ordering between the events in a system run. Constraint (2) requires the domain of π to be partitioned into clusters such that the sum of the probabilities assigned to all events in a cluster equals one. In this way, a cluster Q can be considered to represent a stochastic experiment for which the probability of outcome $e \in Q$ equals $\pi(e)$.

A pes is depicted as follows: Events are denoted by dots. Near the dot the action label is given. $e \sharp e'$ is indicated by a dotted line between e and e'. A bundle $X \mapsto e$ is indicated by an arrow to e and to this arrow each event in X is connected via a line. The probability of an event is depicted near to the event. A cluster is indicated by a shaded surface. When no confusion arises, we identify an event with its action label.

Example 2.2.2. Figure 1 is a pes. It can be understood as a system of message sender with probabilistic information, where $r =$ *read-message*, $a =$ *use-*

channel-1, b = *use-channel*-2, c = *use-channel*-3, s = *send-message*, and the internal actions indicate the system chooses the corresponding channels. The probability of choosing channel-1 is 0.28, the probability of choosing channel-2 is 0.42, and the probability of choosing channel-3 is 0.3.

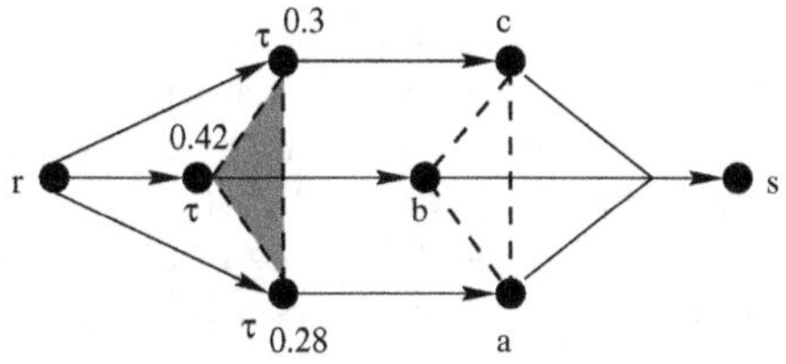

Figure 1. An example pes

By PES we denote the set of pes's. Elements of PES are denoted by Π and Π_k, where $\Pi = (E, \sharp, \mapsto, l, \pi)$ and $\Pi_k = (E_k, \sharp_k, \mapsto_k, l_k, \pi_k)$. When necessary, we also use E_Π, $\sharp_\Pi$, $\mapsto_\Pi$, l_Π and π_Π to represent the components of Π. Without loss of generality, we assume $E_f \cap E_g = \emptyset$ for two different pes's Π_f and Π_g.

We use $init(\Pi)$ to denote the set of initial events (of Π), and $exit(\Pi)$ denotes the set of successful termination events:

$$init(\Pi) = \{e \in E \mid \neg(\exists X \subseteq E : X \mapsto e)\}, exit(\Pi) = \{e \in E \mid l(e) = \surd\}.$$

To describe system runs of a pes Π, we first consider its underlying set of events. Suppose that γ is a finite sequence $e_1, \cdots, e_n$ of events of Π, where e_i and e_j are distinct whenever $i \neq j$. By $E(\gamma)$ we denote the event set $\{e_1, \cdots, e_n\}$. Sequence $\gamma_{i-1} = e_1, \cdots, e_{i-1}$ represents the $(i-1)$-th prefix of γ.

In a system run any two events that occur should not be in conflict, and if a non-initial event happens in a system run then events that cause it should have happened before. Formally

$$en(\gamma_{i-1}) = \{e \in E \setminus E(\gamma_{i-1}) \mid (\forall e_j \in E(\gamma_{i-1}) : \neg(e \sharp e_j)) \wedge \\ (\forall X \mapsto e : X \cap E(\gamma_{i-1}) \neq \emptyset)\}.$$

is the set of events that are enabled after the performance of γ_{i-1}.

Definition 2.2.3 (*Configuration*). If the event sequence $\gamma = e_1, \cdots, e_n$ satisfies the condition $e_i \in en(\gamma_{i-1})$ for $1 \leq i \leq n$, then the set $E(\gamma)$ of all events occurring in γ is called a configuration of Π.

By $C(\Pi)$ we denote the set of all configurations of Π. Let σ be a configuration of Π. Event $e \in \sigma$ is said to be *maximal* in σ if for any bundle-set X and $e' \in E$, if $e \in X$ and $X \mapsto e'$ then $e' \notin \sigma$. We say that σ *successfully terminates* if there exists $e \in \sigma$ such that e is labelled with the successful termination action $\surd$. Π is called *well-labelled* if $\sigma \cap exit(\Pi)$ is empty or a singleton for any configuration σ of Π.

A system run may be represented as an equivalence class of labelled partial orders (pomsets). We define in the following a linear-time equivalence, termed pomset trace equivalence, on PES. A branching-time equivalence, termed history preserving bisimulation equivalence (see e.g. [10]), can be defined similarly, which further records where choices are made. Here we only describe the former equivalence relation in detail.

Assume that σ is a configuration of Π, and $e_i, e_j \in \sigma$. By $e_j \mapsto |_\sigma e_i$ we mean that there exists a bundle-set X such that $e_j \in X$ and $X \mapsto e_i$. Without confusion, we use $\rightarrow |_\sigma$ again to represent the transitive closure of $\rightarrow |_\sigma$. We use $l|_\sigma$ as the restriction of l on σ, i.e. $l|_\sigma(e) = l(e)$ for $e \in \sigma$.

A configuration σ_1 of Π_1 and a configuration σ_2 of Π_2 are said to be *isomorphic*, denoted $\sigma_1 \approx \sigma_2$, if there exists a bijection $h : \sigma_1 \rightarrow \sigma_2$, such that for arbitrary $e, e' \in \sigma_1$,

(1) $e \rightarrow_1 |_{\sigma_1} e'$ iff $h(e) \rightarrow_2 |_{\sigma_2} h(e')$,
(2) $l_1|_{\sigma_1}(e) = l_2|_{\sigma_2}(h(e))$,
(3) $e \in dom(\pi_1)$ and $\pi_1(e) = p$ iff $h(e) \in dom(\pi_2)$ and $\pi_2(h(e)) = p$.

When $\sigma_1 \approx \sigma_2$, σ_1 and σ_2 only differ in the event names, i.e. the action labels, causality relations as well as the probabilities of events remain the same.

By $C(\Pi_1) \approx C(\Pi_2)$ we denote the fact that for any configuration of Π_1, there is an isomorphic configuration of Π_2, and vice versa.

Definition 2.2.4 (*Pomset Trace Equivalence*). Π_1 and Π_2 are pomset trace equivalent, denoted $\Pi_1 \cong_p \Pi_2$, if $C(\Pi_1) \approx C(\Pi_2)$.

A Denotational Semantics. We first describe the operators on PES corresponding to those in the probabilistic process algebra. Here we only present the definition of probabilistic choice $+_p$ in detail. For the definitions of *action prefix* $a_T.$, *abstraction* $\backslash A$, *relabelling* $[\lambda]$, *sequential composition* ;, *parallel composition* $\|_A$ and *choice* $+$, we refer the reader to [5,6,15,16,17].

Probabilistic Choice. $\Pi_1 +_p \Pi_2 = (E_1 \cup E_2, \sharp, \mapsto_1 \cup \mapsto_2, l_1 \cup l_2, \pi)$, where
$\sharp = \sharp_1 \cup \sharp_2 \cup (init(\Pi_1) \times init(\Pi_2)) \cup (init(\Pi_2) \times init(\Pi_1))$,
$\pi = \pi_1|_{(E_1 \setminus init(\Pi_1))} \cup \pi_2|_{(E_2 \setminus init(\Pi_2))} \cup \{(e, p) \mid e \in init(\Pi_1) \setminus dom(\pi_1)\}$
$\cup \{(e, p \cdot \pi_1(e)) \mid e \in init(\Pi_1) \cap dom(\pi_1)\} \cup \{(e, 1-p) \mid e \in init(\Pi_2) \setminus dom(\pi_2)\}$
$\cup \{(e, (1-p) \cdot \pi_2(e)) \mid e \in init(\Pi_2) \cap dom(\pi_2)\}$.

In this definition, we assume events in $init(\Pi_1)$ and $init(\Pi_2)$ are all labelled with internal τ-actions. This assumption guarantees that $\Pi_1 +_p \Pi_2$ remains a pes, and has no influence on providing semantics for the probabilistic process algebra. $\pi_k|_{(E_k \setminus init(\Pi_k))}$ represents the restriction of π_k on $E_k \setminus init(\Pi_k)$ $(k = 1, 2)$.

An expression is called *closed*, if every occurrence of a process variable x is in the range of a μx-operator. A recursion $\mu x.B_1$ is said to be *guarded*, if

B_1 becomes guarded by substituting for a finite number of times the bodies of processes for the process instantiations occurring in B_1 [15].

We want to use the cpo-based denotational semantics as proposed in [15]. For this we have to require that all the recursions occurring in the given closed expression B are guarded [15]. Let s denote this semantics. The semantic model $s(B)$ of expression B is a pes, which is derived as follows:

$$s(0) = (\emptyset, \emptyset, \emptyset, \emptyset, \emptyset),$$
$$s(1) = (\{e\}, \emptyset, \emptyset, \{(e, \surd)\}, \emptyset),$$
$$s(\circ B_1) = \circ s(B_1) \text{ for } \circ \in \{a_T., \backslash A, [\lambda]\},$$
$$s(B_1 \circ B_2) = s(B_1) \circ s(B_2) \text{ for } \circ \in \{;, \|_\surd, +, +_p\},$$

and for guarded recursion $(\mu x.B_1)$, $s(\mu x.B_1)$ is defined as the least upper bound of a set of pes's with a complete partial order (cpo) [1,8,15].

When $s(B)$ is mentioned in the following, we always suppose that expression B is closed and all the recursions occurring in B are guarded.

Example 2.2.5. The semantic model $s(B)$ of the expression B of Example 2.1.3 is pomset trace equivalent to the pes Π of Example 2.2.2 (Figure 1). That is, $s(B) \cong_p \Pi$.

3 Action Refinement

In this section, we discuss syntactic action refinement operation in the probabilistic process algebra. The semantic counterpart is also investigated.

3.1 Syntactic Refinement

Firstly, we have to sort out some 'bound' actions. For a given expression B, let

$$Sort(B) = \cup\{A \mid \backslash A \text{ occurs in } B \text{ or } A = dom(\lambda) \text{ and } [\lambda] \text{ occurs in } B\}.$$

All actions in $Sort(B)$ are not allowed to be refined, since they may lead to a confusion of the communication levels like in the unprobabilistic case [13].

Assume that Ω_0 is a subset of Ω, representing the set of actions that need not or cannot be refined such that $Sort(B) \cup \{\tau, \surd\} \subseteq \Omega_0$. Let $g : \Omega \setminus \Omega_0 \to PPA$ be a function. For convenience, in the following we also use $A_{g(a)}$ to denote the action singleton $\{a\}$ when $a \in \Theta \cap \Omega_0$.

Definition 3.1.1 (*Refinement Function for Expression B*).

g is a refinement function for 0, 1 and x,

g is a refinement function for $a.B_1$ iff g is a refinement function for B_1,

g is a refinement function for $B_1 \circ B_2$ iff g is a refinement function for B_1 and B_2, where $\circ \in \{;, +, +_p\}$,

g is a refinement function for $B_1 \|_A B_2$ iff g is a refinement function for B_1 and B_2, and for any two distinct $a \in A$ and $b \in \Theta$, $A_{g(a)} \cap A_{g(b)} = \emptyset$,

g is a refinement function for $\circ B_1$ iff g is a refinement function for B_1, and for any $a \in \Omega \setminus \Omega_0$, $A_{g(a)} \cap Sort(P) = \emptyset$, where $\circ \in \{\backslash A, [\lambda]\}$,

g is a refinement function for $\mu x.B_1$ iff g is a refinement function for B_1.

We forbid the actions in $Sort(B)$ to occur in $g(a)$, and we pose one more constraint in the definition for parallel composition. Usually, like the unprobabilistic case such constraints cannot be avoided for refinement on syntactic level. Otherwise they may result in a confusion of the communication levels [13].

Example 3.1.2. Suppose that B and B_c are the expressions of Example 2.1.3, and $\Omega_0 = \Omega \setminus \{c\}$. Let $g : c \mapsto B_c$. Then g is a refinement function for B.

Let g be a refinement function for expression B. Now the question is how g can be applied to B to obtain a refined expression. Our basic idea is that the refined expression of $a.B_1$ where $a \notin \Omega_0$ is defined as the sequential composition of $\tau.g(a)$ and the refined expression of B_1. This construction guarantees an equivalence between syntactic and semantic refinement. An intuitive explanation of the prefix τ can be given as follows: We may assume that every process has a start point that executes an internal silent action, the prefix τ here is used to model such a start point of the process $g(a)$ used to substitute action a.

Definition 3.1.3 (*Syntactic Refinement of Expression B*). The refinement $g(B)$ of expression B is defined as follows:

$$
\begin{aligned}
&g(0) = 0, g(1) = 1, g(x) = x,\\
&g(a.B_1) = \begin{cases} a.g(B_1) \text{ if } a \in \Omega_0,\\ (\tau.g(a)); g(B_1) \text{ otherwise,}\end{cases}\\
&g(\circ B_1) = \circ g(B_1) \text{ for } \circ \in \{\backslash A, [\lambda]\},\\
&g(B_1 \circ B_2) = g(B_1) \circ g(B_2) \text{ for } \circ \in \{;, +, +_p\},\\
&g(B_1 \parallel_A B_2) = g(B_1) \parallel_{g(A)} g(B_2) \text{ where } g(A) = \cup_{a \in A} A_{g(a)},\\
&g(\mu x.B_1) = \mu x.g(B_1).
\end{aligned}
$$

Theorem 3.1.4. Suppose that $B \in PPA$ and g is a refinement function for B. Then $g(B) \in PPA$.

Furthermore, $g(B)$ is closed if B is closed. All the recursions occurring in $g(B)$ are guarded if all the recursions occurring in B are guarded.

Example 3.1.5. Consider B and g of Example 3.1.2.
$g(B) = r.((\tau.a.s.0 +_{0.4} \tau.b.s.0) +_{0.7} \tau.((\tau.(\tau.c_1.1 +_{0.8} \tau.c_2.1)); (s.0)))$.

3.2 Semantic Refinement

The refinement methodology developed for various event structures in e.g. [10] does not work for pes's. For example, when we refine action a in the pes

a•- - - -•τ by the pes of Figure 2(a), the resulting Figure 2(b) is no longer a pes because the property of clusters is violated. Our solution is to introduce a special event in each pes used to substitute an action.

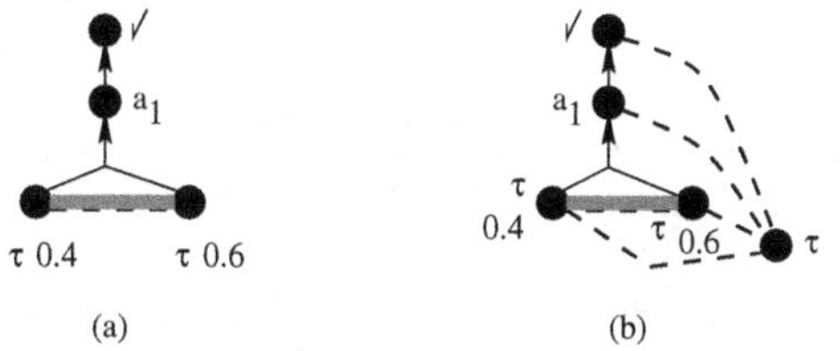

Figure 2. An example showing a refinement methodology does not work

We call $\tau.\Pi$, denoted $r(\Pi)$, the *rooted pes* associated with Π. In $r(\Pi)$, the newly introduced event that corresponds to the prefix τ is called the *start-event* of Π, and is denoted by $o_{r(\Pi)}$. This new event resembles the special events used in [4,9,22,24,25], and can be understood as the start point of system, which is usually supposed to be executing an internal silent action. In the definition of syntactic refinement, we introduced the prefix τ for $g(a)$ with the intention to obtain an 'equivalence result' for syntactic and semantic refinement.

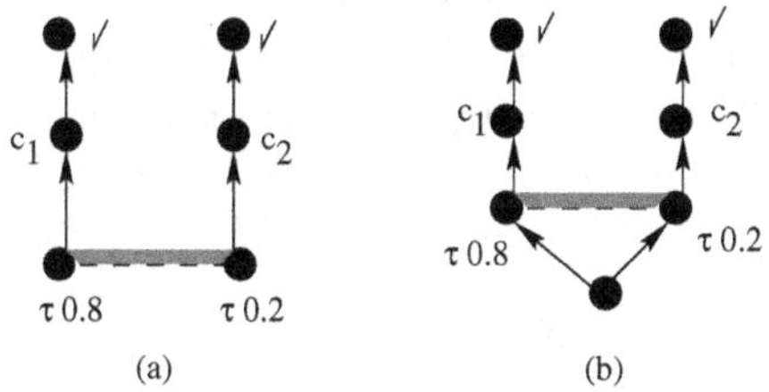

Figure 3. A pes and the rooted pes associated with it

Example 3.2.1. Let Π_c be the pes of Figure 3(a). Then $r(\mathcal{E}_c)$ is the pes of Figure 3(b). The pes Π_c of Figure 3(a) is in fact the semantic model of expression B_c of Example 2.1.3. Namely $s(B_c) = \Pi_c$. In the following, we want to replace action c in the pes of Example 2.2.2 (Figure 1) by Π_c. We may understand that in this example c_1 and c_2 are two new channels that can be used to decompose c. Similarly, the internal actions indicate the system chooses the corresponding channels. The probabilities of choosing channel c_1 and channel c_2 are 0.8 and 0.2, respectively.

The refinement of an action a, say Π_a, should be a pes. Since action $\surd$ only represents successful termination of a system, a system run, if it is not a deadlock, should contain exactly one $\surd$-event. So we require that Π_a is well-labelled.

Definition 3.2.2 (*Semantic Refinement Function*). $f : \Omega \setminus \Omega_0 \to PES$ is called a *refinement function*, if $f(a)$ is well-labelled for any action $a \in \Omega \setminus \Omega_0$.

We say that $f(a)$ is a *refinement of action a*. In practice, it is reasonable to require that $f(a)$ is finite. If so the condition in this definition can be checked and easily enforced.

Example 3.2.3. Suppose $\Omega \setminus \Omega_0 = \{c\}$, and Π_c is the pes of Example 3.2.1. Then $f(c) = \Pi_c$ is a refinement of action c.

Let f be a semantic refinement function. Now we see how f can be used to refine a pes. For simplicity, we use $rfl(e)$ and $rf(a)$ to abbreviate $r(f(l(e)))$ and $r(f(a))$, respectively.

Definition 3.2.4 (*Refinement of a pes*). The *refinement* of Π is defined as $f(\Pi) = (E_f, \sharp_f, \mapsto_f, l_f, \pi_f)$, where

- $E_f = \{(e, e') \mid (e \in E) \wedge (l(e) \notin \Omega_0) \wedge (e' \in E_{rfl(e)})\} \cup$
  $\{(e, e) \mid (e \in E) \wedge (l(e) \in \Omega_0)\}$,

- $\forall (e_1, e_2), (e_1', e_2') \in E_f$, $(e_1, e_2) \sharp_f (e_1', e_2')$ iff
  if $(e_1 = e_1')$ then $(e_2 \sharp_{rfl(e_1)} e_2') \vee (e_2, e_2' \in exit(rfl(e_1)) \wedge e_2 \neq e_2')$,
  if $(e_1 \neq e_1')$ then
  if $(e_2 = e_1) \wedge (e_2' = e_1')$ then $(e_1 \sharp e_1')$,
  if $(e_2 \neq e_1) \wedge (e_2' = e_1')$ then $(e_1 \sharp e_1') \wedge (e_2 \in \{o_{rfl(e_1)}\} \cup exit(rfl(e_1)))$,
  if $(e_2 = e_1) \wedge (e_2' \neq e_1')$ then $(e_1 \sharp e_1') \wedge (e_2' \in \{o_{rfl(e_1')}\} \cup exit(rfl(e_1')))$,
  if $(e_2 \neq e_1) \wedge (e_2' \neq e_1')$ then $(e_1 \sharp e_1') \wedge ((e_2 = o_{rfl(e_1)} \wedge e_2' = o_{rfl(e_1')})$
  $\vee (e_2 \in exit(rfl(e_1)) \wedge e_2' \in exit(rfl(e_1'))))$,

- $\forall X \subseteq E_f, (e_1, e_2) \in E_f$, $X \mapsto_f (e_1, e_2)$ iff
  if $(e_2 \neq e_1) \wedge (e_2 \in E_{rfl(e_1)} \setminus \{o_{rfl(e_1)}\})$ then
  $(\rho_1(X) = \{e_1\}) \wedge (\rho_2(X) \mapsto_{rfl(e_1)} e_2)$,
  if $(e_2 \neq e_1 \wedge e_2 = o_{rfl(e_1)}) \vee (e_2 = e_1)$ then $(\rho_1(X) \mapsto e_1)$
  $\wedge (\rho_2(X) = \cup_{e \in \rho_1(X), l(e) \notin \Omega_0} exit(rfl(e)) \cup (\cup_{e \in \rho_1(X), l(e) \in \Omega_0} \{e\}))$,

- $\forall (e_1, e_2) \in E_f$,
  if $(e_2 \neq e_1)$ then
  if $(e_2 \notin exit(rfl(e_1)))$ then $(l_f(e_1, e_2) = l_{rfl(e_1)}(e_2))$,
  if $(e_2 \in exit(rfl(e_1)))$ then $(l_f(e_1, e_2) = \tau)$,
  if $(e_2 = e_1)$ then $(l_f(e_1, e_2) = l(e_1))$,

- $\forall (e_1, e_2) \in E_f$, $(e_1, e_2) \in dom(\pi_f)$ iff
  $(e_1 \in dom(\pi)) \vee ((e_1 \neq e_2) \wedge (e_2 \in dom(\pi_{rfl(e_1)})))$,
  if $(e_1 \in dom(\pi))$ then $(\pi_f(e_1, e_2) = \pi(e_1))$,
  if $(e_1 \neq e_2) \wedge (e_2 \in dom(\pi_{rfl(e_1)}))$ then $(\pi_f(e_1, e_2) = \pi_{rfl(e_1)}(e_2))$.

Here $\rho_1(X) = \{e \mid (e, e') \in X\}$ and $\rho_2(X) = \{e' \mid (e, e') \in X\}$.

Theorem 3.2.5. Suppose that $\Pi \in PES$ and f is a refinement function. Then $f(\Pi) \in PES$.

Example 3.2.6. Let Π be the pes of Example 2.2.2 (Figure 1) and $f(c)$ be the refinement of action c given in Example 3.2.3 (Figure 3(a)). Then $f(\Pi)$ is the pes of Figure 4.

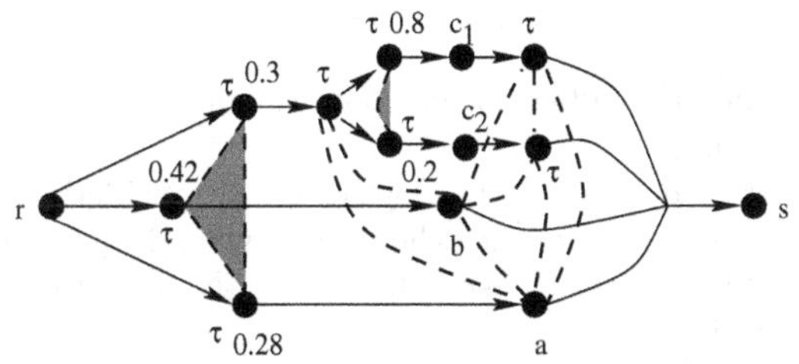

Figure 4. The refined pes

4 Properties

Hereby, suppose that $g : \Omega \setminus \Omega_0 \to PPA$ is a syntactic refinement function for a given expression B, and $f : a \mapsto s(g(a))$ $(a \in \Omega \setminus \Omega_0)$ is a semantic refinement function. Π, Π_1 and Π_2 are pes's, and f_1 and f_2 are two semantic refinement functions.

Let σ be a configuration of Π, $e \in \sigma$ with $l(e) \notin \Omega_0$, and σ_e a configuration of pes $rfl(e)$ which is used to substitute for that action labelled to event e. Furthermore, assume $rfl(e)$ satisfies the condition that σ_e successfully terminates if event e is not maximal in σ. Let

$$\sigma_g = \{(e, e_j) \mid e \in \sigma \text{ and if } l(e) \in \Omega_0 \text{ then } e_j = e \text{ otherwise } e_j \in \sigma_e\}.$$

We call σ_g a *refinement of configuration* σ. It is derived by replacing each event e in configuration σ with $l(e) \notin \Omega_0$ by a configuration σ_e of $rfl(e)$.

Theorem 4.1.1 (*Safety*). $C(f(\Pi)) = \{\sigma_f \mid \sigma_f \text{ is a refinement of } \sigma \in C(\Pi)\}$.

It is natural to use $rfl(e)$ rather than $f(l(e))$ to substitute action $l(e)$ in Π. The start-event we introduced is to imitate the start point of system. Theorem 4.1.1 demonstrates that the refinement of a configuration of Π can be obtained by replacing each event e with $l(e) \notin \Omega_0$ in this configuration by a configuration of $rfl(e)$.

It is straightforward to see that the causality relations in each σ_e remain unchanged in the refinement of σ. The causality relations in σ remain unchanged in the meaning that if e causes e' in σ, then the maximal event in σ_e causes the minimal event in $\sigma_{e'}$. Furthermore, the probability of $(e, e_j) \in \sigma_g$ equals the probability of $e \in \sigma$ when $e \in dom(\pi)$ and equals the probability of $e_j \in \sigma_e$ when $e_j \in dom(\pi_{f(l(e))})$.

The behaviour of a refined system can be thus inferred compositionally from the behaviour of the original system and from the behaviour of those substituted for the actions. Hence, our refinement is safe.

Theorem 4.1.2 (*Congruence Result*). If $\Pi_1 \cong_p \Pi_2$ and $f_1(a) \cong_p f_2(a)$ for any $a \in \Omega \setminus \Omega_0$, then $f_1(\Pi_1) \cong_p f_2(\Pi_2)$.

This theorem indicates that pomset trace equivalence is a congruence under our refinement.

The start-events can be neglected by considering only the observable behaviour of systems. On the other hand, when dealing with the coincidence problem, we have to consider such behaviour. Namely we have to abstract from the internal actions.

Let σ be a configuration of Π. Then $\{e \in \sigma \mid l(e) \neq \tau\}$ is called an *observational configuration* of Π. This observational configuration is said to be derived from configuration σ.

Assume that σ_0 is a configuration of Π, and $e_i, e_j \in \sigma_0$. For the observational configuration σ derived from σ_0, a binary relation $\rightarrow |_\sigma$ on σ is defined as the relation $\mapsto |_{\sigma_0}$ in which the causally necessary internal τ-events that occurs are neglected. That is, $e_j \rightarrow |_\sigma e_i$ iff $e_j \mapsto |_{\sigma_0} e_i$ or there exist $e_{k_p} \in \sigma$ with $l(e_{k_p}) = \tau$ for $p = 1, \cdots, m$, such that $e_j \mapsto |_{\sigma_0} e_{k_1}$, $e_{k_q} \mapsto |_{\sigma_0} e_{k_{q+1}} (q = 1, \cdots, m-1)$ and $e_{k_m} \mapsto |_{\sigma_0} e_i$. It is actually the causality relation that is observable in σ_0. In the sequel, we use $\rightarrow |_\sigma$ again to represent the transitive closure of $\rightarrow |_\sigma$.

As in Section 2.2, we can similarly define that an observational configuration of Π_1 and an observational configuration of Π_2 are *isomorphic*, and we say that Π_1 and Π_2 are *observational pomset trace equivalent*, denoted $\Pi_1 \cong_{op} \Pi_2$, if for any observational configuration of Π_1, Π_2 has an observational configuration isomorphic to it, and vice versa. Clearly, $\Pi_1 \cong_p \Pi_2$ implies $\Pi_1 \cong_{op} \Pi_2$.

Theorem 4.1.3 (*Coincidence Result*). $f(s(B)) \cong_{op} s(g(B))$.

This theorem demonstrates that with respect to the cpo-based denotational semantics our syntactic and semantic refinement operations coincide up to observational pomset trace equivalence.

Please note that all the recursions occurring in expression B should be guarded in Theorem 4.1.3. In addition, Theorem 4.1.3 may not hold if we do not abstract away from τ-events. The main reason is that we can by no means require that some τ-actions can be executed simultaneously. The constraint that actions in $Sort(B)$ are not allowed to be refined and the constraint for refinement of parallel composition are also necessary.

Example 4.1.4. The semantic model of the expression $g(B)$ of Example 3.1.5 is observational pomset trace equivalent to the pes $f(\Pi)$ of Example 3.2.5 (Figure 4). Therefore, up to observational pomset trace equivalence the refinement of pes Π can be used as the semantic model of the refinement of expression B, and on the other hand the refinement of expression B can be used as the specification of the refinement of pes Π.

If we require that the pes Π_a used to replace any observable action a remains 'observable', namely any configuration of Π_a that successfully terminates contains at least one event which is labelled with an observable action, then Theorem 4.1.1 holds for observational configurations and Theorem 4.1.2 holds for observational pomset trace equivalence.

All the conclusions above hold for history preserving bisimulation equivalence and observational history preserving bisimulation equivalence.

5 Concluding Remarks

In this paper, we defined the notions of syntactic and semantic action refinements in a probabilistic process algebra with a true concurrency model. We showed that they coincide with each other up to observational pomset trace equivalence and observational history preserving bisimulation equivalence with respect to a cpo-based denotational semantics. We demonstrated that they are safe and (observational) pomset trace equivalence and history preserving bisimulation equivalence are congruences under them.

This paper is a further development of [10]. The refinement operation defined here can be mapped in a safe refinement operation in the non-deterministic counterpart. Moreover, The results presented in this paper also hold for the metric semantics developed in [19]. This semantics can be easily extended to the probabilistic case.

Appendix

In this part we sketch proofs of our theorems.

Theorems 3.1.4 follows from the fact that $ppa(B) = true$ implies $ppa(g(B)) = true$. Theorem 3.2.5 can be proved by directly checking the two conditions in the definition of pes's. We hereby focus on the remaining statements.

Lemma 1 below shows that introducing the start-event into a pes does not have influence on the observational behaviours. It does not alter the properties of being well-labelled.

Lemma 1. (1) Π is well-labelled iff $r(\Pi)$ is well-labelled,
(2) $\Pi_1 \cong_p \Pi_2$ iff $r(\Pi_1) \cong_p r(\Pi_2)$.

Now suppose that σ_f is a configuration of $f(\Pi)$. Let

$$\rho_1(\sigma_f) = \{e \mid \exists(e, e_j) \in \sigma_f\}.$$

For $e \in \rho_1(\sigma_f)$ with $l(e) \notin \Omega_0$, let

$$\rho_2(\sigma_f, e) = \{e_j \mid \exists(e, e_j) \in \sigma_f\}.$$

$\rho_1(\sigma_f)$ is the projection of σ_f on Π, and $\rho_2(\sigma_f, e)$ the projection of σ_f on $rfl(e)$.

Lemma 2. (1) $\rho_1(\sigma_f) \in C(\Pi)$, (2) $\rho_2(\sigma_f, e) \in C(rfl(e))$ and it successfully terminates if e is not maximal in $\rho_1(\sigma_f)$.

The semantic models of expressions have the following properties:

• σ_1 is a configuration of $s(B_1)$ iff $\sigma = \{e_a\} \cup \sigma_1$ is a configuration of $s(a_T.B_1)$, where event e_a is labelled with action a. The causality relations, action labels and probabilities of events in σ_1 remains unchanged in σ.

• σ is a configuration of $s(B_1; B_2)$ iff either σ is a configuration of $s(B_1)$, or $\sigma = \sigma_1 \cup \sigma_2$, where σ_1 is a configuration of $s(B_1)$ that successfully terminates and σ_2 is a configuration of $s(B_2)$. The causality relations, action labels and probabilities of events in σ_1 and σ_2 remain unchanged in σ except that the event in σ_1 labelled with action $\surd$ is relabelled with action τ in σ.

• σ is a configuration of $s(B_1 + B_2)$ iff it is a configuration of $s(B_1)$ or a configuration of $s(B_2)$. The causality relations, action labels and probabilities of events remains unchanged.

• σ is a configuration of $s(B_1 +_p B_2)$ iff it is a configuration of $s(B_1)$ or a configuration of $s(B_2)$. The causality relations and action labels remains unchanged. If e is an initial event and its probability is q in $s(B_1)$ [$s(B_2)$], then the probability of e in $s(B_1 +_p B_2)$ is $q \cdot p$ [resp. $q \cdot (1-p)$].

• For a configuration σ of $s(B_1 \parallel_A B_2)$, let

$$\rho_1(\sigma) = \{e_1 \mid \exists(e_1, e_2) \in \sigma \wedge e_1 \neq *\} \text{ and}$$
$$\rho_2(\sigma) = \{e_2 \mid \exists(e_1, e_2) \in \sigma \wedge e_2 \neq *\}$$

then $\rho_k(\sigma)$ is a configuration of $s(B_k)$ ($k \in \{1, 2\}$). On the other hand, assume that σ_k is a configuration of $s(B_k)$ which satisfies the conditions:

(1) $e^a_{k1}, \cdots, e^a_{kn}$ are all the events in σ_k labelled with action $a \in A$, and

(2) if $e^a_{1i} \mapsto_1 e^a_{1j}$ then $e^a_{2j} \not\mapsto_2 e^a_{2i}$ and if $e^a_{2i} \mapsto_2 e^a_{2j}$ then $e^a_{1j} \not\mapsto_1 e^a_{1i}$.

Then

$$\begin{array}{l}\{(e_1, *) \mid e_1 \in E(\sigma_1) \wedge \text{ the action labelled to } e_1 \text{ is not in } A\} \cup \\ \{(*, e_2) \mid e_2 \in E(\sigma_2) \wedge \text{ the action labelled to } e_2 \text{ is not in } A\} \cup \\ \cup_{a \in A}\{(e^a_{1i}, e^a_{2i}) \mid i = 1, \cdots, n\}\end{array}$$

is a configuration of $s(B_1 \parallel_A B_2)$. In addition, event (e_1, e_2) causes event (e'_1, e'_2) in $s(B_1 \parallel_A B_2)$ iff event e_1 causes event e'_1 in $s(B_1)$ or event e_2 causes event e'_2 in $s(B_2)$ (See e.g. [15,17] for the usage of $*$-event).

• σ_1 is a configuration of $s(B_1)$ iff it is a configuration of $s(B_1[\lambda])$, but each event $e \in E(\sigma_1)$ is relabelled by λ. The causality relations remain unchanged.

• σ_1 is a configuration of $s(B_1)$ iff it is a configuration of $s(B_1 \backslash A)$, but each event $e \in E(\sigma_1) \cap A$ is relabelled with τ. The causality relations remain unchanged.

• σ is a configuration of $s(\mu x.B_1)$ iff there exists $k \geq 0$ such that σ is a configuration of $v^k_{B_1}(\bot)$. The causality relations remain unchanged. Here pes $\bot = (\emptyset, \emptyset, \emptyset, \emptyset, \emptyset)$, function $v_{B_1} : PES \to PES$ is defined that substitutes a pes for each occurrence of x in B_1, and $v^k_{B_1}(\bot) = v_{B_1}(v^{k-1}_{B_1}(\bot))$.

An immediate corollary of these facts is the following lemma.

Lemma 3. If $s(B_1) \cong_{op} s(B_2)$ and $s(B_3) \cong_{op} s(B_4)$ then
$\circ s(B_1) \cong_{op} \circ s(B_2)$ for $\circ \in \{a_T., \backslash A, [\lambda]\}$,
$s(B_1 \circ B_3) \cong_{op} s(B_2 \circ B_4)$ for $\circ \in \{;, \|_A, +, +_p\}$.

Lemma 4 follows from the above facts, Lemma 3 and the requirement that the actions involved with abstraction and relabelling operators are not allowed to be refined.

Lemma 4. $C(f(a.s(B_1))) = C(a.f(s(B_1)))$ if $a \in \Omega_0$,
$C(f(a.s(B_1))) = C((\tau.f(a)); f(s(B_1)))$ if $a \notin \Omega_0$,
$C(f(\circ s(B_1))) = C(\circ f(s(B_1))) for \circ \in \{\backslash A, [\lambda]\}$,
$C(f(s(B_1) \circ s(B_2))) = C(f(s(B_1)) \circ f(s(B_1)))$ for $\circ \in \{;, +, +_p\}$.

By the facts above, Lemma 4 and the condition that $A_{g(a)} \cap A_{g(b)} = \emptyset$ for two distinct $a \in A$ and $b \in \Theta$, for parallel composition we have the following lemma. Remark that it does not hold if τ-events are not abstracted away.

Lemma 5. $f(s(B_1) \|_A s(B_2)) \cong_{op} f(s(B_1)) \|_{g(A)} f(s(B_2))$.

Theorem 4.1.1 follows from Lemma 1 and Lemma 2. Theorem 4.1.2 follows from Lemma 2 and Theorem 4.1.1. Theorem 4.1.3 follows from Lemmas 3, 4 and 5. The conclusion for recursion follows from the fact that $f(v^k_{B_1}(\bot)) \cong_{op} v^k_{g(B_1)}(\bot)$ for $k \geq 0$. This fact can be proved via the conclusions for the other operators.

Acknowledgements

Thanks to the anonymous referees for their valuable comments and suggestions.

References

1. Abramsky, S., Jung, A., Domain Theory, *Handbook of Logic in Computer Science* (S. Abramsky and Dov M. Gabbay eds.), Clarendon Press, Volume 3: 1 – 168, 1994.
2. Aceto, L, *Action Refinement in Process Algebra*, Cambridge Univ. Press, 1992.

3. Bolognesi, T., Brinksma, E., Introduction to the ISO Specification Language LOTOS, *Comp. Netw. and ISDN Syst.*, 14: 25 – 59, 1987.
4. Bowman, H., Derrick, J., Extending LOTOS with Time: A True Concurrency Perspective, *Lecture Notes in Computer Science*, 1231: 383 – 399, 1997.
5. Bowman, H., Katoen, J.-P., A True Concurrency Semantics for ET-LOTOS, *Proceedings Int. Conference on Applications of Concurrency to System Design*, 228 – 239, 1998.
6. Brinksma, E., Katoen, J.-P., Langerak, R., and Latella, D., Partial-Order Models for Quantitative Extensions of LOTOS, *Computer Networks and ISDN Systems*, 30(9/10): 925 – 950, 1998.
7. D'Argenio, P. R., Hermanns, H., and Katoen, J.-P., On Generative Parallel Composition, *Electronic Notes in Theoretical Computer Science*, 22, 1999.
8. Fecher, H., Majster-Cederbaum, M., and Wu, J., Bundle Event Structures: A Revised Cpo Approach, *Information Processing Letters*, accepted, 2001.
9. Fecher, H., Majster-Cederbaum, M., and Wu, J., Refinement of Actions in a Real-Time Process Algebra with a True Concurrency Model, *REFINE 2002*, accepted, 2002.
10. van Glabbeek, R., Goltz, U., Refinement of Actions and Equivalence Notions for Concurrent Systems, *Acta Informatica*, 37: 229 – 327, 2001.
11. van Glabbeek, R., Smolka, S. A., and Steffen, B., Reactive, Generative and Stratified Models of Probabilistic Processes, *Information and Computation*, 121: 59 – 80, 1995.
12. van Glabbeek, R., Smolka, S. A., Steffen, B., and Tofts, C. M. N., Reactive, Generative and Stratified Models of Probabilistic Processes, *Proc. LICS'90*, IEEE-CS Press, 130 – 141, 1990.
13. Gorrieri, R., Rensink, A., Action Refinement, *Handbook of Process Algebra*, Elsevier Science, 1047 – 1147, 2001.
14. Hoare, C.A.R., *Communicating Sequential Processes*, Prentice-Hall, 1985.
15. Katoen, J.-P., *Quantitative and Qualitative Extensions of Event Structures*, PhD Thesis, University of Twente, 1996.
16. Katoen, J.-P., Baier, C., and Latella, D., Metric Semantics for True Concurrent Real Time, *Th. Comp. Sci.*, 254(1-2): 501 – 542, 2001.
17. Katoen, J-P., Langerak, R., Latella, D., and Brinksma, E., On Specifying Real-Time Systems in a Causality-Based Setting, in: *Formal Techniques in Real-Time and Fault-Tolerant Systems*, Lecture Notes in Computer Science, 1135: 385 – 405, 1996.
18. Langerak, R., *Transformations and Semantics for LOTOS*, PhD Thesis, University of Twente, 1992.
19. Loogn, R., Goltz, U., Modelling Nondeterministic Concurrent Processes with Event Structures, *Fund. Inf.*, 14(1): 39 – 74, 1991.
20. Majster-Cederbaum, M., Salger, F., Correctness by Construction: Towards Verification in Hierarchical System Development, *Lecture Notes in Computer Science*, 1885: 163 – 180, 2000.
21. Majster-Cederbaum, M., Salger, F., and Sorea, M., A Priori Verification of Reactive Systems, *Proc. FORTE/PSTV 2000*, Kluwer Academic Publishers, 35 – 50, 2000.
22. Majster-Cederbaum, M., Wu, J., Action Refinement for True Concurrent Real Time, *Proc. ICECCS 2001*, IEEE Computer Society Press, 58 – 68, 2001.
23. Milner, R., *Communication and Concurrency*, Prentice-Hall, 1989.
24. Murphy, D., Pitt, D., Real-Timed Concurrent Refineable Behaviours, *Lecture Notes in Computer Science*, 571: 529 – 545, 1992.
25. Žic, J., Time-Constrained Buffer Specifications in CSP+T and Timed CSP, *ACM Transactions on Programming and Systems*, 16(6): 1661 – 1674, 1994.

Probabilistic Unfoldings and Partial Order Fairness in Petri Nets

Stefan Haar*

INRIA
IRISA, Campus de Beaulieu
35042 Rennes cedex, France
Stefan.Haar@irisa.fr

Abstract. The article investigates fairness and conspiracy in a probabilistic framework, based on unfoldings of Petri nets. Here, the unfolding semantics uses a new, cluster-based view of local choice. The algorithmic construction of the unfolding proceeds on two levels, *choice* of steps inside conflict clusters, where the choice may be fair or unfair, and the *policy* controlling the order in which clusters may act; this policy may or may not conspire, e.g., against a transition. In the context of an example where conspiracy can hide in the partial order behavior of a life and 1-safe Petri net, we show that, under non-degenerate i.i.d. randomization on both levels, both conspiracy and unfair behavior have probability 0. The probabilistic model, using special Gibbs potentials, is presented here in the context of 1-safe nets, but extends to any Petri net.

1 Introduction

Fairness in concurrent systems and algorithms has long been a subject of extensive studies ([18]). The extension to partial order properties has been studied much less often, see however Best [9], Vogler [32] and Völzer [31]. The subtlety and difficulty of fairness w.r.t. partial order behavior is that *conspiracies* may hide behind concurrency of processes. In our opinion, *Petri nets* provide the most transparent way to study these issues; this article introduces new tools for using Petri nets in a probabilistic model, and apply this model to a problem in partial order fairness.

To start, consider Figure 1. Its left hand side models the *five philosophers problem* studied, e.g., in [9]: let transition x (A, B, C, D) represent philosopher x (A, B, C, D) taking two forks and starting to eat. This is obviously possible in the initial marking, and even if one or both of x's neighbors eat (that is, tokens are absent from u or v or both), the tokens will eventually return (if progress is assumed). However, x's neighbors may alternate their "eating intervals" in such a way that at most one of the two places u and v carries a token at any given moment, but never both of them. At the same time, x is relatively independent of A and B, but their behavior has an *indirect* impact on x via the *direct* conflicts with C and D; we will come back to dependencies of this kind below.

* This work was supported by the MAGDA2 project, RNRT.

H. Hermanns and R. Segala (Eds.): PAPM-PROBMIV 2002, LNCS 2399, pp. 95–114, 2002.

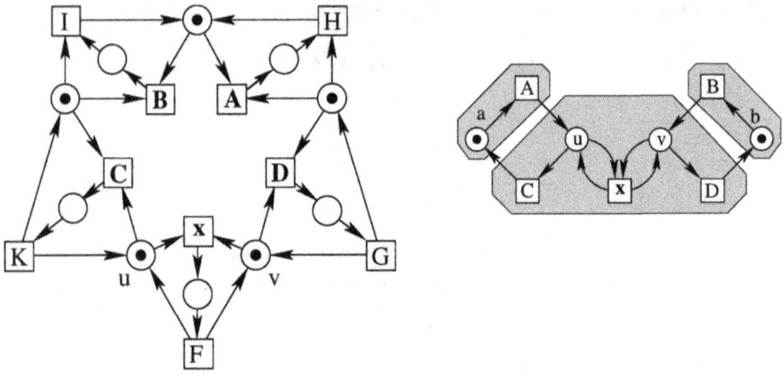

Fig. 1. Left: Five philosophers; Right: Net $\mathcal{N}$ (Clusters shown as shaded areas)

To study the above *conspiracy* phenomenon more closely, consider the Petri net $\mathcal{N}$ on the right hand side in Figure 1; it isolates a local aspect of the *five philosophers problem.* We modified the structure of the net for simplicity; still, focusing on Transition x, we see that its situation vis-à-vis its conflict neighbors C and D is the same as on the left. Which behaviors "treat" transition x "correctly" ? A first approach, which seems "maximal", is to require *strong fairness*: iff places u and v are marked jointly an infinite number of times, then x should fire infinitely often. However, there is a possibility for the net to *"conspire against* x", that does not violate strong fairness and has all other transitions firing infinitely often: take any infinite firing sequence $\sigma \in ((AC)^*(BD)^*)^*$. On any such σ, x will never be enabled; so, from the point of view of x, it cannot even be *noticed* that x is denied its opportunities. This shows that *strong fairness* is not *strong enough* to ensure "just" or "equitable" treatment of possible actions. The stronger notion of ∞-*fairness* introduced in [9] captures some of these cases:

1. a marking M *k-enables* ($k \in \mathbb{N}_0$) transition t iff there is a firing sequence $\mathsf{M} \longrightarrow \mathsf{M}'$ of length at most k to some marking M' that enables t;
2. a firing sequence σ is k-unfair (∞-unfair) w.r.t. t iff t occurs finitely often in σ and is k-enabled (∞-enabled) by infinitely many markings along σ.

In the example, *any* infinite sequence is either strongly fair or ∞-unfair w.r.t. x; in the latter case, x is conspired against, since it is denied any *actual* opportunity to fire infinitely often[1]. However, systems with no concurrency at all may be ∞-unfair as well; since there can be no *conspiracy* in such systems, ∞-fairness seems not fine enough a criterion. In our view, the essential problem lies in the fact that sequences do not reflect concurrency, and that interleaving-based fairness notions do not capture partial order phenomena because of the inherent global state view; hence the need to consider fairness with a partial order viewpoint,

[1] It is interesting to note that at least $\sigma \in ((AC)^+(BD)^+)^*$ as above violates *computable fairness* as introduced by Jaeger [23].

as e.g. in [7,6,9,25]. However, the semantics here will be new; it is particularly adapted to probabilistic reasoning, as we will show below.

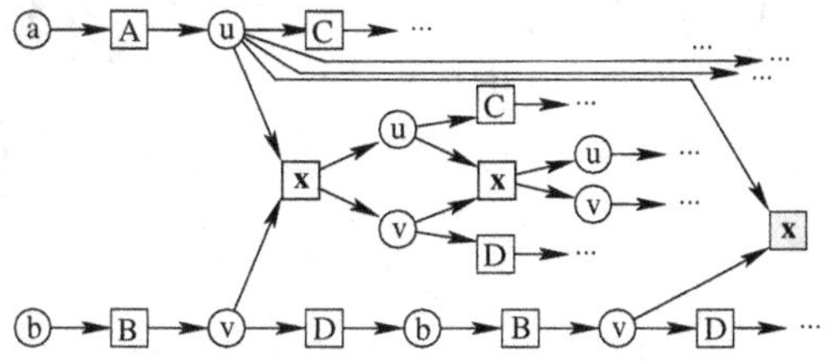

Fig. 2. A branching process for the net of Figure 1

In the example, it is *concurrency*, more precisely the partial order behavior of the net that reveals the opportunities that x "misses". Hence we will consider a partial order semantics given by *concurrent runs.*

Further, reasoning about fairness is strengthened by weighing the possible behavior by an adequate probability measure. In fact, fairness properties can rarely be shown for *all possible* behaviors of distributed systems; it is thus of theoretical as well as practical relevance to have plausible measures of probability for (sets) behaviors that allow to focus on likely rather that exotic cases. In fact, since fairness properties concern the infinite runs only, zero-one-laws apply, and the only relevant probabilities are 0 (*almost surely excluded*) and 1 (*almost surely satisfied*).

However, in the vast majority of the literature on the subject, the probabilistic approach is restricted to sequential models, such as in *Stochastic Transition Systems* (de Alfaro[4,5]) and *probabilistic automata* (Rabin [28]; Segala [30]). Völzer [31] shows that partial order models can also be used for probabilistic verification. His approach is, however, limited to unfoldings for nets where all *probabilistic* choice is made in *Free choice* ([12]) conflicts; other forms of bifurcation, if they arise, are thus left totally or partially *non-deterministic.*

This restriction, while keeping the element of probabilistic routing, is lifted in [8]: all conflicts are randomized, using renormalization whenever conflicts are not resolved locally. Here, combinatorial oddities may prevent the existence of the renormalization and hence of the probability measure (see the discussion of Figure 2 below). The approach is limited to a class that properly includes Free choice nets, yet has to leave out most of the cases interesting for fairness/conspiracy questions.

We argue here that the use of *probabilistic Petri Net unfoldings* is a key to the probabilistic analysis of concurrent systems described by Petri Nets. Still, the approach of [8] has to be modified in a central point. Consider the branching process [27,15] for the above net, displayed in Figure 2. Recall that condition elements stand for tokens on places, and events for occurrences of transitions; we

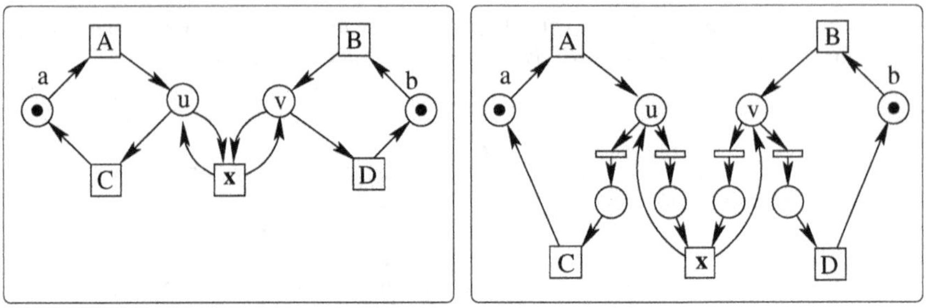

Fig. 3. The use of pre-selection (see [8])

have labeled all condition and place nodes with the name of the corresponding node in Figure 1.

The conflicts to be randomized are *(i)* between x and C at u and *(ii)* between x and D at v. Now, a token on u may wait to be consumed by x, while B and D fire an arbitrarily large number of times; hence, the conditions labeled u in Figure 2 have an infinite number of outgoing arcs to different x-labeled events. Randomizing these choices is thus a problematic task: how should the probability mass for x be distributed over the instances of x in the post-set of any u-condition ? And how can this possibly be done with the knowledge of only a *finite* prefix ?

An (apparently) obvious way to push randomization in such situations was suggested and discussed in [8], namely, *pre-selection*. This technique transforms any conflict that resembles the one in place u above into a Free choice, by splitting the outgoing arcs as illustrated in Figure 3. This allows to probabilize just the initial choice between C and "eventually x, that is, *waiting* for the corresponding choice to be made in v.

However, this auxiliary technique will not work in general, as the example in Figure 4 shows: for the live and safe net on the left hand side, pre-selection (center) introduces the possibility of new blocked markings, impossible in the original net: on the right hand side, a prefix of the branching process unfolding under pre-selection is shown; the shaded places correspond to a new, dead marking, made possible by the pre-selection.

We think these problems are intrinsic to the standard unfolding semantics based on *branching processes*. In our view, that semantics does not sufficiently reflect the interdependence between local conflicts, and hence of the random variables involved in the conflicts. In order to treat concurrent runs with an adequate probabilistic model, we propose to use a different semantics. Below, we will present a *cluster-oriented* partial order semantics on Petri nets, introduced in [21], that allows to capture conspiracy as a measurable set of nodes, and moreover allows to show that in the above example this conspiracy is almost surely excluded, for all non-degenerate probability measures adapted to that semantics in the way we described below. We will outline some elements of the probabilistic model here; more on the theory can be found in [22]. Section 2

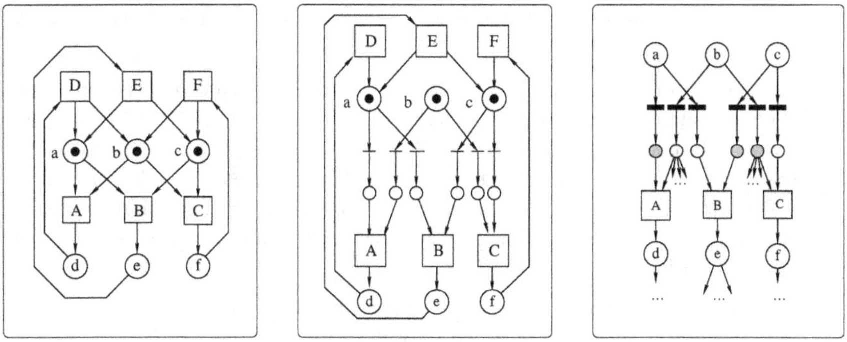

Fig. 4. A net (left) for which pre-selection falsifies the behavior (center, right; see [8])

develops the unfolding under the cluster viewpoint; the duality of local actions and their scheduling in Petri nets leads to the use of the *cluster net* explained in Section 3. The probabilistic model developed in Section 4 will be applied in Section 5 to the above example, proving that conspiracy has probability 0.

2 Petri Nets and their Unfoldings

To fix ideas, let us first review definitions. A *Petri net* or *(PN)* is a tuple of the form $\mathcal{N} = (\mathcal{P}, \mathcal{T}, \mathsf{F}, \mathsf{M})$, with $\mathcal{P} = \mathcal{P}(\mathcal{N})$ a set of places and $\mathcal{T} = \mathcal{T}(\mathcal{N})$ a set of transitions such that $\mathcal{P} \cap \mathcal{T} = \emptyset$ and $\mathcal{P} \cup \mathcal{T}$ is finite; $\mathsf{F} \subseteq [(\mathcal{P} \times \mathcal{T}) \cup (\mathcal{T} \times \mathcal{P})]$ is a set of *arcs*. For $\mathsf{x} \in (\mathcal{P} \cup \mathcal{T})$, denote pre-and postset of x as ${}^\bullet\mathsf{x} \stackrel{\mathsf{def}}{=} \{\mathsf{x}' \mid (\mathsf{x}', \mathsf{x}) \in \mathsf{F}\}$ and $\mathsf{x}^\bullet \stackrel{\mathsf{def}}{=} \{\mathsf{x}' \mid (\mathsf{x}, \mathsf{x}') \in \mathsf{F}\}$, respectively; by extension, write for $\mathcal{X} \subseteq (\mathcal{P} \cup \mathcal{T})$,

$$ {}^\bullet\mathcal{X} \stackrel{\mathsf{def}}{=} \bigcup_{\mathsf{x} \in \mathcal{X}} [{}^\bullet\mathsf{x}] \text{ and } \mathcal{X}^\bullet \stackrel{\mathsf{def}}{=} \bigcup_{\mathsf{x} \in \mathcal{X}} [\mathsf{x}^\bullet] . $$

For simplicity, we will only consider 1-safe nets here; the extension to the general case is possible, see [22]. A *marking* of $\mathcal{N}$ is a set of places; $\mathsf{M}_0 \subseteq \mathcal{P}$ is the *initial marking* of $\mathcal{N}$. Transitions may fire one by one or in sets; any transition set $\xi \subseteq \mathcal{T}$ is called a *step*. In particular, denote as λ the *empty step*; the set of steps of $\mathcal{N}$ is denoted $\mathfrak{S}(\mathcal{N})$. A step ξ is *enabled* in a marking M, denoted $\mathsf{M} \stackrel{\xi}{\longrightarrow}$, iff

$$ {}^\bullet\xi \subseteq \mathsf{M}. \tag{1} $$

Step ξ *transforms* marking M into M' , denoted $\mathsf{M} \stackrel{\xi}{\longrightarrow} \mathsf{M}'$, iff $\mathsf{M} \stackrel{\xi}{\longrightarrow}$ and

$$ \mathsf{M}'(\mathsf{p}) = [\mathsf{M}(\mathsf{p}) \backslash {}^\bullet\xi] \cup \xi^\bullet. \tag{2} $$

A marking M is *reachable* from M_0, denoted $\mathsf{M}_0 \xrightarrow{*} \mathsf{M}$, iff: (i) $\mathsf{M} = \mathsf{M}_0$, or (ii) there exists a *firing sequence* $\mathsf{M}_0 \xrightarrow{\xi_1} \mathsf{M}_1 \xrightarrow{\xi_2} \dots \xrightarrow{\xi_n} \mathsf{M}_{n+1} = \mathsf{M}$. The semantic domain for *branching unfoldings* of Petri Nets are *Occurrence nets*; some preparations before defining them. Set $< := \mathsf{F}^+$ and $\leqslant = \mathsf{F}^*$. For $\mathsf{t}_1, \mathsf{t}_2 \in \mathcal{T}$, write $\mathsf{t}_1 \ \mathsf{ic} \ \mathsf{t}_2$ iff $\mathsf{t}_1 \neq \mathsf{t}_2$ and ${}^\bullet\mathsf{t}_1 \cap {}^\bullet\mathsf{t}_2 \neq \emptyset$. The *conflict* relation $\#$ is given as follows: For $\mathsf{x}, \mathsf{x}' \in \mathcal{P} \cup \mathcal{T}$, write $\mathsf{x} \# \mathsf{x}'$ iff there exist $\mathsf{t}, \mathsf{t}' \in \mathcal{T}$: **(i)** $\mathsf{t} \ \mathsf{ic} \ \mathsf{t}'$, **(ii)** $\mathsf{t} \leqslant \mathsf{x}$, and **(iii)** $\mathsf{t}' \leqslant \mathsf{x}'$. A tuple of three sets $\overline{\mathsf{N}} = (\mathsf{B}, \mathsf{E}, \overline{\mathsf{F}})$ that has a *PN-like structure*[2], i.e. $\mathsf{B} \cap \mathsf{E} = \emptyset$, and $\overline{\mathsf{F}} \subseteq [(\mathsf{B} \times \mathsf{E}) \cup (\mathsf{E} \times \mathsf{B})]$, is called a *pre-occurrence net* iff it satisfies:

1. *no backward branching*: $|{}^\bullet\mathsf{b}| \leqslant 1$ for all $\mathsf{b} \in \mathsf{B}$, with ${}^\bullet(\cdot)$ given by $\overline{\mathsf{F}}$;
2. *Acyclicity*: With $<$ as above, $\neg(x < x)$ for all $x \in (\mathcal{P} \cup \mathcal{T})$;
3. *absence of auto-conflict*: $\neg(x \# x)$ for all $x \in (\mathcal{P} \cup \mathcal{T})$;
4. With $\mathbf{c}_0 = min_<$, $\overline{\mathsf{N}}$ is *place-initialized*, i.e. $\mathbf{c}_0 \subseteq \mathsf{B}$.

Hence $< \stackrel{\text{def}}{=} \overline{\mathsf{F}}^+$ as above yields a partial order in pre-occurrence nets. An *occurrence net* or ON is a pre-occurrence net that, in addition, is **(i)** *well-founded*, i.e. there exists no infinitely $<$-decreasing sequence, and **(ii)** *place-bordered*: all maximal nodes are in B. The *place-bordered* requirement is non-standard but inessential; any net meeting all other requirements can be extended into a place-bordered one, without changing its other properties.

The objects we will need below to build our probability space are *configurations*. For this, we first introduce the following auxiliary relations: Set $\mathsf{id} := \{(x, x) : x \in \mathsf{B} \cup \mathsf{E}\}$; the *causal dependence* is $\mathsf{li} := < \cup <^{-1}$, and *elementary concurrency* is $\mathsf{co} := (\mathsf{B} \cup \mathsf{E})^2 - (\mathsf{li} \cup \# \cup \mathsf{id})$. A *prefix* of $\overline{\mathsf{N}} = (\mathsf{B}, \mathsf{E}, \overline{\mathsf{F}})$ is any subnet spanned by a place-bordered set $\mathcal{U} \subseteq (\mathsf{B} \cup \mathsf{E})$ that is *causally closed*: If $\mathsf{x}_1 \in \mathcal{U}$ and $\mathsf{x}_2 \leqslant \mathsf{x}_1$, then $\mathsf{x}_2 \in \mathcal{U}$. Obviously, every prefix is *initialized*, that is, $\mathbf{c}_0 \subseteq \mathcal{U}$. The set of prefixes of $\overline{\mathsf{N}}$ is denoted Pref. A *configuration* is a conflict-free prefix $\mathbf{C}$, i.e. such that $\mathsf{x} \# \mathsf{x}'$ never holds for $\mathsf{x}, \mathsf{x}' \in \mathbf{C}$. Denote the set of configurations of $\overline{\mathsf{N}}$ as $\mathbb{C}(\overline{\mathsf{N}})$, and the set of *maximal* configurations as Ω; any $\omega \in \Omega$ is called a *run*. Runs correspond to maximal concurrent executions of a net, under some process semantics. Concerning these semantics, we saw, in Figure 2, some difficulties with branching processes as in [15,16,17,26,27,31]; instead, we will use the *cluster semantics* of [21,22]. It gives the same roles to cuts and configurations, but its local mechanism is based on the fact that the net $\mathcal{N}$ is naturally partitioned into subnets that are "minimally closed under conflicts"; the unfoldings are successively created from local choices of steps. These subnets are the *clusters* of $\mathcal{N}$ according to the following definition (see [12]):

The **cluster** $\gamma(\mathsf{x})$ *of* $\mathsf{x} \in (\mathcal{P} \cup \mathcal{T})$ *is the smallest set containing* x *such that* :

$$\forall\ \mathsf{t} \in \mathcal{T}: \qquad {}^\bullet\mathsf{t} \cap \gamma(\mathsf{x}) \neq \emptyset \Rightarrow \mathsf{t} \in \gamma(\mathsf{x}); \tag{3}$$

$$\forall\ \mathsf{p} \in \mathcal{P}: \qquad \mathsf{p}^\bullet \cap \gamma(\mathsf{x}) \neq \emptyset \Rightarrow \mathsf{p} \in \gamma(\mathsf{x}) \tag{4}$$

By extension, we call *cluster* any set $\gamma \subseteq (\mathcal{P} \cup \mathcal{T})$ for which there exist $x \in (\mathcal{P} \cup \mathcal{T})$ such that $\gamma = \gamma(x)$; we denote the set of clusters of $\mathcal{N}$ as $\Gamma(\mathcal{N})$. In Figure

[2] Note that we cannot require finiteness here, as we did in the definition of Petri nets, since the occurrence nets obtained from unfoldings of finite nets are typically infinite.

1, $\mathcal{N}$ has three clusters: $\gamma(A) = \{a, A\}$, $\gamma(B) = \{b, B\}$, and $\gamma(\mathsf{x}) = \{u, v, \mathsf{x}, C, D\}$; the three clusters are shown in shaded areas. Of course, $\gamma(\mathsf{x}) = \gamma(C) = \gamma(D)$.

Clusters are *closed under local conflicts*: the minimal sets that contain, with any node x, all elements (places, competing transitions) pertaining to the resolution of any given conflict that involves x.

Remark 1. *In [1,2], conflicts between individual untimed transitions in GSPN are randomized using static priorities and transition weights: the probability of firing* t *is the weight of* t*, divided by the sum of the weights of* all *transitions in the* conflict set*. These conflict sets coincide with the* clusters *here, in the class of Petri Nets where both definitions are applicable (in particular, where all transitions have equal priority). There, single firing of transitions is required; so, steps of transitions with a common pre-place are not considered as enabled, and auto-concurrency of transitions is excluded. With the obvious generalization of the weight approach to steps, this* local *randomization is included as a special case in the approach here (cf. the cluster measures below). Note, however, that the framework of [2] is in a* global time *setting, and the non-determinism between clusters that we will encounter below is not resolved; hence the two models are compatible only locally,* inside *clusters (of untimed transitions).*

Clusters are of a very simple form in Free-Choice-Nets: all transitions of a given cluster have the same, and only, pre-place. This means that any transition is enabled iff all other transitions of its cluster are; and the choice between these transitions can be made locally by the common pre-place, call it p, by *routing* any token arriving on p to some transition. This will not work in general clusters: transitions may depend on several places that may not all be marked, or whose routers do not agree with one another, and conflicts are no longer *locally* decidable.

We will unfold $\mathcal{N}$ in such a way that the events of the unfolding represent instances not necessarily of *single* transitions but of *steps*. So, *joint* firing will be reflected, in all of the above cases, by a single event. However, using *global* steps would be unwise from a computational point of view, and also ignore the actual independence of events at a great distance (in terms of causal influence) from one another. So, rather than looking at *all* steps enabled in some global state of the net, examine the *local steps* (subset of $\mathcal{T}_\gamma \stackrel{\mathsf{def}}{=} \gamma \cap \mathcal{T}$) within each *cluster*; once enabled, they can be fired irrespective of the behavior of other clusters, and allow to calculate the global steps as their combinations. This is the *cluster viewpoint*, which can be called "semi-local".

To formalize, let $\mathfrak{S}(\gamma)$ be the set of γ-steps, and $\mathfrak{S}_\Gamma(\mathcal{N})$ the set of all γ-steps of $\mathcal{N}$. We will denote as $\mathbf{in}(\gamma) \stackrel{\mathsf{def}}{=} \gamma \cap \mathcal{P} = {}^\bullet\gamma$ and $\mathbf{out}(\gamma) \stackrel{\mathsf{def}}{=} \{p \in \mathcal{P}\backslash\gamma \mid {}^\bullet p \cap \gamma \neq \emptyset\}$ the sets of *input* and *output* places of a cluster γ, respectively. Note that $\mathbf{out}(\gamma)$ is, in general, a proper subset of $\gamma^\bullet$. For $\gamma \in \Gamma(\mathcal{N})$, a *$\gamma$-step* is a subset of $\mathcal{T}(\gamma)$ and thus a step of $\mathcal{N}$; denote as $\mathfrak{S}(\gamma)$ the set of γ-steps, and set $\mathfrak{S}_\Gamma(\mathcal{N}) := \bigcup_{\gamma\in\Gamma(\mathcal{N})} \mathfrak{S}(\gamma)$. Thus, $\mathfrak{S}_\Gamma(\mathcal{N}) \subseteq \mathfrak{S}(\mathcal{N})$, and the inclusion is proper in general.

Cluster processes associate an occurrence net to $\mathcal{N}$, reflecting place markings by conditions (circles in the figures) and cluster steps by *events* (rectangles):

Definition 1. *Let* $\mathcal{N} = (\mathcal{P}, \mathcal{T}, \mathsf{F}, \mathsf{M}_0)$ *be a net and* $\overline{\mathsf{N}} = (\mathsf{B}, \mathsf{E}, \overline{\mathsf{F}}, \mathbf{c}_0)$ *an* ON, *and let* $\pi : \mathsf{B} \to \mathcal{P}$, $\mu : \mathsf{B} \to \{0, 1\}$ *and* $\beta : \mathsf{E} \to \mathfrak{S}_\Gamma(\mathcal{N})$ *be mappings.* $\Pi = (\overline{\mathsf{N}}, \pi, \beta, \mu)$ *is a* **cluster process** *iff:*

1. *for all* $\mathsf{b} \in \mathbf{c}_0$, $\mu(\mathsf{b}) = \mathsf{M}_0(\pi(\mathsf{b}))$;
2. *for all* $\mathsf{e}_1, \mathsf{e}_2 \in \mathsf{E}$, $[{}^\bullet\mathsf{e}_1 = {}^\bullet\mathsf{e}_2 \wedge \beta(\mathsf{e}_1) = \beta(\mathsf{e}_2)] \Rightarrow \mathsf{e}_1 = \mathsf{e}_2$;
3. *for all* $\mathsf{b}_1, \mathsf{b}_2 \in \mathsf{B}$: $[\mathsf{b}_1 \ \mathsf{co} \ \mathsf{b}_2] \Rightarrow \pi(\mathsf{b}_1) \neq \pi(\mathsf{b}_2)$;
4. *for all* $\mathsf{e} \in \mathsf{E}$ *and any* $\mathsf{p} \in [\mathbf{in}(\beta(\mathsf{e})) \cup \mathbf{out}(\beta(\mathsf{e}))]$, *there exist* $\mathsf{p}_{in}, \mathsf{p}_{out} \in \mathcal{P}$ *such that:* **(i)** ${}^\bullet\xi \cap \pi^{-1}(\{p\}) = \{\mathsf{p}_{in}\}$ *and* $\xi^\bullet \cap \pi^{-1}(\{p\}) = \{\mathsf{p}_{out}\}$, **(ii)** $\mu(\mathsf{p}_{in}) \geqslant \sum_{\mathsf{t} \in \xi} \mathbb{1}_{({}^\bullet\mathsf{t})}(\mathsf{p})$, *and* **(iii)** $\mu(\mathsf{p}_{out}) = [\mu(\mathsf{p}_{in}) - \sum_{\mathsf{t} \in \xi} \mathbb{1}_{({}^\bullet\mathsf{t})}(\mathsf{p})] + \sum_{\mathsf{t} \in \xi} \mathbb{1}_{(\mathsf{t}^\bullet)}(\mathsf{p})$, *where* $\xi := \beta(\mathsf{e})$; *compare (1) and (2).*

The mappings π and β correspond to the structural unfolding, taking conditions to places and events to steps; note in particular that the empty step, which we denote as λ, is always enabled; hence, an empty event will appear in all unfoldings. Since conditions represent *states* of places, we also need the mapping μ to assign token numbers to conditions. The only values of μ that we are dealing with here are 0 and 1. We cannot, as is done in the branching process semantics, discard all conditions in $\mu^{-1}(0)$: the choice, in cluster γ, of a step to be fired is made based on a complete knowledge of the tokens that can be used by γ; absence of a condition for place p must therefore be considered as lack of information on the state of p, and blocks the decision process of γ.

Remark 2. *In order to extend Definition 1 to Petri nets that are not 1-safe, it suffices to modify Condition 4: allow the steps* ξ *to be multi-sets of transitions, and arc-weights greater than 1. One obtains conditions of the form*

$$\mu(\mathsf{b}_{in}) \geqslant \langle \pi(\mathsf{b})^{\odot}, \delta \rangle \ \textit{and} \ \mu(\mathsf{b}_{out}) = (\mu(\mathsf{b}_{in}) - \langle \pi(\mathsf{b})^{\odot}, \delta \rangle) + \langle {}^{\odot}\mathsf{b}, \delta \rangle, \tag{5}$$

where $\langle \bullet, \bullet \rangle$ *denotes inner product of the appropriate* $\mathbb{Z}$*-module, and* ${}^{\odot}(\bullet)$, $(\bullet)^{\odot}$ *the vectors of in-and outgoing arc weights for the respective place. The probabilistic models below both extend to the general, not 1-safe case, with analogous adjustments as for the semantics here.*

For the net $\mathcal{N}$ of Figure 1, some processes are shown in Figure 6. Consider the following sequences of actions:

1. Suppose $\gamma(\mathsf{x})$ acts first; then nothing happens, so we may ignore this case.
2. If $\gamma(A)$ acts and then $\gamma(\mathsf{x})$, we obtain the center part of Figure 6.
3. A symmetric scenario in the right hand part of Figure 6: $\gamma(B)$ before $\gamma(\mathsf{x})$.
4. If $\gamma(A)$ and $\gamma(B)$ precede $\gamma(\mathsf{x})$, we obtain the process $\hat{\pi}$ on the left hand side. $\gamma(A)$ and $\gamma(B)$ act *concurrently* here, their ordering is irrelevant.

All processes of $\mathcal{N}$ are composed of these three basic pieces, glued together at the cuts indicated by oblong boxes and corresponding to the initial marking.

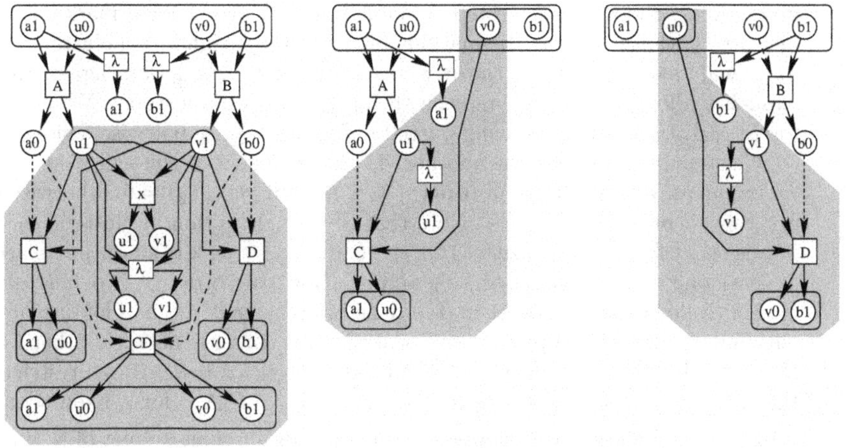

Fig. 5. Three different processes for Figure 1; $\hat{\pi}$ shown on the left

We will see how non-determinism arises in the algorithm that constructs processes as defined in Definition 1. Intuitively, the procedure is the following: The initial marking is represented by a set of conditions b, such that $\pi(\mathsf{b})$ is a place and $\mu(\mathsf{b})$ the number of tokens present. Choose a cluster γ and determine the set of γ-steps enabled under γ's current local marking (i.e., the global marking M restricted to the places p of γ). For each of these steps ξ, create an event e (i.e. $\beta(\mathsf{e}) = \xi$); draw an $\overline{\mathsf{F}}$-arc to e from all b in the initial cut such that $\pi(\mathsf{b})$ belongs to γ *or is an output place of* ξ; put a copy b' of each pre-condition b of e behind e, i.e. with an $\overline{\mathsf{F}}$-arc from e to b'; and set $\mu(\mathsf{b}')$ equal to $\mu(\mathsf{b})$ plus/minus the effect of ξ (for a formal definition and the extension to the general, not 1-bounded case see [22].) The construction extends, in each round, every branch of the process by a tile associated to the cluster γ selected there: this tile contains an event for every possible local step. Denote any maximal process obtained in this way as *ND-unfolding*. Then, there exists a unique *maximal ND-unfolding*, see below.

3 Scheduling of Clusters: The Cluster Net

We now study more closely the choice of clusters in the n-th round. In fact, it is here that conspiracies are possible: as described above in the context of Figure 1, it is by suitably scheduling the cluster actions that x was denied any enabling. To formalize this, we will refine the scheduling model from cluster sequences to more "parallel" orderings below. Assume first that clusters are inspected in the order given by some infinite *sequence* $\mathbf{d} = \mathbf{d}_1, \mathbf{d}_2, \mathbf{d}_3, \ldots$, where every $\mathbf{d}_i$ stands for the choice of some cluster γ of $\mathcal{N}$. Following the control theoretic terminology, we will speak of *designs* for such sequences. In the n-th action, the cluster γ_n

given by $\mathbf{d}_n$ chooses one γ_n-step ξ enabled in the current local marking on γ_n, and fires ξ, thus changing the marking on γ_n and its neighboring clusters. Then, γ_{n+1} will take its turn, in the new situation after γ_n's action; again, we will explicitly allow the choice – and the "firing"– of the empty step λ. We note that for *fixed* $\mathbf{d}$, tiles in the resulting partial unfolding $\Pi_{\mathbf{d}}$ are pairwise disjoint.

Clusters interact with one another via places; thus, two clusters γ_1 and γ_2 are *independent*, written $\gamma_1 \mathbb{I} \gamma_2$, iff $(\mathbf{out}(\gamma_1) \cup \gamma_1) \cap (\mathbf{out}(\gamma_2) \cup \gamma_2) = \emptyset$; otherwise they are *dependent*, written $\gamma_1 \mathbb{D} \gamma_2$. Note that, by the definition of clusters, $\gamma_1 \neq \gamma_2$ implies that for any $\mathsf{x} \in (\mathbf{out}(\gamma_1) \cup \gamma_1) \cap (\mathbf{out}(\gamma_2) \cup \gamma_2)$, x is an output place for at least one of the two clusters γ_1 and γ_2. Now, the order of two *independent* clusters following one another in $\mathbf{d}$ may be interchanged in the unfolding process without changing $\Pi_{\mathbf{d}}$ (up to an isomorphism of occurrence nets). In other words, the unfolding depends on $\mathbf{d}$ only via its *Mazurkiewicz trace*. If one restricts, in the nth round, the new nodes to a $\mathbb{I}$-clique $\Delta_n \subset \Gamma(\mathcal{N})$; for a fixed sequence $(\Delta_n)_{n \in \mathbb{N}}$ of $\mathbb{I}$-cliques, called a *policy*, one obtains a unique subnet of $\hat{\overline{\mathsf{N}}}$, and the associated process is the Δ*-unfolding.*

So, for any $\mathbb{I}$-clique $\mathcal{X} \subseteq \Gamma$, all clusters of $\mathcal{X}$ may be considered *simultaneously* in the above algorithm, yielding the same unfolding (up to an isomorphism of marked graphs) as any ordering of $\mathcal{X}$; in the five philosophers' case, this corresponds to two non-neighboring philosophers eating in parallel. We will thus use sequences $\Delta_1 \Delta_2 \ldots$ of such $\mathbb{I}$-cliques; call these sequences *policies.*

Policies can be generated in a distributed way with a maximum of parallelization, using an abstract Petri Net. This *cluster net* $\mathcal{N}^\Gamma$ is not itself included in the structure of $\mathcal{N}$, which remains unchanged; rather, it provides a formal model producing all parallelized policies possible on $\mathcal{N}$, with in addition a natural randomization of these policies. Thus, we can use a unifying model on several levels; in particular, the probability measures we will introduce below for the local *choice inside a cluster* can also, on the higher level, schedule the global order in which clusters act. This will allow a unified parallelized and randomized unfolding. $\mathcal{N}^\Gamma$ has as place set a copy $\mathcal{P}^\Gamma$ of $\mathcal{P}$, $\Gamma(\mathcal{N})$ as set of transition, and

$$\mathsf{F}^\Gamma \overset{\text{def}}{=} \{(\gamma, \mathsf{p}), (\mathsf{p}, \gamma) \mid \mathsf{p} \in \gamma \ \wedge \ \exists \mathsf{t} \in \gamma : \ \mathsf{t} \ \mathsf{F} \ \mathsf{p}\}.$$

With $\mathsf{M}^\Gamma(\mathsf{p}) \overset{\text{def}}{=} 1$ for all $\mathsf{p} \in \mathcal{P}$, call $\mathcal{N}^\Gamma \overset{\text{def}}{=} (\mathcal{P}^\Gamma, \Gamma, \mathsf{F}^\Gamma, \mathsf{M}^\Gamma)$ the **cluster net** of $\mathcal{N}$. Figure 6 shows the cluster net obtained for the running example from Figure 1. So the cluster net of $\mathcal{N}$ has one transition t_γ for every cluster γ of $\mathcal{N}$, and a copy for every place p from $\mathcal{N}$; that copy, which we also denote as p, is either **(i)** connected to t_γ by an in- and an outgoing arc if $\mathsf{p} \in (\gamma \cup \mathbf{out}(\gamma))$, or **(ii)** no arc at all otherwise. Every place p in $\mathcal{N}^\Gamma$ contains one token to indicate that p can be "used" by its own cluster, or some of its neighbors that may put tokens on the p in the bottom layer $\mathcal{N}$. Since all arcs of $\mathcal{N}^\Gamma$ go in both directions, all transitions are *individually* enabled initially, and M^Γ is reproduced by every firing, so M^Γ is the only reachable marking in $\mathcal{N}^\Gamma$. Note that, if $\mathcal{N}$ is connected, $\mathcal{N}^\Gamma$ consists of a single cluster.

Any design over $\mathcal{N}$ is a single transition firing sequence of $\mathcal{N}^\Gamma$; and a step ξ of $\mathcal{N}^\Gamma$ consists of single firings of mutually independent $\mathcal{N}^\Gamma$-transitions. The

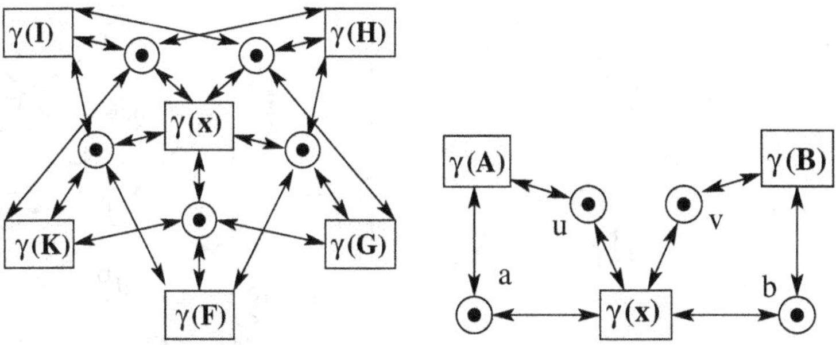

Fig. 6. The cluster nets for Figure 1; double arrows for the symmetric arc relation F^{Γ}

policies for $\mathcal{N}$ are thus exactly the step firing sequences of $\mathcal{N}^{\Gamma}$. Fix a policy Δ; in the n-th round of the ND-unfolding procedure, append one tile for each cluster $\gamma \in \Delta_n^{\Gamma}$; the result is the **$\Delta$-directed unfolding** or **Δ-unfolding** .

Figure 6 shows the cluster nets obtained from the Petri Nets of Figure 1. The set of policies on the right hand side is $\{\{\gamma(A)\}, \{\gamma(B)\}, \{\gamma(A), \gamma(B)\}\{\gamma(\mathsf{x})\}\}^{\omega}$, where $\mathcal{X}^{\omega}$ denotes the set of infinite words over $\mathcal{X}$; there are no independent clusters. By contrast, on the left hand side, we have $\gamma(F)\mathbb{I}\gamma(I)$, $\gamma(F)\mathbb{I}\gamma(K)$, etc; the policies for this larger net can thus allow parallel actions for such pairs of clusters.

On the other hand, merging inductively the tiles added by *different* policies, one obtains a series of prefixes $\hat{\mathcal{U}}_n$, $n \in \mathbb{N}$, each of which contains all behaviors possible with n successive rounds of cluster actions. In Figure 7, the original net in Part (A) has two different policy-unfoldings, cf. Parts (C) and (D). The occurrence net in Part (B) contains both of them; it represents $\hat{\mathcal{U}}_2$, containing all possible behaviors after two policy steps. The limit of the $\hat{\mathcal{U}}_n$ as $n \to \infty$ is a unique maximal process $\hat{\Pi} = (\hat{\overline{\mathsf{N}}}, \hat{\pi}, \hat{\beta}, \hat{\mu})$, called the **full unfolding** of $\mathcal{N}$.

4 The Probabilistic Model

Our probabilistic model needs to measure sets of *runs*. As we saw above, the run realized by $\mathcal{N}$ is determined by a policy Δ and the subsequent choices of steps in the clusters selected by Δ. Let Ω_1 be the set of possible policies, and $\mathbb{P}^1$ a probability on the $\mathbb{I}$-cliques of $\mathcal{N}^{\Gamma}$; let Δ_n be i.i.d., and for $\mathcal{X} \subseteq \Gamma$, let $\mathbb{P}^1\{\mathcal{X}\}$ be the probability that $\Delta_1 = \mathcal{X}$. So we identify ω_1 with the unique policy Δ it generates; the associated filtration is $(\mathcal{F}_n^1)_{n \in \mathbb{N}}$, where $\mathcal{F}_n^1$ is the σ-algebra generated by the random variables $\Delta_1, \ldots, \Delta_n$.

As we saw above, every policy Δ is associated to a unique unfolding $\Pi = \Pi_{\Delta}$. Now, consider the set of clusters $\Delta_n = \Delta_n(\omega_1) \subseteq \Gamma$; by assumption, Δ is a clique of $\mathbb{I}$. Thus all $\gamma \in \Delta_n$ can choose their step independently of one another, given the local marking that has been reached after all decisions in $\Delta_1, ..., \Delta_{n-1}$. For

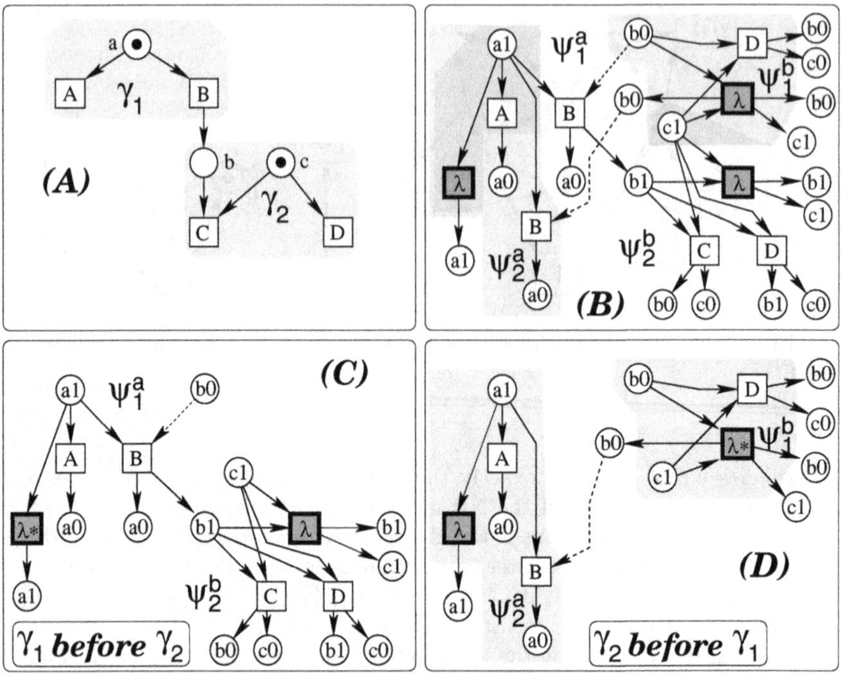

Fig. 7. The non-determinism between clusters

$\gamma \in \Gamma(\mathcal{N})$ and a local marking $\mathsf{M}\gamma \subseteq \mathcal{P}_\gamma \overset{\mathsf{def}}{=} \gamma \cap \mathcal{P}$, write $\mathsf{Enabled}_{\mathsf{M},\gamma}$ for the set of γ-steps enabled under M_γ; finally, identify Ω_2 with the sample space of the subsequent choices of the clusters.

4.1 Cluster Measures

We now consider a single cluster γ of $\mathcal{N}$ (note that the same approach will work applied to the cluster net $\mathcal{N}^\Gamma$ and hence to the policy). All steps enabled under the restriction $\mathsf{M}_{|\gamma}$ of a marking M to γ belong to $\mathfrak{S}(\gamma)$. We postulate that, to be coherent with Petri net dynamics, a probability measure for the choice of the step to be fired should depend in a functional way on $\mathsf{M}_{|\gamma}$. As we saw earlier, there is no *choice* involved in the *creation* of tokens: a step that fires is assured to produce its tokens on its post-places. By contrast, it is at the *beginning* of a step that choices are made, i.e. the enabled steps compete for the *incoming* tokens. Therefore, we have to equip with probabilities the input, and not the output of tokens: given the choice of a step, the number and places of tokens to be created is fully determined.

We will introduce some examples of cluster measures arising in a "natural" way from locally randomized behavior. In the following, let ξ be the random element of $\mathfrak{S}(\gamma)$ to be selected, and $\mathcal{F}$ the σ-algebra generated by the sets $\mathcal{X}_t \stackrel{\text{def}}{=} \{t \in \xi\}$, where t runs over $\mathcal{T}_\gamma$.

In a cluster, all transitions influence one another by a direct conflict, or indirectly through a chain of conflicts. However, suppose two transitions $t_1, t_2 \in \mathcal{T}_\gamma$ are *structurally independent*, i.e. $({}^\bullet t_1 \cup t_1{}^\bullet) \cap {}^\bullet t_2 = \emptyset$ and $({}^\bullet t_2 \cup t_2{}^\bullet) \cap {}^\bullet t_1 = \emptyset$; denote this as "$t_1$ **ind** t_2". Then, whether or not t_1 fires should not depend on t_2, given the the behavior of the other transitions of γ. Of course, there will be some influence of t_2 on the possibilities of t_1's opponents; what we mean is that once the behavior of the rest of γ is fixed and known, the conditional laws for t_1 and t_2 are independent.

The direct approach to constructing probability measures for the choice of a cluster step is the *Gibbs measure construction*, which we will present next, following the presentation of Markov field theory in [24]. Fix a cluster γ and a local marking $M_\gamma \subseteq \mathcal{P}_\gamma$. The graph $\mathcal{G}_\gamma$ has edges between transitions that are neighbors in some conflict under M_γ; i.e. let $\mathcal{V}_\gamma \stackrel{\text{def}}{=} \mathcal{T}_\gamma$, and

$$\mathcal{E}_\gamma \stackrel{\text{def}}{=} \{(t_1, t_2) \mid ({}^\bullet t_1 \cup {}^\bullet t_2) \subseteq M_\gamma \ \wedge \ {}^\bullet t_1 \cap {}^\bullet t_2 \neq \emptyset\};$$

then $\mathcal{G}_\gamma = (\mathcal{V}_\gamma, \mathcal{E}_\gamma)$. Obviously, $\mathcal{G}_\gamma$ is not directed: if $(t_1, t_2) \in \mathcal{E}_\gamma$, then also $(t_2, t_1) \in \mathcal{E}_\gamma$. On this graph, we want to establish a probability $\mathbb{P}$ for a random choice ξ of transitions jointly firable under M_γ; in other words, a probability on $\mathfrak{P}(\mathcal{T}_\gamma)$. Denote the restriction of a γ-step ξ to $\mathcal{X} \subseteq \mathcal{T}_\gamma$ as $\xi_\mathcal{X}$, and for a set $\mathcal{X} \subseteq \mathcal{T}_\gamma$, denote the *boundary* of $\mathcal{X}$ as

$$\partial\mathcal{X} \stackrel{\text{def}}{=} \{t \in \mathcal{T}_\gamma \backslash \mathcal{X} \mid \exists\, t' \in \mathcal{X} : \ (t, t') \in \mathcal{E}_\gamma\}; \tag{6}$$

then $\mathbb{P}$ is called (see [24]) a **Markov field** iff for all $\mathcal{X} \subseteq \mathcal{T}_\gamma$,

$$\mathbb{P}\left(\xi_\mathcal{X} \mid \xi_{\mathcal{T}_\gamma \backslash \mathcal{X}}\right) = \mathbb{P}\left(\xi_\mathcal{X} \mid \xi_{\partial\mathcal{X}}\right). \tag{7}$$

This requirement is natural in the light of Petri net semantics: the behaviors of two (sets of) transitions that are not close neighbors are independent once it is known whether or not their immediate conflict opponents fire. So, Markov fields are a natural and important class of measures for cluster choices. It is well known (see [24]) that the following construction based on *potentials* yields Markov fields; below, we will look at some potentials that reflect Petri net semantics, and thus lead to natural cluster measures.

A *Gibbs potential* on γ is a mapping D that associates to every subset $\mathcal{X}$ of $\mathcal{T}_\gamma$ and every γ-step ξ, a number $D_\mathcal{X}(\xi) = D(\xi_\mathcal{X})$, such that $D_\mathcal{X}(\xi) = 0$ unless $\mathcal{X}$ is a clique of $\mathcal{G}_\gamma$. The *energy* U of ξ is

$$U(\xi) \stackrel{\text{def}}{=} - \sum_{\mathcal{X} \subseteq \mathcal{T}_\gamma} D_\mathcal{X}(\xi); \tag{8}$$

obviously, the sum needs only be taken over the $\mathcal{X}$ that are cliques of $\mathcal{G}_\gamma$. Now, no ξ that contains more than one transition from any given clique is admissible in the 1-safe Petri net dynamics. We thus introduce a subclass of Gibbs potentials D on $\mathcal{G}_\gamma$ that satisfy, for all cliques $\mathcal{X}$ of $\mathcal{G}_\gamma$:

$$\forall\, \xi \subseteq \mathcal{T}_\gamma :\ |\xi \cap \mathcal{X}| > 1 \Rightarrow D_{\mathcal{X}}(\xi) = 0. \tag{9}$$

Remark 3. *Again, the approach extends to the non-1-safe case, with (9) replaced analogously to (5), and the choice measures giving probabilities for* **multi-sets** *of transitions rather than sets.*

We call every Gibbs potential D on $\mathcal{G}_\gamma$ that satisfies (9) a *1-PN potential.* Now, as usual ([24]), define the *partition function* (which will serve as a renormalization constant below) as

$$Z_{\mathsf{M}}^{\gamma} \stackrel{\mathsf{def}}{=} \sum_{\xi \subseteq \mathcal{T}_\gamma} \mathsf{e}^{-U(\xi)}; \tag{10}$$

then the associated *Gibbs measure* is

$$\mathbb{P}_{\mathsf{M}}^{\gamma}(\xi) \stackrel{\mathsf{def}}{=} \frac{\mathsf{e}^{-U(\xi)}}{Z}. \tag{11}$$

So, the essential part in specifying a Markov field is to find the right potential; we will give two examples now.

Token Routing. Routing probabilities were used in [8] and, as *coin flips*, in [31]; neither work, however, studies clusters. The idea is to assign every token that arrives in a place to a randomly chosen post-transition of that place. In Free choice conflicts, this immediately enables that transition; in the general case, enabling depends on more than one place, so its probability is not given by the router of one place: witness the situation of x in the running example.

To formalize, let us first add, for every place p of γ, a new transition $\mathsf{t_p}$ *"looped around* p*"*, i.e. ${}^\bullet\mathsf{t_p} = \mathsf{t_p}^\bullet = \{\mathsf{p}\}$. These $\mathsf{t_p}$'s model *idling* on place p, i.e. if a step $\xi \in \mathfrak{S}(\gamma)$ does not consume a token on place p, we regard ξ as containing a firing of $\mathsf{t_p}$; note that this leaves the marking process unchanged. In this way, we only need to consider steps that use all tokens on $\mathbf{in}(\gamma)$. Write

$$\hat{\mathcal{T}}_\gamma \stackrel{\mathsf{def}}{=} \mathcal{T}_\gamma \cup \{\mathsf{t_p} \mid \mathsf{p} \in \mathcal{P}_\gamma\}.$$

Then, fix routing probabilities ν_{p} on $\mathsf{p}^\bullet \cup \{\mathsf{t_p}\}$. The 1-PN potential D is then given by: for every clique $\mathcal{X} \subseteq \hat{\mathcal{T}}\gamma$ of $\mathcal{G}_\gamma$ and every step $\xi \subseteq \hat{\mathcal{T}}_\gamma$,

$$D_{\mathcal{X}}(\xi) \stackrel{\mathsf{def}}{=} \mathbb{1}_{\{|\xi\cap\mathcal{X}|=1\}} \cdot \left\{ \sum_{\mathsf{t}\in(\mathcal{X}\cap\xi)} \sum_{\mathsf{p}\in{}^\bullet\mathsf{t}} \ln\left[\nu_{\mathsf{p}}(\mathsf{t})\right] \right\} + \mathbb{1}_{\{|\xi\cap\mathcal{X}|\neq 1\}} \cdot (-\infty).$$

Applied to the cluster $\gamma(\mathsf{x})$ of the running example, with the reachable local marking $\mathsf{M} \stackrel{\text{def}}{=} \{u, v\}$ on $\gamma(\mathsf{x})$, this yields, e.g.:

$$\mathbb{P}_{\mathsf{M}}^{\gamma(x)} = \frac{\nu_u(\mathsf{x}) \cdot \nu_v(\mathsf{x})}{\begin{pmatrix} \nu_u(\mathsf{x}) \cdot \nu_v(\mathsf{x}) + \nu_u(\mathsf{t}_u) \cdot \nu_v(\mathsf{t}_v) \\ +\nu_u(C) \cdot \nu_v(\mathsf{t}_v) + \nu_u(\mathsf{t}_u) \cdot \nu_v(D) + \nu_u(C) \cdot \nu_v(D) \end{pmatrix}}.$$

That is, the probability of firing x is the probability of both place routers independently choosing x, re-normalized by the probabilities of all other routings that *agree* on some step.

Transition Coin Toss. Here, it is not the tokens that choose but the *transitions:* Assume every transition $\mathsf{t} \in \gamma$ chooses whether or not to fire – that is, $\mathsf{t} \in \xi$ or $\mathsf{t} \notin \xi$ — by a *"coin toss"* decision $\eta_{\mathsf{t}} \in \{0, 1\}$; then, discard the combinations that do not yield an enabled step. Here, we have *no* idling transitions; the probability that a transition t chooses "yes" is denoted η_{t}; the probability of "no" as $\overline{\eta}_{\mathsf{t}}$. Then, we define a 1-PN potential D by setting, for every clique $\mathcal{X} \subseteq \mathcal{T}_\gamma$ and every step $\xi \subseteq \mathcal{T}_\gamma$,

$$\begin{aligned} D_{\mathcal{X}}(\xi) \stackrel{\text{def}}{=}\ & \mathbb{1}_{\{|\xi \cap \mathcal{X}| = 1\}} \cdot \left[\sum_{\mathsf{t} \in \mathcal{X}} \ln(\eta_{\mathsf{t}})\, \mathbb{1}_{\xi}(\mathsf{t}) + \ln(\overline{\eta}_{\mathsf{t}})\, (1 - \mathbb{1}_{\xi}(\mathsf{t})) \right] \\ & + \mathbb{1}_{\{|\xi \cap \mathcal{X}| \neq 1\}} \cdot (-\infty) \end{aligned}$$

With $\gamma(\mathsf{x})$ and $\mathsf{M} \stackrel{\text{def}}{=} \{u, v\}$ as above, we have:

$$\mathbb{P}_{\mathsf{M}}^{\gamma(x)} = \frac{\eta_{\mathsf{x}} \cdot \overline{\eta}_C \cdot \overline{\eta}_D}{\begin{pmatrix} \eta_{\mathsf{x}} \cdot \overline{\eta}_C \cdot \overline{\eta}_D + \overline{\eta}_{\mathsf{x}} \cdot \overline{\eta}_C \cdot \overline{\eta}_D \\ +\overline{\eta}_{\mathsf{x}} \cdot \eta_C \cdot \overline{\eta}_D + \overline{\eta}_{\mathsf{x}} \cdot \overline{\eta}_C \cdot \eta_D + \overline{\eta}_{\mathsf{x}} \cdot \eta_C \cdot \eta_D \end{pmatrix}}.$$

So, there exist cluster measures that have, by virtue of being Markov fields, the conditional independence property (7)

Obviously, there are countless other Markov fields possible as cluster measures; the two given above seem, to us, important and natural from a Petri net viewpoint, hence adequate for a probability on partially ordered runs.

4.2 The Joint Probability Space for Scheduling and Choice

Let us turn now from the randomization of local choice to the global probabilistic model. The following construction of the filtration follows the one used in [8] and [22], with a probability space containing all *runs* and filtered by increasing prefixes. This is also the idea for *cones* in Völzer's model [31], except for the non-randomized non-deterministic parts there.

Let $\mathcal{N} = (\mathcal{P}, \mathcal{T}, \mathsf{F}, \mathsf{M})$ be a Petri Net, and $\Pi = (\overline{\mathsf{N}}, \pi, \beta, \mu)$ the full cluster unfolding of $\mathcal{N}$; here, $\overline{\mathsf{N}} = (\mathsf{B}, \mathsf{E}, \overline{\mathsf{F}})$, with initial cut $\mathbf{c}_0$. Recall that all processes are *prefixes* of the unfolding; we denote a prefix generically as $\mathcal{U}$, and the set of prefixes of $\mathcal{N}$ is $\mathsf{Pref}(\mathcal{N})$. For $\omega \in \Omega$ and a prefix $\mathcal{U}$, define the *projection of* ω *on* $\mathcal{U}$ as the configuration $\omega_{\mathcal{U}} \overset{\text{def}}{=} \omega \cap \mathcal{U}$. For fixed $\mathcal{U}$, the relation $\sim_{\mathcal{U}}$ given by

$$\omega \sim_{\mathcal{U}} \omega' \overset{\text{def}}{\Longleftrightarrow} \omega_{\mathcal{U}} = \omega'_{\mathcal{U}}, \tag{12}$$

is an equivalence on Ω; its extension to $\mathbb{C}$ is an equivalence as well. Let $\mathcal{F}_{\mathcal{U}}$ be the σ-algebra given by:

$$A \in \mathcal{F}_{\mathcal{U}} \overset{\text{def}}{\Longleftrightarrow} \left[\left. \begin{matrix} \omega \in A \\ \omega' \sim_{\mathcal{U}} \omega \end{matrix} \right\} \Rightarrow \omega' \in A \right] \tag{13}$$

Thus $(\mathcal{F}_{\mathcal{U}})_{\mathcal{U} \in \mathsf{Pref}}$ is a filtration over Ω. For a node x, set $\mathcal{A}_{\mathsf{x}} \overset{\text{def}}{=} \{\omega \mid \mathsf{x} \in \omega\}$; by extension, for a set $\mathcal{X}$ of nodes, write $\mathcal{A}_{\mathcal{X}} \overset{\text{def}}{=} \{\omega \mid \mathcal{X} \subseteq \omega\}$. If no confusion can occur, we just write x for the set $\mathcal{A}_{\mathsf{x}}$.

As above, write $\mathcal{U}_n$ for the prefix of the full unfolding obtained after n steps of Δ, and $\mathcal{F}_n$ for $\mathcal{F}_{\mathcal{U}_n}$; then $(\mathcal{F}_n)_{n \in \mathbb{N}}$ is a filtration of $\Omega \overset{\text{def}}{=} \Omega_1 \times \Omega_2$.

For $n \in \mathbb{N}$, let $\mathbf{C}_n = \mathbf{C}_n(\omega)$, $\mathbf{c}_n = \mathbf{c}_n(\omega)$ and M_n as above. That is, in the section $\mathcal{U}_n(\omega_1)$ of prefix $\mathcal{U}_n$ given by $\Delta_1(\omega_1), \ldots, \Delta_n(\omega_1)$, the cluster choices governed by ω_2 have let to the configuration $\mathbf{C}_n$, ending in condition cut $\mathbf{c}_n$ which represents marking M_n. Further, let ψ be a tile such that $\mathbf{in}(\psi) \subseteq \mathbf{c}_n$, and e an event from ψ; denote as ξ the step $\beta(\mathsf{e})$, and the cluster corresponding to ψ by γ. Then, define inductively: a family $\mathbf{C}_n$ of configuration- valued random variables; the associated probability measures $\mathbf{P}_n$, such that $\mathbf{P}_0$ and $\mathbf{C}_0$ satisfy $\mathbf{P}_0(\mathbf{C}_0 = \mathbf{c}_0) = 1$; and

$$\mathbf{P}_{n+1}(\mathsf{e} \in \mathbf{C}_{n+1} \mid \mathbf{C}_n) = \mathbf{P}^1(\gamma \in \Delta_{n+1}) \cdot \mathbf{P}^{\gamma}_{\mathsf{M}_n}(\mathsf{e} \in \mathbf{C}_{n+1}). \tag{14}$$

Recursion (14) yields a unique ([22]) limit probability that we denote by $\mathbf{P}$.

5 Almost Sure Fairness and Non-conspiracy

Definition 2. *Let* $\mathcal{N} = (\mathcal{P}, \mathcal{T}, \mathsf{F}, \mathsf{M}_0)$ *be a Petri net,* $\mathsf{t} \in \mathcal{T}$*, and* ω *a run of* $\mathcal{N}$*.*

1. ω conspires *against* t *iff*
   (a) ω *is* ∞*-fair w.r.t.* t*, and*
   (b) ω denies t*, i.e. contains only finitely many co-cliques labeled by* ${}^\bullet\mathsf{t}$*.*
2. ω *is* choice-unfair *w.r.t.* t *iff* ω *does not deny* t *but contains only finitely many copies of* t*.*

Note that if ω is choice-unfair w.r.t. t, it may nonetheless have strongly fair (w.r.t. t) interleavings, namely those in which t is only enabled a finite number of times. On the other hand, if ω is not choice-unfair w.r.t. t, then all interleavings of ω are strongly fair w.r.t. t.

The questions of *Fairness* and *Conspiracy* with respect to a given transition concern two levels of the behavior, the policy and the local choice. The examples show that one can often not expect to have all *possible* runs fair and free of conspiracy; the probabilistic model given together with the cluster unfoldings permits, however, to show that *essentially all* runs have these properties, under mild assumptions.

We call a cluster measure $\mathbb{P}^{\gamma}_{\mathsf{M}}$ *non-degenerate* iff every γ-step enabled in M has positive probability under $\mathbb{P}^{\gamma}_{\mathsf{M}}$; the family $(\mathbb{P}^{\gamma}_{\mathsf{M}})_{\gamma,\mathsf{M}}$ is i.i.d. iff

(a) on the same γ, the different choices made at different occasions are independent of the previous choices, but governed by the same marking-indexed family $(\mathbb{P}^{\gamma}_{\mathsf{M}})_{\gamma\in\Gamma,\mathsf{M}\subseteq\mathcal{P}_\gamma}$ of probability measures, and
(b) decisions on cluster γ are made independently of all $\gamma' \neq \gamma$.

Informally, an infinite prefix of the net $\mathcal{N}$ of Figure 1 *conspires* against x iff it contains only finitely many copies of $\hat{\pi}$ in Figure 6. Then we obtain:

- *Let all cluster measures for $\mathcal{N}$ be non-degenerate and i.i.d. Then*
  1. *Conspiracy against* x *has probability 0; and*
  2. x *occurs almost surely infinitely often.*

Proof: Note first that there are two clusters with two possible markings and one, namely $\gamma(\mathsf{x})$, with four ($\emptyset$, $\{u\}$, $\{v\}$, $\{u,v\}$). $\mathcal{N}$ has four reachable markings: $\{a,b\}$, $\{u,b\}$, $\{v,a\}$, $\{u,v\}$; they are in 1-1 correspondence with their restrictions $\emptyset$, $\{u\}$, $\{v\}$, and $\{u,v\}$ to $\gamma(\mathsf{x})$. Since the cluster measure for $\mathcal{N}^{\Gamma}$ is non-degenerate and the policy is i.i.d., $\gamma(\mathsf{x})$ is selected infinitely often with probability 1 by the *divergence part of the Borel-Cantelli Lemma (dBC)* ([10], Thm. 4.2.4). But that means that, if conspiracy is assumed, either *(a)* $\{a,v\}$ is reached infinitely many times, and then $\gamma(\mathsf{x})$ selected consistently except for finitely many instances, or *(b)* the same as *(a)* with $\{b,u\}$ instead of $\{a,v\}$. This has probability 0 by dBC.

For the second assertion, recall that $\{u,v\}$ is reached infinitely often; the i.i.d. assumption and dBC then lead to conclude that x occurs infinitely often in all runs, excluding a set of probability 0. □

This example shows how the above probabilistic model helps identify – and, under "reasonable" circumstances, exclude almost surely – possible conspiracies and choice-unfair behaviors. The treatment of the above example generalizes to many cases; note in particular that one would have obtained the same result if the net had possible blocking behaviors. In fact, cutting the edges leading back from x to u and v makes the *choice-unfair runs* the only infinite ones. We can repeat the above argument but with some cumbersome technical adjustments, concluding that this modified net will eventually be clean (i.e. have no tokens at all) almost surely.

One obtains more general results using the strong Markov property, proven in [22]. We will use it to show that the above case generalizes to an important class of systems; first, we have to introduce *stopping times*. These are particular prefixes of the full unfolding, with the following additional property:

Definition 3. *Let $\mathcal{N}$ and $\hat{\mathcal{U}}_n$ as above. Then a* stopping time *of $\mathcal{N}$ is a prefix τ satisfying*

$$\forall\, n \in \mathbb{N}: \ \left\{\omega \mid \omega_\tau \subseteq \hat{\mathcal{U}}_n\right\} \in \mathcal{F}_{\hat{\mathcal{U}}_n}. \tag{15}$$

Denote the set of stopping times as $\mathfrak{T}$; we will often write $\tau(\omega)$ for ω_τ.

Note that, if $\mathcal{N}$ is an S-net (contains no branching transitions), the stopping times as defined here coincide with the usual ones in linear times. The simplest examples of stopping times are *constants*, i.e. prefixes of the form $\hat{\mathcal{U}}_n$. A less trivial class, and arguably the most important one, is formed by *hitting times*: Let $\mathfrak{U}$ be a set of markings for $\mathcal{N}$, and $\mathfrak{C}_\mathfrak{U} \subseteq \mathfrak{C}(\overline{\mathsf{N}})$ be the set $\mathfrak{C}_\mathfrak{U} \stackrel{\mathsf{def}}{=} \{\mathbf{c} \in \mathfrak{C} \mid \exists\, \mathsf{M}_\mathbf{c} \in \mathfrak{U} : \ \mu_{|\mathbf{c}} = \mathsf{M}_\mathbf{c} \circ \pi_{|\mathbf{c}}\}$ of the corresponding cuts. For $\omega \in \Omega(\mathcal{N})$, let $\mathbf{c}_{\mathsf{hit}_\mathfrak{U}}$ be the earliest (with respect to the partial order on ω) cut from $\mathfrak{C}_\mathfrak{U}$, if such a cut exists on ω. Let the random variable $\mathbf{C}_{\mathsf{hit}_\mathfrak{U}}$ be the configuration

$$\mathbf{C}_{\mathsf{hit}_\mathfrak{U}}(\omega) \stackrel{\mathsf{def}}{=} \begin{cases} (\mathbf{c}_{\mathsf{hit}_\mathfrak{U}}(\omega))^{\Downarrow} & \textit{if } \mathbf{c}_{\mathsf{hit}_\mathfrak{U}} \textit{ is defined, and} \\ \omega & \textit{otherwise.} \end{cases} \tag{16}$$

Then the *(first) hitting time* for $\mathfrak{U}$ is the prefix

$$\rho_\mathfrak{U} \stackrel{\mathsf{def}}{=} \bigcup_{\omega \in \Omega} \mathbf{C}_{\mathsf{hit}_\mathfrak{U}}(\omega).$$

For any set $\mathfrak{U}$ of markings, the hitting time $\tau_{\mathsf{hit}_\mathfrak{U}}$ is a stopping time (cf. [22]).

To reason about the future evolution, we need another family of σ-algebras that abstracts in an appropriate way from the "pre-history" of the process.

Definition 4. *Fix $\tau \in \mathfrak{T}$. For $\mathcal{E} \subseteq (\mathsf{B} \cup \mathsf{E})$ and $\omega, \omega' \in \Omega$, let $\omega \sim_\mathcal{E} \omega'$ iff $\omega \cap \mathcal{E}$ and $\omega' \cap \mathcal{E}$ are isomorphic as labeled graphs (i.e. the isomorphism commutes with π, μ and β); note the weakening of "$\sim$" compared to (12). We obtain a σ-algebra $\mathcal{F}_\mathcal{E}$ on Ω by letting $\mathcal{X} \in \mathcal{F}_\mathcal{E}$ iff:*

$$\left.\begin{array}{r} \omega \in \mathcal{A} \\ \omega' \sim_\mathcal{E} \omega \end{array}\right\} \Rightarrow \omega' \in \mathcal{A};$$

so the construction is analogous to (13) but based on a weaker equivalence.

Note that, if $\mathcal{E}$ is a prefix, both definitions for $\mathcal{F}_\mathcal{E}$ are equivalent. Let the *future* of τ be the subnet $\mathfrak{Z}_\tau$ of $\overline{\mathsf{N}}$ spanned by $\mathcal{OH}([\mathsf{B} \cup \mathsf{E}] \backslash \tau)$, and the σ-algebra $\mathcal{F}_{\mathfrak{Z}_\tau}$ is defined according to Definition 4. This allows to state

Theorem 1. (Strong Markov Property, from [22]) *For all $\tau \in \mathfrak{T}$ finite and $\mathcal{A} \in \mathcal{F}_{\mathfrak{Z}_\tau}$,*

$$\mathbb{P}(\mathcal{A} \mid \mathcal{F}_\tau) = \mathbb{P}(\mathcal{A} \mid \mathsf{M}_\tau). \tag{17}$$

Here, $\mathsf{M}_\tau = \mathsf{M}_\tau(\omega)$ is the marking that corresponds to the cut given by configuration $\mathbf{C}_\tau(\omega)$

Proof: [22] □

We can thus consider a Petri net that is currently in a marking M as equivalent to one that *starts* in M; its history has no effect on the future. To make use of this *re-starting*, recall that a *home marking* M of a Petri net $\mathcal{N} = (\mathcal{P}, \mathcal{T}, \mathsf{F}, \mathsf{M}_0)$ is a marking that is reachable *from* any marking M' that is reachable from M_0. Then:

Theorem 2. *Let* $\mathcal{N} = (\mathcal{P}, \mathcal{T}, \mathsf{F}, \mathsf{M}_0)$ *be a live and safe Petri net, such that* M_0 *is a home marking. Further, assume that all cluster measures are non-degenerate. Then for all transitions* $\mathsf{t} \in \mathcal{T}$*, the probability that* t *occurs infinitely often is* 1.

Sketch of proof: Note first that *all* reachable markings are home markings under the hypothesis. For reachable markings M_1 and M_2, denote as $\mathbb{P}(\mathsf{M}_1 \longrightarrow \mathsf{M}_2)$ the probability that, starting in M_1, eventually reaches M_2. Then non-degeneracy implies $\mathbb{P}(\mathsf{M}_1 \longrightarrow \mathsf{M}_2) > 0$ for any choice of M_1 and M_2. $\mathcal{N}$ thus induces an irreducible Markov chain on its reachability graph $\mathcal{R}$. Since $\mathcal{R}$ is finite by assumption, all markings in it are recurrent, i.e. will almost surely be reached an infinite number of times. By assumption, there exists a reachable marking M_t such that $\mathsf{M}_\mathsf{t} \xrightarrow{\mathsf{t}}$. Let q be the probability that t is not selected by $\mathbb{P}^c lun_\mathsf{M}$, where γ is the cluster of t. Then non-degeneracy implies $q < 1$, so, using Theorem 1 and the above, the probability that t fires only finitely often is bounded above by $\lim_{n\to\infty} q^n = 0$, so must be 0. □

The last argument in this proof generalizes to the *convergence* part of the Borel-Cantelli Lemma [10]. One sees that Theorem 2 shows, for almost all runs, non-conspiracy *and* fairness w.r.t. all transitions, in the particularly well-behaved class of nets considered. This contains both example nets from Figure 1 For more general cases, the work still needs to be done.

Acknowledgments

I wish to thank, for many discussions and remarks, A. Benveniste, H. Völzer, and E. Fabre. Special thanks to four anonymous referees whose comments helped greatly to improve on an earlier version of this paper.

References

1. M. Ajmone Marsan, G. Balbo, G. Conte, S. Donatelli, and G. Franceschinis. *Modeling with Generalized Stochastic Petri Nets.* Wiley, 1995.
2. M. Ajmone Marsan, G. Balbo, G. Chiola, and G. Conte. *Generalized Stochastic Petri Nets Revisited: Random Switches and Priorities.* In: *Proceedings of PNPM'87*, IEEE-CS Press, pp. 44–53.
3. A. Aghasaryan, E. Fabre, A. Benveniste, R. Boubour, and C. Jard. Fault detection and diagnosis in distributed systems: An approach by partially stochastic petri nets. *Discrete event dynamic systems* **8**:203–231, 1998.
4. L. de Alfaro. From Fairness to Chance. In: *Electronic Notes on Theoretical Computer Science* **22**, Elsevier, 2000.

5. L. de Alfaro. Stochastic Transition Systems. In: *CONCUR'98*, LNCS **1466**:423-438, Springer-Verlag, 1998.
6. K.R. Apt, N. Francez, and S. Katz. Appraising fairness in languages for distributed programming. *Distributed Computing* **2**:226–241,1988.
7. K.R. Attie, N. Francez, and O. Grumberg. Fairness and hyperfairness in multi-parti interactions. *Distributed Computing* **6**:245–254,1993.
8. A. Benveniste, S. Haar, and E. Fabre. Markov Nets: probabilistic Models for Distributed and Concurrent Systems. INRIA *Report* **4235**, 2001.
9. E. Best. Fairness and Conspiracies. *Information Processing Letters* **18**(4):215–220, 1984. Erratum ibidem **19**:162.
10. K. L. Chung. *A Course in Probability Theory.* Academic Press, 1974.
11. C. Derman. *Finite State Markovian Decision Processes.* Acad. Press, 1970.
12. J. Desel and J. Esparza. *Free Choice Petri Nets.* Camb. Univ. Press, 1995.
13. R. Dobrushin. The description of a random field by means of conditional probabilities and conditions of its regularity. *Th. Prob. Appl.* **13**:197-224, 1968.
14. S. Dolev, A. Israeli, and S. Moran. Analyzing expected time by scheduler-luck games. *IEEE Trans on Par. and Dist. Systems*, **8**(4):424–440, 1997.
15. J. Engelfriet. Branching Processes of Petri Nets. *Acta Inf.* **28**:575-591, 1991.
16. J. Esparza, S. Römer, and W. Vogler. An Improvement of McMillan's Unfolding Algorithm. In: *TACAS'96*, LNCS **1055** :87–106, Springer 1996. Extended version to appear in *Formal Methods in System Design.*
17. J. Esparza. Model Checking Using Net Unfoldings. *Science of Computer Programming* **23**:151–195, 1994.
18. N. Francez. *Fairness.* Springer, 1986.
19. S. Haar. Branching Processes of general S/T-Systems. *Electronic Notes in Theoretical Computer Science* **18**, 1998.
20. S. Haar. Occurrence Net Logics. *Fund. Informaticae* **43**:105-127, 2000.
21. S. Haar. Clusters, confusion and unfoldings. *Fund.Informaticae* **47**(3-4):259-270, 2001.
22. S. Haar. Probabilistic Cluster Unfoldings for Petri Nets. INRIA *Research Report* **4426**, 2002; `http://www.inria.fr/rrrt/rr-4426.html`; `ftp://ftp.inria.fr/INRIA/publication/RR/RR-4426.ps.gz`
23. M. Jaeger. Fairness, Computable Fairness and Randomness. *PROBMIV'99*, Tech. Report CSR-99-8, School of CS, Univ. of Birmingham 1999.
24. R. Kindermann and J. L. Snell. *Markov Random Fields and their Applications.* AMS Contemporary Mathematics Vol. **1**, Providence 1980.
25. M. Kwiatkowska. Event fairness and non-interleaving concurrency. *Formal Aspects of Computing* **1**:213–228, 1989.
26. K. McMillan. Using Unfoldings to avoid the state explosion problem in the verification of asynchronous circuits. *4th CAV*, pp. 164–174, 1992.
27. M. Nielsen, G. Plotkin, and G. Winskel. Petri nets, event structures, and domains. Part I. *Theoretical Computer Science* **13**:85–108, 1981.
28. M.O. Rabin. Probabilistic Automata. *Inf. and Control* **6**:230–245, 1963.
29. G. Rozenberg and J. Engelfriet. Elementary Net Systems. In: *Lectures on Petri Nets I: Basic Models. LNCS* **1491**, pp. 12–121, Springer, 1998.
30. R. Segala. Verification of Randomized Distributed Algorithms. In: LNCS **2090**:232-260, Springer 2001.
31. H. Völzer. Randomized non-sequential processes. In: *Proc. CONCUR 2001.* LNCS **2154**, Springer, August 2001.
32. W. Vogler. Fairness and Partial Order Semantics. *Inf. Proc. Letters* **55**(1): 33-39, 1995.

Possibilistic and Probabilistic Abstraction-Based Model Checking

Michael Huth

Department of Computing
Imperial College of Science, Technology and Medicine
180 Queen's Gate, London, SW7 2BZ, United Kingdom
huth@cis.ksu.edu

Abstract. We present a framework for the specification of abstract models whose verification results transfer to the abstracted models for a logic with unrestricted use of negation and quantification. This framework is novel in that its models have quantitative or probabilistic observables and state transitions. Properties of a quantitative temporal logic have measurable denotations in these models. For probabilistic models such denotations approximate the probabilistic semantics of full LTL. We show how predicate-based abstractions specify abstract quantitative and probabilistic models with finite state space.

1 Introduction

Probabilistic models of concurrent systems [45] are important for the quantitative design and analysis of safety-critical systems [24]. Such models are also indispensable for the analysis of quantitative behaviour in a wide variety of systems, e.g. through the computation of performance measures [26]. Formal analysis of probabilistic models does not scale well. Model checking LTL formulas on concurrent labelled Markov chains is polynomial in the size of models and doubly exponential in the size of formulas [12]. These complexity bounds inflate the effects of the state-explosion problem, that the number of states of a composed model is often exponential in the number of its components. Therefore, probabilistic verification of realistic models requires the use of aggressive abstraction techniques.

Model checks $\mathcal{M} \models \phi$ can be abstracted by simplifying the model $\mathcal{M}$ [10], the property ϕ [25], or the satisfaction relation $\models$. To be effective, such simplifications need to ensure that they render sound, and hopefully useful, analysis results. For qualitative systems, we present instances for all of these simplifications, prove their soundness, and discuss their utility for analysing probabilistic systems.

These instances are realised by transferring work on three-valued model checking [33,5,6,29,22] to the realm of probabilistic verification. Three-valued models allow specifiers to state under-determinacy in non-deterministic choices: if "There are possible delays on the Bakerloo Line." is the only available information, then it should not mean "For all other lines, there are no delays." This

H. Hermanns and R. Segala (Eds.): PAPM-PROBMIV 2002, LNCS 2399, pp. 115–134, 2002.

additional expressiveness is also critically needed for the computation of abstract models whose verified properties carry over to the models they abstract, where properties range over a full logic with negation and quantification. Such a range is required, for example, if one mixes abstraction-based checks with simple fairness assumptions [3,28] or for the verification of properties that combine safety and liveness aspects. We offer this transfer also for systems whose quantities are specified in any partial order (cost, total energy, weighted sums, etc).

The resulting relational calculus for specifying and computing sound abstract models works well for abstractions based on finite, measurable partitions of the underlying state space, making it a powerful tool for practical verification tasks. However, models require that basic observables be measurable. This limits the freedom of specifying models as well as the scope of the applicability of our relational abstraction calculus from a foundational point of view. It is hoped that a more general theory will emerge from this paper that generalises its results to abstractions that are continuous-state concurrent labelled Markov chains. Markov kernels [21], as outlined in [38,18,19], are a likely candidate for such a theory.

Since this paper works with a branching-time logic, we are able to give a coordinated approximation of the satisfaction relation and thresholds of probabilistic LTL formulas. Although we cannot yet comment on the practical utility of this abstraction, it has apparent connections to bounded model checking techniques [9].

Outline of paper. In Section 2 we survey existing work on abstraction of probabilistic systems. Section 3 generalises labelled concurrent Markov chains to modal quantitative structures and develops our notions of possibilistic abstraction and refinement. A possibilistic property semantics for a quantitative mu-calculus is given in Section 4, its consistency and soundness with respect to refinement is proved and a path lemma (for CTL* formulations) is shown. In Section 5, we prove that our possibilistic property semantics is an abstraction of the usual branching-time probabilistic logic PCTL of Hansson [24] for modal probabilistic systems. A systematic way of specifying abstract models through an abstraction relation on states is presented in Section 6 and its soundness and compositionality proved. We show that a modal version of probabilistic simulations [30] is the operational equivalent of our possibilistic refinement for functional and discrete abstractions. Finally, Section 7 concludes.

2 Related Work

Di Pierro and Wiklicky [41] use the Moore-Penrose pseudo-inverse of linear operators to re-cast Galois connections, which require orderings, in the setting of vector spaces; this allows for a re-formulation of soundness and optimality principles for abstract interpretations in linear spaces. Monniaux [36] systematically develops abstract interpretations of infinite-state concurrent Markov chains [45]; these analyses may use state-space partitioning.

Jonsson and Larsen [30] generalise probabilistic bisimulation [32] to a satisfaction relation between probabilistic specifications — multi-set versions of probabilistic transition systems — and probabilistic transition systems; two notions and algorithms for refinement between probabilistic specifications are presented. D'Argenio et al. [15] define simulations between concurrent Markov chains that are based on a discrimination criterion and the co-inductive existence of distributions. Such simulations allow for the sound verification of safety properties and incremental refinement of abstractions driven by refutation evidence.

Clark et al. [8] present a program analysis of probabilistic idealised Algol that collects possibilistic information flow between high and low security variables; this (abstract) analysis is shown to be sound for the probabilistic non-interference of Sabelfeld and Sands [43]. Di Pierro et al. [40] provide a quantitative version of identity confinement for probabilistic concurrent constraint programming (without non-determinism), using a probabilistic version of the widening operator [13] for a safe abstraction of their concrete collection semantics.

In the framework of probabilistic automata, Segala and Lynch present and investigate several notions of probabilistic simulations with respect to compositionality — where these notions fare well — and the preservation of properties written in probabilistic CTL [24] — where these notions fare poorly.

Desharnais et al. [17] approximate continuous-state Markov processes by a family of finite-state labelled Markov chains; they define a notion of (probabilistic) simulation and prove its soundness with respect to a fragment of probabilistic propositional modal logic. Vardi (e.g. [46]) shows that properties of the form "with probability 1 satisfies ϕ" can be expressed as an ergodic analysis and therefore checked through automata-theoretic means.

Morgan et al. [37] study a probabilistic version of the process algebra CSP and show that probabilistic choice distributes through all other operators; a failure/divergence semantics [27] supplies a refinement notion between such processes. MvIver [34] generalises stationary distributions of Markov processes to models of probabilistic programs that include non-determinism (abstraction) with support for Dijkstra-style reasoning.

For a simple but practically important fragment of temporal logic, Zuck [47] replaces probabilistic assumptions with strong fairness assumptions and thereby reduces P-validity checks on parameterised probabilistic systems to validity checks over non-probabilistic programs.

Andova and Baeten [1] define a branching probabilistic bisimulation for a probabilistic process algebra without non-deterministic choice whose rooted branching variant is a congruence with respect to sequential composition and probabilistic choice; abstractions operate on internal actions.

3 Modal Quantitative Systems

We present modal versions of *quantitative* models for abstraction-based model checking. Labeled concurrent Markov chains [16,45] and their modal abstractions turn out to be a special instance of such models. In a partial order $(P, \leq)$, we write $\geq$ for the relational inverse $\{(r, r') \in P \times P \mid r' \leq r\}$ of $\leq$. The relation $<$ is obtained by removing from $\leq$ the diagonal $\{(r, r') \in P \times P \mid r = r'\}$ of P; as customary, its inverse is denoted by $>$.

Definition 1 (Modal Quantitative Kripke Structures).

1. *Let $\mathcal{F}$ be a σ-algebra [23] over a state set Σ, $(P, \leq)$ a partial order of quantities, and $[\mathcal{F} \to P]$ the set of monotone (total) functions of type $(\mathcal{F}, \subseteq) \to (P, \leq)$; elements of $[\mathcal{F} \to P]$ are* quantitative measures.
2. *Given a set* AP *of state observables, a* quantitative Kripke structure $\mathcal{K}$ with signature $(\mathtt{AP}, \mathcal{F}, P)$ *is a triple (Σ, R, L), where $L \colon \Sigma \to \mathcal{P}(\mathtt{AP})$ is a labelling function and $R \subseteq \Sigma \times [\mathcal{F} \to P]$ a transition relation such that for all $A \in \mathcal{F}$, $\sqsupseteq \in \{\geq, >\}$, $r \in P$, and $p \in \mathtt{AP}$*

$$\mathrm{pre}_{\sqsupseteq r}(A) \stackrel{\mathrm{def}}{=} \{s \in \Sigma \mid \exists (s, \mu) \in R,\ \mu(A) \sqsupseteq r\} \in \mathcal{F} \tag{1}$$
$$\{s \in \Sigma \mid p \in L(s)\} \in \mathcal{F};$$

3. *a* modal quantitative Kripke structure $\mathcal{M}$ with signature $(\mathtt{AP}, \mathcal{F}, P)$ *is a pair $(\mathcal{M}^{\mathrm{a}}, \mathcal{M}^{\mathrm{c}})$ of quantitative Kripke structures $\mathcal{M}^{\mathrm{m}} = (\Sigma, R^{\mathrm{m}}, L^{\mathrm{m}})$, $\mathrm{m} \in \{\mathrm{a}, \mathrm{c}\}$, with the same signature such that $R^{\mathrm{a}} \subseteq R^{\mathrm{c}}$ and $L^{\mathrm{a}}(s) \subseteq L^{\mathrm{c}}(s)$ for all $s \in \Sigma$.*

Throughout this paper, we assume that modal quantitative Kripke structures are *finitely branching*: for all $\mathrm{m} \in \{\mathrm{a}, \mathrm{c}\}$ and $s \in \Sigma$, the set $\{\mu \in [\mathcal{F} \to P] \mid (s, \mu) \in R^{\mathrm{m}}\}$ is finite. Since every probability measure is also a quantitative measure [23], our models are generalisations of established systems that combine non-determinism and probabilistic transitions. Our abstractions of probability measures in Section 6 turn out to be probability measures if boolean or Cartesian abstraction is used.

Example 1. **Neural systems.** Given a Kripke structure with finite state set Σ and state observables AP, we transform it into a quantitative Kripke structure for the partial order $[0, \infty)$. Let $\mathcal{F}$ be $\mathcal{P}(\Sigma)$. We endow each state $s \in \Sigma$ with a stimulus $k_s \in [0, \infty)$. Each vector of weights $(w_s)_{s \in \Sigma}$ with $w_s \in [0, \infty)$ then determines an element $\mu \in [\mathcal{F} \to [0, \infty)]$ given by

$$\mu(A) = \sum_{a \in A} w_a \cdot k_a. \tag{2}$$

To complete the model, we specify a labelling function L and transitions of the form (s, μ) such that all μ can be represented in the form of (2). The semantics of modal operators is then similar to the effects of information propagation in neural networks [42]. Let A be the set of all states that satisfy $p \in L(s)$. Checking whether $\mathsf{EX}_{\sqsupseteq r}\, p$ holds at s_0 — as specified in Section 4 — amounts to verifying whether there is a transition $(s_0, \mu) \in R$ such that the weighted sum in (2) is $\sqsupseteq r$. In our framework of quantitative systems, the stimuli k_s are static and the monotonicity of μ demands that all weights w_s and stimuli k_s be non-negative.

Concurrent Labelled Markov Chains. Finite-state modal quantitative Kripke structures that satisfy $\mathcal{M}^{a} = \mathcal{M}^{c}$ and $\mathcal{F} = \mathcal{P}(\Sigma)$ and whose quantitative measures are probability measures are essentially labelled concurrent Markov chains [45] (without designated fair states).

An Abstraction. Figure 1 depicts a concurrent labelled Markov chain (left) and its abstraction (right), which we re-visit in Section 6.

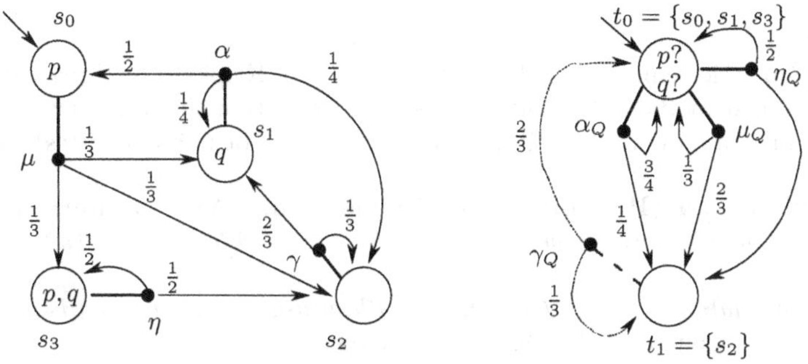

Fig. 1. Left: A graphical representation of a concurrent, labelled Markov chain. Right: Predicate abstraction, along the predicate $p \vee q$, of the model on the left. The only R^{a}-transition is γ_Q. We write $p?$ for $p \in L^{c}(t_0) \setminus L^{a}(t_0)$ etc.

In this paper, we base the specification of abstract models on relations. For $Q \subseteq \Sigma_1 \times \Sigma_2$, $A \subseteq \Sigma_1$, and $B \subseteq \Sigma_2$, we recall the standard relational navigation $A.Q \stackrel{\text{def}}{=} \{t \in \Sigma_2 \mid \exists a \in A : \ (a,t) \in Q\}$ and $Q.B \stackrel{\text{def}}{=} \{s \in \Sigma_1 \mid \exists b \in B : (s,b) \in Q\}$. In measure theory, it is well known that relational navigation does not mix well with the preservation or reflection of measurable sets. Nonetheless, this paper uses relational navigation as a means for specifying abstract models, since important abstraction techniques can be carried out successfully with such a pedestrian approach. Thus, we offer a direct means of specifying boolean and Cartesian abstractions for practitioners, but we encounter limitations if more general abstractions are desired. A general theory for abstraction of probabilistic systems, with Markov kernels as the prime candidates, is desirable.

Definition 2 (Measurable Navigation). *Given two σ-algebras* $(\Sigma_1, \mathcal{F}_1)$ *and* $(\Sigma_2, \mathcal{F}_2)$*, a relation* $Q \subseteq \Sigma_1 \times \Sigma_2$ has measurable navigation *iff for all* $A \in \mathcal{F}_1$ *and* $B \in \mathcal{F}_2$*,* $A.Q \in \mathcal{F}_2$ *and* $Q.B \in \mathcal{F}_1$*.*

We emphasise that relations do not have measurable navigation in general, but we show in Section 6 that important abstraction relations do enjoy that property. For that, condition (1) is needed to ensure that all properties of our logic denote measurable sets; this is a non-condition for finite systems.

Before we present a property semantics for modal quantitative systems, we discuss their co-inductive definition of possibilistic refinement and abstraction. These notions require a lift of relations over state spaces to relations over quantitative measures. Our definitions assume a fixed partial order P and our results hold for all partial orders.

Definition 3 (Lifting Relations to Quantitative Measures). *For every $Q \subseteq (\Sigma, \mathcal{F}) \times (\Sigma, \mathcal{F})$ with measurable navigation, we define $Q^{\text{ps}} \subseteq [\mathcal{F} \to P] \times [\mathcal{F} \to P]$ by $(\mu, \eta) \in Q^{\text{ps}}$ iff for all $A, B \in \mathcal{F}$, $\eta(A.Q) \geq \mu(A)$ and $\mu(Q.B) \geq \eta(B)$.*

Possibilistic refinements of modal quantitative Kripke structures are defined as for Kripke modal transition systems [29], except that the co-inductive constraints are put on pairs of quantitative measures (and not pairs of states) via Q^{ps}.

Definition 4 (Possibilistic Refinement and Abstraction). *Let $\mathcal{M}$ be a modal quantitative Kripke structure $(\mathcal{M}^{\text{a}}, \mathcal{M}^{\text{c}})$ with signature $(\texttt{AP}, \mathcal{F}, P)$.*

1. *A relation $Q \subseteq (\Sigma, \mathcal{F}) \times (\Sigma, \mathcal{F})$ with measurable navigation is a* possibilistic refinement *in $\mathcal{M}$ iff $(s, t) \in Q$ implies*
   (a) whenever $(t, \eta) \in R^{\text{a}}$, there is some $(s, \mu) \in R^{\text{a}}$ such that $(\mu, \eta) \in Q^{\text{ps}}$;
   (b) whenever $(s, \mu) \in R^{\text{c}}$, there is some $(t, \eta) \in R^{\text{c}}$ such that $(\mu, \eta) \in Q^{\text{ps}}$;
   (c) $L^{\text{a}}(t) \subseteq L^{\text{a}}(s)$ and $L^{\text{c}}(s) \subseteq L^{\text{c}}(t)$.
2. *The relational inverse Q^{-1} of a refinement Q is a* possibilistic abstraction.

Given modal quantitative Kripke structures $\mathcal{M}_i = ((\Sigma_i, R_i{}^{\text{a}}, L_i{}^{\text{a}}), (\Sigma_i, R_i{}^{\text{c}}, L_i{}^{\text{c}}))$ with $i = 1, 2$ and signatures over the same partial order $(P, \leq)$, we can define their sum $\mathcal{M}_1 + \mathcal{M}_2$, whose state space, state observables, and σ-algebra are the sum of the respective two structures. Thus, Definition 4 also applies *between* such models (with initial state) in the usual manner. If all measures of $\mathcal{M}_1$ and $\mathcal{M}_2$ are probabilistic, then that is also the case for their sum.

Co-inductive, monotone definitions over complete lattices have a greatest fixed point [39,35]. Since σ-algebras are not complete lattices in general, we need to ensure that the computation of such a greatest fixed point resides within a given σ-algebra. This is guaranteed for finite-state systems.

Proposition 1 (Greatest Possibilistic Refinement). *For every modal quantitative Kripke structure $\mathcal{M}$ with signature $(\texttt{AP}, \mathcal{F}, P)$, possibilistic refinements in $\mathcal{M}$ are closed under the diagonal $\{(s, s) \mid s \in \Sigma\}$ of Σ, relational composition, and countable unions. In particular, for finite Σ the greatest possibilistic refinement $\prec_{\mathcal{M}}$ exists in $\mathcal{M}$ and is a preorder on Σ.*

Proof. The diagonal of Σ is a possibilistic refinement, since the order on P is reflexive.

Possibilistic refinements Q_1 and Q_2 are closed under composition because the transitivity of $\geq$ in P implies $(Q_1)^{\text{ps}}; (Q_2)^{\text{ps}} \subseteq (Q_1; Q_2)^{\text{ps}}$.

Possibilistic refinements Q_i are closed under countable unions, since $A \mapsto A.Q$ and $B \mapsto Q.B$ preserve all unions, so $(Q_{i_0})^{\text{ps}} \subseteq (\bigcup_{i \in I} Q_i)^{\text{ps}}$ for all i_0 in the countable set I.

For finite Σ, the union of all possibilistic refinements is finite which, by the above, is reflexive and transitive. ∎

4 Possibilistic Property Semantics

We consider the logic $\mathcal{L}_{\text{pr}}$, a quantitative mu-calculus, defined by $\phi ::= \mathtt{tt} \mid p \mid Z \mid \neg\phi \mid \phi\wedge\phi \mid \mathsf{EX}_{\sqsupseteq r}\phi \mid \mu Z.\phi$, where $Z \in \text{var}$ for a countable set of recursion variables var, $p \in \mathtt{AP}$, r ranges over elements of a partial order $(P, \leq)$, $\sqsupseteq$ equals $\geq$ or $>$, and all ϕ are formally monotone in $\mu Z.\phi$. For $\rho = (\rho^{\text{a}}, \rho^{\text{c}})$ with $\rho^{\text{m}} \colon \text{var} \to \mathcal{F}$, $\text{m} \in \{\text{a}, \text{c}\}$, we write $s \models^{\text{a}}_{\rho} \phi$ (positive assertion check) and $s \models^{\text{c}}_{\rho} \phi$ (positive consistency check) iff $s \in [\![\phi]\!]^{\text{a}}_{\rho}$ and $s \in [\![\phi]\!]^{\text{c}}_{\rho}$, respectively.

Remark 1. Our models are *under-specified* in that they have more than one "complete" refinement, models that satisfy $\mathcal{M}^{\text{a}} = \mathcal{M}^{\text{c}}$. Positive assertion checks state that the property in question holds for all (complete) refinements. Positive consistency checks state that there is a (complete) refinement with that property.

The possibilistic property semantics $[\![\cdot]\!]^{\text{m}}_{\cdot}$ is defined in Figure 2, where $\neg\text{a} \stackrel{\text{def}}{=} \text{c}$, $\neg\text{c} \stackrel{\text{def}}{=} \text{a}$, and $\text{pre}^{\text{m}}_{\sqsupseteq r}(A) \stackrel{\text{def}}{=} \{s \in \Sigma \mid \exists (s, \eta) \in R^{\text{m}} : \eta(A) \sqsupseteq r\}$ for $A \in \mathcal{F}$, is the modal quantitative version of the usual pre-image operator. This operator computes the set of states s from which there is a quantitative R^{m}-transition (s, η) that reaches set A with a quantity $\sqsupseteq r$. Although least fixed points are computed in $(\mathcal{P}(\Sigma), \subseteq)$, condition (1) maintains that all denotations are elements of the underlying σ-algebra — as shown below. Note the special treatment of negation: to evaluate $\neg\phi$ in mode m, first evaluate ϕ in mode $\neg$m and then negate that result [31]. Since models are finitely branching, we have $s \models^{\text{m}} \mu Z.\phi$ iff for some $k \geq 0$, $s \models^{\text{m}} \mu_k Z.\phi$, where $\mu_k Z.\phi$ is the standard syntactic approximation of $\mu Z.\phi$ defined by $\mu_0 Z.\phi \stackrel{\text{def}}{=} \neg\mathtt{tt}$ and $\mu_{k+1} Z.\phi \stackrel{\text{def}}{=} \phi[Z \mapsto \mu_k Z.\phi]$ for all $k \geq 0$. (As customary, $\phi[Z \mapsto \mu_k Z.\phi]$ denotes the formula obtained by replacing in ϕ all free occurrences of Z with $\mu_k Z.\phi$.) Incidentally, our treatment of $\mathtt{tt}$, negation,

$$
\begin{aligned}
[\![\mathtt{tt}]\!]^{\text{m}}_{\rho} &\stackrel{\text{def}}{=} \Sigma & [\![\neg\phi]\!]^{\text{m}}_{\rho} &\stackrel{\text{def}}{=} \Sigma \setminus [\![\phi]\!]^{\neg\text{m}}_{\rho} \\
[\![p]\!]^{\text{m}}_{\rho} &\stackrel{\text{def}}{=} \{s \in \Sigma \mid p \in L^{\text{m}}(s)\} & [\![\phi_1 \wedge \phi_2]\!]^{\text{m}}_{\rho} &\stackrel{\text{def}}{=} [\![\phi_1]\!]^{\text{m}}_{\rho} \cap [\![\phi_2]\!]^{\text{m}}_{\rho} \\
[\![Z]\!]^{\text{m}}_{\rho} &\stackrel{\text{def}}{=} \rho^{\text{m}}(Z) & [\![\mathsf{EX}_{\sqsupseteq r}\phi]\!]^{\text{m}}_{\rho} &\stackrel{\text{def}}{=} \text{pre}^{\text{m}}_{\sqsupseteq r}([\![\phi]\!]^{\text{m}}_{\rho}) \\
[\![\mu Z.\phi]\!]^{\text{m}}_{\rho} &\stackrel{\text{def}}{=} \text{lfp}\, F^{\text{m}}; & \text{where } F^{\text{m}}(A) &\stackrel{\text{def}}{=} [\![\phi]\!]^{\text{m}}_{\rho^{\text{m}}[Z \mapsto A]}.
\end{aligned}
$$

Fig. 2. Semantics for modal quantitative Kripke structures in mode $\text{m} \in \{\text{a}, \text{c}\}$

conjunction, and fixed points in Figure 2 means that the set of denotations has to be a σ-algebra *even if the model is a quantitative, non-probabilistic one.*

Disjunction, implication, formulas $\mathsf{AX}_{\sqsubseteq r}\phi$, and greatest fixed points are definable in this logic. However, this is not the case for checks of the form $s\models^{\mathrm{m}}\mathsf{AX}_{\sqsupseteq r}\phi$ which are positive if for all $R^{\neg\mathrm{m}}$-transitions (s,μ), $\mu(\llbracket \phi \rrbracket^{\mathrm{m}}) \sqsupseteq r$.

Our property semantics for modal quantitative Kripke structures is sound with respect to possibilistic refinement and abstraction for *all* closed formulas of $\mathcal{L}_{\mathrm{pr}}$: positive assertion checks $t\models^{\mathrm{a}}\phi$ remain valid for all possibilistic refinements of t; dually, positive consistency checks $t\models^{\mathrm{c}}\phi$ remain consistent for all possibilistic abstractions of t.

Theorem 1 (Soundness). *Let $\mathcal{M}$ be a modal quantitative Kripke structure, ρ an environment, $\phi \in \mathcal{L}_{\mathrm{pr}}$ a closed formula, and $\mathrm{m} \in \{\mathrm{a},\mathrm{c}\}$. Then $\llbracket \phi \rrbracket^{\mathrm{m}}_{\rho} \in \mathcal{F}$. Moreover for any possibilistic refinement Q, $(s,t) \in Q$ entails that (1) $t\models^{\mathrm{a}}_{\rho}\phi$ implies $s\models^{\mathrm{a}}_{\rho}\phi$; and (2) $s\models^{\mathrm{c}}_{\rho}\phi$ implies $t\models^{\mathrm{c}}_{\rho}\phi$.*

Proof. For measurability, the closure properties of σ-algebras take care of the clauses `tt`, negation, conjunction, and fixed points (since models are finitely branching such fixed points are the countable union of their unfoldings). Condition (1) takes care of the clauses p and $\mathsf{EX}_{\sqsupseteq r}$.

We show items 1 and 2 for the key clauses of $\mathsf{EX}_{\sqsupseteq r}$ and fixed points. Let $t\models^{\mathrm{a}}_{\rho}\mathsf{EX}_{\sqsupseteq r}\phi$. Then there exists some $(t,\eta) \in R^{\mathrm{a}}$ such that $\eta(\llbracket \phi \rrbracket^{\mathrm{a}}_{\rho}) \sqsupseteq r$, where $\llbracket \phi \rrbracket^{\mathrm{a}}_{\rho} \in \mathcal{F}$ and item 1 holds for ϕ by induction. Since Q is a possibilistic refinement, $(s,t) \in Q$ implies that there exists some $(s,\mu) \in R^{\mathrm{a}}$ such that $(\mu,\eta) \in Q^{\mathrm{ps}}$. In particular, $Q.\llbracket \phi \rrbracket^{\mathrm{a}}_{\rho} \in \mathcal{F}$. By induction, $Q.\llbracket \phi \rrbracket^{\mathrm{a}}_{\rho} \subseteq \llbracket \phi \rrbracket^{\mathrm{a}}_{\rho}$, so $\mu(\llbracket \phi \rrbracket^{\mathrm{a}}_{\rho}) \geq \mu(Q.\llbracket \phi \rrbracket^{\mathrm{a}}_{\rho}) \geq \eta(\llbracket \phi \rrbracket^{\mathrm{a}}_{\rho})$ — the first inequality follows from the monotonicity of μ, the second one follows from $(\mu,\eta) \in Q^{\mathrm{ps}}$. But $\eta(\llbracket \phi \rrbracket^{\mathrm{a}}_{\rho}) \sqsupseteq r$ then implies $\mu(\llbracket \phi \rrbracket^{\mathrm{a}}_{\rho}) \sqsupseteq r$, since $\geq$ is transitive and since $\geq \circ >$ and $> \circ \geq$ are contained in $>$. Thus, $(s,\mu) \in R^{\mathrm{a}}$ implies $s\models^{\mathrm{a}}_{\rho}\mathsf{EX}_{\sqsupseteq r}\phi$.

Let $s\models^{\mathrm{c}}_{\rho}\mathsf{EX}_{\sqsupseteq r}\phi$. Then there exists some $(s,\mu) \in R^{\mathrm{c}}$ such that $\mu(\llbracket \phi \rrbracket^{\mathrm{c}}_{\rho}) \sqsupseteq r$, where $\llbracket \phi \rrbracket^{\mathrm{c}}_{\rho} \in \mathcal{F}$ and item 2 holds for ϕ by induction. Since Q is a possibilistic refinement and $(s,t) \in Q$, there exists some $(t,\eta) \in R^{\mathrm{c}}$ such that $(\mu,\eta) \in Q^{\mathrm{ps}}$. In particular, $\llbracket \phi \rrbracket^{\mathrm{c}}_{\rho}.Q \in \mathcal{F}$. By induction, $\llbracket \phi \rrbracket^{\mathrm{c}}_{\rho}.Q \subseteq \llbracket \phi \rrbracket^{\mathrm{c}}_{\rho}$, so $\eta(\llbracket \phi \rrbracket^{\mathrm{c}}_{\rho}) \geq \eta(\llbracket \phi \rrbracket^{\mathrm{c}}_{\rho}.Q) \geq \mu(\llbracket \phi \rrbracket^{\mathrm{c}}_{\rho})$ — the first inequality follows from the monotonicity of η, the second one follows from $(\mu,\eta) \in Q^{\mathrm{ps}}$. But $\mu(\llbracket \phi \rrbracket^{\mathrm{c}}_{\rho}) \sqsupseteq r$ then implies $\eta(\llbracket \phi \rrbracket^{\mathrm{c}}_{\rho}) \sqsupseteq r$, rendering $t\models^{\mathrm{c}}_{\rho}\mathsf{EX}_{\sqsupseteq r}\phi$ since $(t,\eta) \in R^{\mathrm{c}}$.

By induction and the previous item, function F^{a} restricts to type $\mathsf{L}[\prec_{\mathcal{M}}] \to \mathsf{L}[\prec_{\mathcal{M}}]$, where $\mathsf{L}[Q] \stackrel{\mathrm{def}}{=} \{L \in \mathcal{F} \mid \text{ for all } (s,l) \in Q :\ l \in L \Rightarrow s \in L\}$, and so its least fixed point is an element of $\mathsf{L}[\prec_{\mathcal{M}}]$; dually, by induction, function F^{c} restricts to type $\mathsf{U}[\prec_{\mathcal{M}}] \to \mathsf{U}[\prec_{\mathcal{M}}]$, where $\mathsf{U}[Q] \stackrel{\mathrm{def}}{=} \{U \in \mathcal{F} \mid \text{ for all } (u,t) \in Q : u \in U \Rightarrow t \in U\}$, and so its least fixed point is an element of $\mathsf{U}[\prec_{\mathcal{M}}]$. ∎

We emphasise that these results are also valid when the model $\mathcal{M}$ is a sum $\mathcal{M}_1 + \mathcal{M}_2$, where $\mathcal{M}_1$ is a (more) concrete model and $\mathcal{M}_2$ is its possibilistic abstraction. Also note that $\prec_{\mathcal{M}}$ exists and has measurable navigation in finite-state models.

Our property semantics loses precision in two places: the interpretation of disjunction in the assertion mode a, and the interpretation of conjunction in the consistency checking mode c. For example, for $p \in L^{c}(s) \setminus L^{a}(s)$ our semantics computes $s \models^{c}_{\rho} p \wedge \neg p$ and $s \not\models^{a}_{\rho} p \vee \neg p$. The meaning of these checks, as expressed in Remark 1, reveals that positive consistency checks are *over-approximations* (not all such checks are truthful) and positive assertion checks are *under-approximations* (not all such checks discover the truth). Fortunately, the subtle treatment of negation [31] guarantees that the interplay of these approximations is sound.

Theorem 2 (Consistency). *Let $\mathcal{M}$ be a modal quantitative Kripke structure. For all closed $\phi \in \mathcal{L}_{\text{pr}}$ and environments ρ, we have (i) $\| \phi \wedge \neg\phi \|^{a}_{\rho} = \{\}$ and, equivalently, (ii) $\| \phi \|^{a}_{\rho} \subseteq \| \phi \|^{c}_{\rho}$.*

Proof. Items (i) and (ii) are equivalent. We show item 2 for the clause $\mathsf{EX}_{\sqsupseteq r}\phi$. Let $s \in \| \mathsf{EX}_{\sqsupseteq r}\phi \|^{a}_{\rho}$. Then there exists some $(s, \mu) \in R^{a}$ such that $\mu(\| \phi \|^{a}_{\rho}) \sqsupseteq r$. By induction, $\| \phi \|^{a}_{\rho} \subseteq \| \phi \|^{c}_{\rho} \in \mathcal{F}$ and so — using the monotonicity of μ — $\mu(\| \phi \|^{c}_{\rho}) \geq \mu(\| \phi \|^{a}_{\rho}) \sqsupseteq r$ which implies $\mu(\| \phi \|^{c}_{\rho}) \sqsupseteq r$. But $(s, \mu) \in R^{a} \subseteq R^{c}$ renders $s \in \| \mathsf{EX}_{\sqsupseteq r}\phi \|^{c}_{\rho}$. ∎

The loss of precision in $\models^{m}$ may severely impact the quality of an analysis. Various techniques exist for obtaining more precise interpretations in *qualitative* models, although at a significant increase in complexity. We mention the focus operation of Ball et al. [4] and Bruns and Godefroid's generalised model checking [6]; it would be of interest to investigate their quantitative analogues. As we saw above, the loss of precision in our property semantics does not compromise the validity of positive assertion checks.

For a class of quantitative measures that subsumes probability measures, one can show that possibilistic refinements lift to computation paths.

Proposition 2 (Matching Computation Paths). *Let $\mathcal{M}$ be a finite-state modal quantitative Kripke structure with state set Σ and $\mathcal{F} = \mathcal{P}(\Sigma)$ such that its partial order P has a least element 0, and all its quantitative measures μ satisfy that $\mu(X) > 0$ implies $\mu(\{x\}) > 0$ for some $x \in X$. Let Q be a possibilistic refinement in $\mathcal{M}$ with $(s_0, t_0) \in Q$. If the path $\pi = (t_0, \eta_0)(t_1, \eta_1) \ldots$ is such that $(t_i, \eta_i) \in R^{a}$ and $\eta_i(\{t_{i+1}\}) > 0$ for all $i \geq 0$, then there exists a matching path $\kappa = (s_0, \mu_0)(s_1, \mu_1) \ldots$ such that $(s_i, \mu_i) \in R^{a}$, $\mu_i(\{s_{i+1}\}) > 0$, $(\mu_i, \eta_i) \in Q^{ps}$, and $(s_i, t_i) \in Q$ for all $i \geq 0$. A dual property holds for R^{c}-paths beginning in s_0.*

Proof. Since $(t_0, \eta_0) \in R^{a}$ and $(s_0, t_0) \in Q$, there is some $(s_0, \mu_0) \in R^{a}$ such that $(\mu_0, \eta_0) \in Q^{ps}$. The set $\{t_1\}$ is in $\mathcal{F}$ by assumption. Therefore, $\mu_0(Q.\{t_1\}) \geq \eta_0(\{t_1\})$. But $\eta_0(\{t_1\}) > 0$ by assumption and so $\mu_0(Q.\{t_1\}) > 0$ implies that $\mu_0(\{s_1\}) > 0$ for some $s_1 \in Q.\{t_1\}$, i.e. $(s_1, t_1) \in Q$. Proceeding by induction, we construct a path κ with the desired properties. The proof for the second statement is dual. ∎

Our possibilistic property semantics can be implemented with a conventional labelling algorithm such that the only changes are in the treatment of negation — leading to a-labels and c-labels on states that may be represented with two BDDs [7] — and in the computation of successor sets, based on $\mathrm{pre}^{\mathrm{m}}_{\exists_r}(A)$. In particular, in finite-state systems fixed points always converge with the size of the state space as upper bound on the number of necessary unfoldings. It would be of interest to represent this labelling algorithm symbolically, as done for the standard semantics with MTBBDs in Baier et al. [2].

We conclude this section with a discussion of how our possibilistic refinement relates to the established notion of probabilistic bisimulation [32].

Theorem 3 (Probabilistic Bisimulation). *Let $\mathcal{M}$ be a finite modal quantitative Kripke structure $((\Sigma, R^{\mathrm{a}}, L^{\mathrm{a}}), (\Sigma, R^{\mathrm{c}}, L^{\mathrm{c}}))$ with signature $(\mathtt{AP}, \mathcal{P}(\Sigma), P)$ such that $\mathcal{M}^{\mathrm{a}}$ equals $\mathcal{M}^{\mathrm{c}}$.*

1. *For all closed $\phi \in \mathcal{L}_{\mathrm{pr}}$ and all environments ρ: $\llbracket \phi \rrbracket^{\mathrm{a}}_{\rho}$ equals $\llbracket \phi \rrbracket^{\mathrm{c}}_{\rho}$.*
2. *If P equals $[0,1]$ and every quantitative measure is probabilistic, then the probabilistic bisimulations in $\mathcal{M}^{\mathrm{a}}$ are exactly the possibilistic refinements in $\mathcal{M}$ which are equivalence relations. In particular, the greatest probabilistic bisimulation of $\mathcal{M}^{\mathrm{a}}$ is contained in $\prec_{\mathcal{M}} \cap (\prec^{-1}_{\mathcal{M}})$.*

Proof. Since $\mathcal{M}^{\mathrm{a}} = \mathcal{M}^{\mathrm{c}}$, $\llbracket \cdot \rrbracket^{\mathrm{a}}_{\rho}$ and $\llbracket \cdot \rrbracket^{\mathrm{c}}_{\rho}$ are identical for closed formulas.

Let Q be an equivalence relation and a possibilistic refinement with $(s,t) \in Q$. (i) Definition 4 implies $L^{\mathrm{a}}(s) = L^{\mathrm{a}}(t)$ (since $L^{\mathrm{a}} = L^{\mathrm{c}}$). (ii) Let $(t,\eta) \in R^{\mathrm{a}}$. Then there is some $(s,\mu) \in R^{\mathrm{a}}$ such that $(\mu,\eta) \in Q^{\mathrm{ps}}$. But for every union of a collection of equivalence classes X of Q we have $X.Q = Q.X = X$, so $\eta(X) = \eta(X.Q) \geq \mu(X)$ and $\mu(X) = \mu(Q.X) \geq \eta(X)$. Thus, $\mu(X) = \eta(X)$ for all such X, since the order on P is anti-symmetric. (ii) Let $(s,\mu) \in R^{\mathrm{a}}(= R^{\mathrm{c}})$. Then we reason dually. Thus, Q is a probabilistic bisimulation in $\mathcal{M}^{\mathrm{a}}$.

Conversely, let $\mathcal{R}$ be a probabilistic bisimulation in $\mathcal{M}^{\mathrm{a}}$. Given $A, B \in \mathcal{F}$, the sets $A.\mathcal{R}$ and $\mathcal{R}.B$ are in $\mathcal{F} = \mathcal{P}(\Sigma)$, are unions of a collection of equivalence classes of $\mathcal{R}$, and contain A and B (respectively). Let $(s,t) \in \mathcal{R}$. Then $L^{\mathrm{m}}(s) = L^{\mathrm{m}}(t)$ (i.e. $L^{\mathrm{a}}(t) \subseteq L^{\mathrm{a}}(s)$ and $L^{\mathrm{c}}(s) \subseteq L^{\mathrm{c}}(t)$) since $\mathcal{R}$ is a probabilistic bisimulation. (i) If $(t,\eta) \in R^{\mathrm{a}}$, then there is some $(s,\mu) \in R^{\mathrm{a}}$ such that $\mu(X) = \eta(X)$ for all unions of equivalence classes X of $\mathcal{R}$. But then the reflexivity of $\mathcal{R}$ renders $\eta(A.\mathcal{R}) = \mu(A.\mathcal{R}) \geq \mu(A)$ and $\mu(\mathcal{R}.B) = \eta(\mathcal{R}.B) \geq \eta(B)$, establishing $(\mu,\eta) \in \mathcal{R}^{\mathrm{ps}}$. (ii) If $(s,\mu) \in R^{\mathrm{c}}(= R^{\mathrm{a}})$, we reason dually. Thus, $\mathcal{R}$ is a possibilistic refinement and therefore contained in $\prec_{\mathcal{M}}$. Since $\mathcal{R}$ is symmetric, it is contained in $\prec_{\mathcal{M}} \cap (\prec_{\mathcal{M}})^{-1}$. ∎

5 Approximating Probabilistic Logics

In this section we assume that quantitative models are probabilistic: the partial order P equals $[0,1]$ and all quantitative measures are probabilistic. This assumption enables a direct comparison of possibilistic refinement and property semantics to their probabilistic versions.

For probabilistic models, the possibilistic refinement notion seems suitable since probabilistic bisimulations are then possibilistic refinements that are equivalence classes (Theorem 3). Our possibilistic property semantics, however, is quite different in spirit from established probabilistic logics which are variants or fragments of Hansson's logic PCTL [24]. The latter abstracts a linear semantics that assigns probabilities to sets of traces to a branching semantics through quantification over suitable adversaries. Our semantics is also branching as its models are (probabilistic) computation trees but it abstracts the adversary abstraction of the linear semantics in [24]. (For qualitative models, Cousot and Cousot [14] showed how one may systematically abstract a trace semantics into several branching-time semantics; see [44] for a corresponding abstraction of trace sets to modal transition systems.)

This abstraction is caused by a "memory-less" way of computing successor states for fixed points. Along computation paths, the threshold in $\mathsf{EX}_{\sqsupseteq r}$ is applied to each state in isolation. This is akin to the use of possibility and necessity measures in artificial intelligence [20] — whence the name *possibilistic* refinement — especially if quantitative measures compute maxima: $\mu(A) = \max_{s\in A}\mu(\{a\})$. Given this memory-less treatment of probabilities, it is therefore intuitive that the thresholds for our fixed-point semantics turn into exponential thresholds for the standard probabilistic semantics (see Proposition 3).

To enable a comparison to PCTL, we restrict fixed points to those of the form $\mu Z.\psi \stackrel{\mathrm{def}}{=} \mu Z.q \vee (p \wedge \mathsf{EX}_{\sqsupseteq r}(Z))$. (We assume $p, q \in \mathtt{AP}$ to simplify this discussion.) Since the underlying modal probabilistic Kripke structure $\mathcal{M}$ is finitely-branching, we have $s{\models}^{\mathrm{m}}\mu Z.\psi$ iff $s{\models}^{\mathrm{m}}\mu_k Z.\psi$ for some $k > 0$. For each $\mathrm{m} \in \{\mathrm{a}, \mathrm{c}\}$, we define $f^{\mathrm{m}}\colon \Sigma \times \mathbb{N} \to \{0, 1\}$, parametric in r, through

$$
\begin{aligned}
f^{\mathrm{m}}(s,k) &\stackrel{\mathrm{def}}{=} 1; && \text{if } s{\models}^{\mathrm{m}} q \\
f^{\mathrm{m}}(s,k+1) &\stackrel{\mathrm{def}}{=} 1; && \text{if } s{\models}^{\mathrm{m}} p,\ s{\not\models}^{\mathrm{m}} q \\
& && \text{and } \exists\,(s,\mu) \in R^{\mathrm{m}}\colon \mu(\{t \mid f^{\mathrm{m}}(t,k) = 1\}) \sqsupseteq r \\
f^{\mathrm{m}}(s,k) &\stackrel{\mathrm{def}}{=} 0; && \text{otherwise.}
\end{aligned}
\tag{3}
$$

Lemma 1 (Computing a Possibilistic EU). *For all $0 < r \leq 1$, $\mathrm{m} \in \{\mathrm{a}, \mathrm{c}\}$, and all $k \geq 0$, the function f^{m} is well defined and $s{\models}^{\mathrm{m}}\mu_{k+1}Z.\psi$ iff $f^{\mathrm{m}}(s,k) = 1$.*

Proof. By Theorem 1, induction on k ensures that $f^{\mathrm{m}}(\cdot,\cdot)$ is well defined, since the argument of μ in the second clause of (3) is then in $\mathcal{F}$. For $k = 0$, we have $f^{\mathrm{m}}(s,0) = 1$ iff $s{\models}^{\mathrm{m}}q$ iff — since $r > 0$ implies $s\,{\not\models}^{\mathrm{m}}\mathsf{EX}_{\sqsupseteq r}\neg\mathtt{tt}$ — $s{\models}^{\mathrm{m}}\mu_1 Z.\psi$. Assume that the statement holds for k. Then $s{\models}^{\mathrm{m}}\mu_{k+2}Z.\psi$ iff $s{\models}^{\mathrm{m}}q$ or ($s{\models}^{\mathrm{m}}p$ and $s{\models}^{\mathrm{m}}\mathsf{EX}_{\sqsupseteq r}\mu_{k+1}Z.\psi$) iff $s{\models}^{\mathrm{m}}q$ or ($s{\models}^{\mathrm{m}}p$ and — by induction — there is some $(s,\mu) \in R^{\mathrm{m}}$ such that $\mu(\{t \mid f^{\mathrm{m}}(t,k) = 1\}) \sqsupseteq r$) iff $s{\models}^{\mathrm{m}}q$ or ($s{\models}^{\mathrm{m}}p$ and $s{\not\models}^{\mathrm{m}}q$ and $f^{\mathrm{m}}(s,k+1) = 1$) iff $f^{\mathrm{m}}(s,k+1) = 1$. ∎

A similar specification may be given for an AF connective, rendering a labelling algorithm for our semantics which covers full PCTL. We now compare our semantics to that of a probabilistic Until $[p\,\mathsf{EU}\,q]_{\sqsupseteq r}$ whose meaning may be computed via a function $g^{\mathrm{m}}\colon \Sigma \times \mathbb{N} \to \{0,1\}$:

$$g^{\mathrm{m}}(s,k) \stackrel{\mathrm{def}}{=} 1; \qquad \text{if } s \models^{\mathrm{m}} q \tag{4}$$
$$g^{\mathrm{m}}(s,k+1) \stackrel{\mathrm{def}}{=} \max_{(s,\mu)\in R^{\mathrm{m}}} \{\sum_{t\in\Sigma} \mu(\{t\}) \cdot g^{\mathrm{m}}(t,k)\}; \qquad \text{if } s \models^{\mathrm{m}} p,\ s \not\models^{\mathrm{m}} q;$$
$$g^{\mathrm{m}}(s,k) \stackrel{\mathrm{def}}{=} 0; \qquad \text{otherwise.}$$

Note that this function is well defined only if all singletons are measurable. One may then specify that $s \models^{\mathrm{m}} [p\,\mathsf{EU}\,q]_{\sqsupseteq r}$ holds iff $g(s,k) \sqsupseteq r$ for some $k \geq 0$. In general, our possibilistic property semantics is an abstraction of the probabilistic one in that possibilistic *positive* EU checks imply positive checks in the probabilistic interpretation, alas with an exponential penalty in the threshold.

Proposition 3 (Sound Approximation). *Let all singletons in Σ be measurable. For all $\mathrm{m} \in \{\mathrm{a},\mathrm{c}\}$, $s \in \Sigma$, and $k \geq 0$, $f^{\mathrm{m}}(s,k) = 1$ implies $g^{\mathrm{m}}(s,k) \sqsupseteq r^{k+1}$.*

Proof. For $k = 0$, $f^{\mathrm{m}}(s,0) = 1$ iff $s \models^{\mathrm{m}} q$ iff $g^{\mathrm{m}}(s,0) = 1$, which implies $g^{\mathrm{m}}(s,0) \sqsupseteq r^{0+1}$.

For $k+1$, let $f^{\mathrm{m}}(s,k+1) = 1$. If $s \models^{\mathrm{m}} q$, then $g^{\mathrm{m}}(s,k+1) = 1$ and we are done. Otherwise, $f^{\mathrm{m}}(s,k+1) = 1$ implies $s \models^{\mathrm{m}} p$, $s \not\models^{\mathrm{m}} q$, and the existence of some $(s,\eta) \in R^{\mathrm{m}}$ such that $\sum_{f^{\mathrm{m}}(t,k)=1} \eta(\{t\}) \sqsupseteq r$ since η is a probability measure. For each instance of $f^{\mathrm{m}}(t,k) = 1$ we have $g^{\mathrm{m}}(t,k) \sqsupseteq r^{k+1}$ by induction. Since $s \models^{\mathrm{m}} p$ and $s \not\models^{\mathrm{m}} q$, we therefore obtain $g^{\mathrm{m}}(s,k+1) = \max_{(s,\mu)\in R^{\mathrm{m}}} \{\sum_t \mu(\{t\}) \cdot g^{\mathrm{m}}(t,k)\} \geq \sum_t \eta(\{t\}) \cdot g^{\mathrm{m}}(t,k) \geq \sum_{f^{\mathrm{m}}(t,k)=1} \eta(\{t\}) \cdot g^{\mathrm{m}}(t,k) \sqsupseteq \sum_{f^{\mathrm{m}}(t,k)=1} \eta(\{t\}) \cdot r^{k+1} = r^{k+1} \cdot \sum_{f^{\mathrm{m}}(t,k)=1} \eta(\{t\}) \sqsupseteq r^{k+1} \cdot r = r^{k+2}$. ∎

To decide whether $s \models^{\mathrm{m}} [p\,\mathsf{EU}\,q]_{\sqsupseteq r}$, one may therefore choose some $k \geq 0$, compute τ_k as the $k+1$th square root of r, and then determine $f^{\mathrm{m}}(s,k)$ for parameter value τ_k; if $f^{\mathrm{m}}(s,k)$ equals 1, we know for certain that $s \models^{\mathrm{m}} [p\,\mathsf{EU}\,q]_{\sqsupseteq r}$ holds. Thus, for each $l \geq 2$ the formula $\bigvee_{k=2}^{l} \mu_{k+1} Z.\psi_k$ provides a sound approximation of the probabilistic Until if interpreted in the possibilistic semantics, where all thresholds in ψ_k are τ_k. Unfortunately, the loss of precision in $r \mapsto r^k$ means that possibilistic checks may rarely give insight into true probabilistic behaviour.

How does our possibilistic refinement notion fare with the preservation of properties if the Untils are interpreted via g^{m}? Possibilistic refinements clearly takes care of the clause $\mathsf{EX}_{\sqsupseteq r}$, but fails to secure the fixed-point clause for EU in general, since relational navigation is too coarse grained to reason about the sum expression in (4). Of course, this is where probabilistic simulations [30] do their work to complete satisfaction.

To define *probabilistic* refinements, we keep Definition 4 as is, except for replacing all occurrences of Q^{ps} with Q^{pr}; the latter is the relation $\sqsubseteq_R$ of [15] where Q plays the role of the discriminating criterion C: $(\mu,\eta) \in Q^{\mathrm{pr}}$ iff there is some probability measure δ on $\mathcal{F} \times \mathcal{F}$ such that for all $s,t \in \Sigma$,

$$\mu(\{s\}) = \delta(\{s\} \times \Sigma),\ \eta(\{t\}) = \delta(\Sigma \times \{t\}),\ \delta(\{s\} \times \{t\}) > 0 \Rightarrow (s,t) \in Q. \tag{5}$$

Note that this definition applies to finite models with $\mathcal{F} = \mathcal{P}(\Sigma)$. We can extend the results of [15] to a full logic in re-proving Theorem 1 for the *probabilistic* semantics — which changes the possibilistic semantics of Figure 2 for the computation of Untils by using g^{m} instead of f^{m} — and *probabilistic* refinement. However, this inductive proof works only if denotations of subformulas are measurable. Thus, we encounter potential problems in properties that nest Untils. Specifically, we then owe a proof that $\{s \in \Sigma \mid \exists (s,\mu) \in R^{\mathrm{m}} \colon \sum \mu(\{t\}) \cdot g^{\mathrm{m}}(t,k) \sqsupseteq r\} \in \mathcal{F}$ for all $k \geq 0$ and $r \in [0,1]$. With the help of condition (1), it may be possible to show the inductive step for such a claim, but we did not yet investigate that in sufficient detail.

Theorem 4 (Soundness of Probabilistic Abstraction). *Theorem 1 applies to finite modal probabilistic models with measurable singletons, provided that least fixed points are restricted to Untils whose probabilistic semantics is based on (4) and the refinement is the probabilistic one.*

Proof. Let $(s,t) \in Q$. The proof only changes for the clauses $\mathsf{EX}_{\sqsupseteq r}$ and Until.

Given $t \models^{\mathrm{a}} \mathsf{EX}_{\sqsupseteq r}\phi$, there exists some $(t,\eta) \in R^{\mathrm{a}}$ such that $\eta(\llbracket \phi \rrbracket^{\mathrm{a}}) \sqsupseteq r$. Since Q is a probabilistic refinement, $(s,t) \in Q$ implies that there exists some $(s,\mu) \in R^{\mathrm{a}}$ such that $(\mu,\eta) \in Q^{\mathrm{pr}}$. Let δ be the corresponding witness. Then $\eta(\llbracket \phi \rrbracket^{\mathrm{a}}) = \sum_{t' \in \llbracket \phi \rrbracket^{\mathrm{a}}} \eta(\{t'\})$ which equals $\sum_{t' \in \llbracket \phi \rrbracket^{\mathrm{a}}} \delta(\Sigma \times \{t'\})$. Since $\delta(\{u\} \times \{v\}) > 0$ implies $(u,v) \in Q$, the latter equals $\sum_{t' \in \llbracket \phi \rrbracket^{\mathrm{a}}} \sum_{s' \mid (s',t') \in Q} \delta(\{s'\} \times \{t'\})$ which, by induction, is less than or equal to $\sum_{s' \in \llbracket \phi \rrbracket^{\mathrm{a}}} \sum_{t' \in \llbracket \phi \rrbracket^{\mathrm{a}}} \delta(\{s'\} \times \{t'\}) \leq \sum_{s' \in \llbracket \phi \rrbracket^{\mathrm{a}}} \delta(\{s'\} \times \Sigma)$ which equals $\sum_{s' \in \llbracket \phi \rrbracket^{\mathrm{a}}} \mu(\{s'\}) = \mu(\llbracket \phi \rrbracket^{\mathrm{a}})$. But $\eta(\llbracket \phi \rrbracket^{\mathrm{a}}_{\rho}) \sqsupseteq r$ then implies $\mu(\llbracket \phi \rrbracket^{\mathrm{a}}_{\rho}) \sqsupseteq r$. Thus, $(s,\mu) \in R^{\mathrm{a}}$ entails $s \models^{\mathrm{a}}_{\rho} \mathsf{EX}_{\sqsupseteq r}\phi$. The proof that $s \models^{\mathrm{c}}_{\rho} \mathsf{EX}_{\sqsupseteq r}\phi$ implies $t \models^{\mathrm{c}}_{\rho} \mathsf{EX}_{\sqsupseteq r}\phi$ is dual and omitted.

Given $t \models^{\mathrm{a}} [\phi_1 \, \mathsf{EU} \, \phi_2]_{\sqsupseteq r}$, there is some $k \geq 0$ such that $g^{\mathrm{a}}(t,k) \sqsupseteq r$. Thus, it suffices to show that $g^{\mathrm{a}}(s,k) \geq g^{\mathrm{a}}(t,k)$ for all $(s,t) \in Q$ and $k \geq 0$. For $k = 0$, this follows by induction on ϕ_2. For the inductive step $k+1$, only the second clause of (4) contains a non-trivial proof obligation. Let $(t,\eta) \in R^{\mathrm{a}}$ be the witness for the computation of the maximum value in that clause. Since $(s,t) \in Q$, there is some $(s,\mu) \in R^{\mathrm{a}}$ such that $(\mu,\eta) \in Q^{\mathrm{pr}}$. Let δ be the corresponding witness. We claim that the second clause of (4) applies to $g^{\mathrm{a}}(s,k+1)$ as well: by induction on ϕ_1, we obtain $s \models^{\mathrm{a}} \phi_1$ and if $s \models^{\mathrm{a}} \phi_2$ is the case there is nothing to show as then $g^{\mathrm{a}}(s,k+1) = 1$. Therefore, $g^{\mathrm{a}}(s,k+1) \geq \sum_{s' \in \Sigma} \mu(\{s'\}) \cdot g^{\mathrm{a}}(s',k)$ which equals $\sum_{s' \in \Sigma} \delta(\{s'\} \times \Sigma) \cdot g^{\mathrm{a}}(s',k)$. The latter equals $\sum_{s' \in \Sigma} \sum_{t' \mid (s',t') \in Q} \delta(\{s'\} \times \{t'\}) \cdot g^{\mathrm{a}}(s',k)$ which is greater or equal to $\sum_{s' \in \Sigma} \sum_{t' \mid (s',t') \in Q} \delta(\{s'\} \times \{t'\}) \cdot g^{\mathrm{a}}(t',k) = \sum_{s' \in \Sigma} \sum_{t' \in \Sigma} \delta(\{s'\} \times \{t'\}) \cdot g^{\mathrm{a}}(t',k)$ by induction on k. But the latter equals $\sum_{t' \in \Sigma} \delta(\Sigma \times \{t'\}) \cdot g^{\mathrm{a}}(t',k) = g^{\mathrm{a}}(t,k+1)$. To show that $s \models^{\mathrm{c}} [\phi_1 \, \mathsf{EU} \, \phi_2]_{\sqsupseteq r}$ implies $t \models^{\mathrm{c}} [\phi_1 \, \mathsf{EU} \, \phi_2]_{\sqsupseteq r}$ it suffices to prove that $g^{\mathrm{c}}(t,k) \geq g^{\mathrm{a}}(s,k)$ for all $(s,t) \in Q$ and $k \geq 0$; this proof is dual to the one of the previous item. ∎

Although the scope of Theorem 4 is sufficiently wide for practical abstraction-based model checking, a proper extension of this result to the full scope of models and the logic is desirable.

6 Specifying Abstractions

We defined refinement and abstraction as notions *within* quantitative models. Since these models have sums (which restrict to probabilistic systems as well), such an approach did not compromise any generality. In this section, we avoid the notational overhead of castings into sum types and express abstractions between models directly. We transfer the results of Godefroid et al. [22], where a compositional calculus for specifying and computing relational abstractions of modal transition systems is developed, to modal quantitative structures.

Definition 5 (Specifying Abstractions). *Let* $\mathcal{M} = (\Sigma, R^{\rm a}, L^{\rm a}), (\Sigma, R^{\rm c}, L^{\rm c}))$ *be a modal quantitative Kripke structure with signature* $(\mathtt{AP}, \mathcal{F}, P)$ *and* $Q \subseteq (\Sigma, \mathcal{F}) \times (\Sigma_Q, \mathcal{F}_Q)$ *a left-total and right-total relation with measurable navigation. We define a* possibilistic relational abstraction $\mathcal{M}_{Q^{\rm ps}}$ *(and a* probabilistic relational abstraction $\mathcal{M}_{Q^{\rm pr}}$ *if* $\mathcal{M}$ *happens to be probabilistic) of* $\mathcal{M}$ *via* Q*. Let* $\star \in \{{\rm ps}, {\rm pr}\}$*:*

- *both state sets equal* Σ_Q *and their* σ*-algebra is* $\mathcal{F}_Q$*;*
- *for each* $t \in \Sigma_Q$*, their labelling functions are given by* $p \in L^{\rm a}_{Q^\star}(t)$ *iff for all* $(s', t) \in Q$*,* $p \in L^{\rm a}(s')$*; and* $p \in L^{\rm c}_{Q^\star}(t)$ *iff there exists some* $(s', t) \in Q$ *such that* $p \in L^{\rm c}(s')$*;*
- *as for their transition relations, we have* $(t, \eta) \in R^{\rm a}_{Q^\star}$ *iff for all* $s \in Q.\{t\}$ *there is some* $(s, \mu) \in R^{\rm a}$ *such that* $(\mu, \eta) \in Q^\star$*; dually,* $(t, \eta) \in R^{\rm c}_{Q^\star}$ *iff there is some* $s \in Q.\{t\}$ *and some* $(s, \mu) \in R^{\rm c}$ *such that* $(\mu, \eta) \in Q^\star$*.*

Since $B \mapsto Q.B$ is monotone, every $\mu \in [\mathcal{F} \to P]$ determines a $\mu_Q \in [\mathcal{F}_Q \to P]$ such that $\mu_Q(B) \stackrel{\rm def}{=} \mu(Q.B)$ for all $B \in \mathcal{F}_Q$. We note that $\mu \mapsto \mu_Q$ does not preserve the property of being a probability measure in general, although we show below that this is the case for predicate-based abstractions. To show that the specification above renders relational abstractions, we need to gain a better understanding of the relations $Q^{\rm ps}$ and $Q^{\rm pr}$.

Theorem 5. *Let* $Q \subseteq (\Sigma, \mathcal{F}) \times (\Sigma_Q, \mathcal{F}_Q)$ *be a right-total relation with measurable navigation.*

1. *Q is the graph of a function $f \colon \Sigma \to \Sigma_Q$ iff $B = (Q.B).Q$ for all $B \subseteq \Sigma_Q$.*
2. *For every $\mu \in [\mathcal{F} \to P]$, $(\mu, \mu_Q) \in Q^{\rm ps}$.*
3. *The relation $(\mu, \eta) \in Q^{\rm ps}$ implies $\eta(B) \leq \mu_Q(B)$ for all $B \in \mathcal{F}_Q$.*
4. *Let Q be the graph of a function. Then*
   (a) $(\mu, \eta) \in Q^{\rm ps}$ iff $\eta = \mu_Q$; and
   (b) if μ and η are probabilistic, $\mathcal{F} = \mathcal{P}(\Sigma)$, and $\mathcal{F}_Q = \mathcal{P}(\Sigma_Q)$, then $(\mu, \eta) \in Q^{\rm pr}$ iff $(\mu, \eta) \in Q^{\rm ps}$.

Proof. 1. This is straightforward, noting that Q is right-total.

2. Let $A \in \mathcal{F}$. Then $\mu_Q(A.Q) = \mu(Q.(A.Q))$ by definition. Since Q is right-total, we have $A \subseteq Q.(A.Q)$ and so — using the monotonicity of μ — $\mu_Q(A.Q) = \mu(Q.(A.Q)) \geq \mu(A)$. Let $B \in \mathcal{F}_Q$. Then $\mu(Q.B) = \mu_Q(B)$. Thus, $(\mu, \mu_Q) \in Q^{\rm ps}$.

3. Let $B \in \mathcal{F}_Q$. Then $\mu_Q(B) = \mu(Q.B) \geq \eta(B)$.

4a. $B = (Q.B).Q$ implies $\eta(B) = \eta((Q.B).Q) \geq \mu(Q.B) = \mu_Q(B)$.

4b. Let $(\mu, \eta) \in Q^{\mathrm{ps}}$, i.e. $\eta = \mu_Q$. We show that (5) is met for $\delta(\{s\} \times \{t\}) = \mu(\{s\} \cap Q.\{t\})$. Since Q is the graph of a function, $\sum_t \delta(\{s\} \times \{t\}) = \sum_t \mu(\{s\} \cap Q.\{t\}) = \mu(\{s\})$. But $\sum_s \delta(\{s\} \times \{t\}) = \sum_s \mu(\{s\} \cap Q.\{t\}) = \sum_s \{\mu(\{s\}) \mid (s,t) \in Q\} = \mu(Q.\{t\}) = \mu_Q(\{t\})$. If $\delta(\{s\} \times \{t\}) > 0$, then $\{s\} \cap Q.\{t\}$ is non-empty (as $\mu(\{\}) = 0$), so $(s,t) \in Q$. Finally, $\Sigma \times \Sigma_Q$ has measure 1, for $\sum_s \sum_t \delta(\{s\} \times \{t\}) = \sum_s \mu(\{s\}) = 1$. Thus, $(\mu, \eta) \in Q^{\mathrm{pr}}$.

Conversely, let $(\mu, \eta) \in Q^{\mathrm{pr}}$. It suffices to show that $\eta = \mu_Q$. Let κ be the witness to $(\mu, \eta) \in Q^{\mathrm{pr}}$. By assumption, for every $s \in Q$ there is a unique $t_s \in \Sigma_Q$ with $(s, t_s) \in Q$. But then $\kappa(\{s\} \times \{t\}) = 0 = \delta(\{s\} \times \{t\})$ for all $t \neq t_s$. Thus, $\delta(\{s\} \times \{t_s\}) = \sum_t \delta(\{s\} \times \{t\}) = \mu(\{s\}) = \sum_t \kappa(\{s\} \times \{t\}) = \kappa(\{s\} \times \{t_s\})$ shows $\delta = \kappa$ which implies $\eta = \mu_Q$. ■

Corollary 1 (Abstractions of Finite-State Probabilistic Models). *Possibilistic and probabilistic abstractions coincide if the underlying relation is left-total and right-total, models are finite-state, and all measures are probabilistic.*

Left-total and right-total relations with measurable navigation specify abstractions.

Theorem 6 (Soundness of Specifications). *Let $Q \subseteq (\Sigma, \mathcal{F}) \times (\Sigma_Q, \mathcal{F}_Q)$ be a left-total and right-total relation with measurable navigation.*

1. *The transformations $\mathcal{M} \mapsto \mathcal{M}_{Q^{\mathrm{ps}}}$ and $\mathcal{M} \mapsto \mathcal{M}_{Q^{\mathrm{pr}}}$ preserve the property of being a modal quantitative and probabilistic Kripke structure (respectively). These transformations are equal for functional abstractions of finite-state systems.*
2. *The model $\mathcal{M}$ is a possibilistic refinement of $\mathcal{M}_{Q^{\mathrm{ps}}}$; under the assumptions of Theorem 5.4(b), $\mathcal{M}$ is a probabilistic refinement of $\mathcal{M}_{Q^{\mathrm{pr}}}$.*

Proof. Let $\star \in \{\mathrm{ps}, \mathrm{pr}\}$. 1. For $t \in \Sigma_Q$, $L^{\mathrm{a}}_{Q^\star}(t) \stackrel{\mathrm{def}}{=} \bigcap_{(s',t) \in Q} L^{\mathrm{a}}(s')$ is contained in $\bigcup_{(s',t) \in Q} L^{\mathrm{a}}(s')$. Since $L^{\mathrm{c}}(s') \subseteq L^{\mathrm{a}}(s')$ for all $s' \in \Sigma$, $\bigcup_{(s',t) \in Q} L^{\mathrm{a}}(s')$ is contained in $\bigcup_{(s',t) \in Q} L^{\mathrm{c}}(s')$ which equals $L^{\mathrm{c}}_{Q^\star}(t)$. Similarly, we show $R^{\mathrm{a}}_{Q^\star} \subseteq R^{\mathrm{c}}_{Q^\star}$ using $R^{\mathrm{c}} \subseteq R^{\mathrm{a}}$ and the fact that Q is left-total.

2. For $(s,t) \in Q$, let $(t, \eta) \in R^{\mathrm{a}}_{Q^\star}$. By definition of $R^{\mathrm{a}}_{Q^\star}$, $(s,t) \in Q$ implies the existence of some $(s, \mu) \in R^{\mathrm{a}}$ such that $(\mu, \eta) \in Q^\star$. Let $(s, \mu) \in R^{\mathrm{c}}$. By Theorem 5.(2 or 4(b)), $(\mu, \mu_Q) \in Q^\star$ and so $(t, \mu_Q) \in R^{\mathrm{c}}_Q$. Since $(s,t) \in Q$, we get $L^{\mathrm{a}}_{Q^\star}(t) = \bigcap_{(s',t) \in Q} L^{\mathrm{a}}(s') \subseteq L^{\mathrm{a}}(s)$ and $L^{\mathrm{c}}(s) \subseteq \bigcup_{(s',t) \in Q} L^{\mathrm{c}}(s')$ which equals $L^{\mathrm{c}}_{Q^\star}(t)$. ■

If our abstractions compute only the component $\mathcal{M}^{\mathrm{c}}_{Q^{\mathrm{ps}}}$, then such abstractions are fully compositional in that we may first specify an abstraction via Q_1 and then specify further abstractions on the first abstraction (via some Q_2). However, the computation of $\mathcal{M}^{\mathrm{a}}_{Q^{\mathrm{ps}}}$ requires that we specify an abstract model through the composite $Q_1; Q_2$ — as is the case for qualitative systems [22]. Thus, we owe a proof that abstraction relations are compositional.

Proposition 4 (Compositionality of Possibilistic Abstractions). *The composition of left-total, right-total relations with measurable navigation is left-total, right-total, and has measurable navigation.*

6.1 Predicate Abstraction

Given a finite set $\{\psi_1, \psi_2, \ldots, \psi_n\}$ of closed formulas in $\mathcal{L}_{\mathrm{pr}}$ and a concurrent labelled Markov chain $\mathcal{M}$ — a modal probabilistic Kripke structure $\mathcal{M}$ with $\mathcal{M}^{\mathrm{a}} = \mathcal{M}^{\mathrm{c}}$ — an equivalence relation $\equiv$ may be defined on its set of states Σ by

$$s \equiv s' \qquad \text{iff} \qquad \{i \mid s \models \psi_i\} = \{i \mid s' \models \psi_i\}, \tag{6}$$

where Theorem 3.1 justifies the use of the notation $\models$. On Σ_Q, the finite set of bit vectors of length n, we define $(s,t) \in Q$ iff for all i, $t_i = 1$ iff $s \models \psi_i$. The functional relation Q is left-total and right-total.

The relation Q is also measurable where $\mathcal{F}_Q = \mathcal{P}(\Sigma_Q)$, *regardless of the nature of the underlying σ-algebra $\mathcal{F}$ of the model $\mathcal{M}$*. Given $B \subseteq \Sigma_Q$ and $t \in B$, let ϕ_t be the conjunction of all ψ_i such that $t_i = 1$ and all $\neg\psi_i$ such that $t_i = 0$; define ϕ_B as the disjunction of all such ϕ_t with $t \in B$. Then $Q.B = \llbracket \phi_B \rrbracket_\rho \in \mathcal{F}$ by Theorem 1 and 3. Although this works for every $\mathcal{F}$, any choice of $\mathcal{F}$ severely restricts the class of legitimate models $\mathcal{M}$ through condition (1). By Theorem 5, Q^{ps} relates μ to μ_Q only.

Example 2 (Predicate Abstraction). If we abstract the model in Figure 1(left) with one predicate only, $\psi_1 = p \vee q$, we obtain the model of Figure 1(right). We verify $t_0 \models^{\mathrm{a}} \neg\mathsf{EX}_{>\frac{3}{4}}\neg\mathsf{EX}_{>\frac{3}{10}}\neg p$, i.e. $t_0 \not\models^{\mathrm{c}} \mathsf{EX}_{>\frac{3}{4}}\neg\mathsf{EX}_{>\frac{3}{10}}\neg p$. The latter is equivalent to "for all $(t_0, \kappa) \in R^{\mathrm{c}}_{Q^{\mathrm{pr}}}$, $\kappa(\llbracket \neg\mathsf{EX}_{>\frac{3}{10}}\neg p \rrbracket^{\mathrm{c}}) \leq \frac{3}{4}$", where $\kappa \in \{\alpha_Q, \mu_Q, \eta_Q\}$. But $\llbracket \neg\mathsf{EX}_{>\frac{3}{10}}\neg p \rrbracket^{\mathrm{c}} = \Sigma_Q \setminus \llbracket \mathsf{EX}_{>\frac{3}{10}}\neg p \rrbracket^{\mathrm{a}})$. Now $\llbracket \mathsf{EX}_{>\frac{3}{10}}\neg p \rrbracket^{\mathrm{a}} = \{t \mid \exists (t,\kappa) \in R^{\mathrm{a}}_{Q^{\mathrm{pr}}} : \kappa(\llbracket \neg p \rrbracket^{\mathrm{a}}) > \frac{3}{10}\}$ equals $\{t_1\}$, since (t_1, γ_Q) is the only $R^{\mathrm{a}}_{Q^{\mathrm{pr}}}$-transition and $\gamma_Q(\{t_1\}) = \frac{1}{3} > \frac{3}{10}$. Thus, $\llbracket \neg\mathsf{EX}_{>\frac{3}{10}}\neg p \rrbracket^{\mathrm{c}} = \{t_0\}$ from which we infer $\alpha_Q(\{t_0\}) = \frac{2}{3} \leq \frac{3}{4}$, $\mu_Q(\{t_0\}) = \frac{3}{4} \leq \frac{3}{4}$, and $\eta_Q(\{t_0\}) = \frac{1}{2} \leq \frac{3}{4}$. By Theorem 1, we now know that s_0, s_1, and s_3 satisfy $\neg\mathsf{EX}_{>\frac{3}{4}}\neg\mathsf{EX}_{>\frac{3}{10}}\neg p$. (Note that $\neg\mathsf{EX}_{>\frac{3}{4}}\neg$ may be abbreviated as $\mathsf{AX}_{\leq\frac{3}{4}}$.)

In the paragraph above, we proved that predicate abstraction always gives rise to an abstraction relation that has measurable navigation.

Theorem 7 (Predicate Abstraction Has Measurable Navigation). *Let $\mathcal{M}$ be a concurrent labelled Markov chain whose σ-algebra $(\Sigma, \mathcal{F})$ satisfies condition (1). For $\equiv$ defined as in (6), assume that its set of equivalence classes, Σ_Q, is finite and contained in $\mathcal{F}$. Then $Q = \{(s,t) \in \Sigma \times \Sigma_Q \mid s \in t\}$ has measurable navigation, where the σ-algebra on Σ_Q is discrete.*

The proof of this fact works as specified above, except that the disjunctive normalform has elements of $\mathcal{F}$ as literals.

Although we presented a compositional sound formalism for the specification of relational abstractions, the computation of such abstractions may prove to be expensive. This is indeed the case for qualitative systems, where the tradeoff between size and precision has been anticipated by Cleaveland et al. in [11]. For $\star \in \{\mathrm{ps}, \mathrm{pr}\}$, one seeks abstraction relations $Q \subseteq \Sigma \times \Sigma_Q$ such that $\mathcal{M}^{\mathrm{c}}_{Q\star}$ is as small and $\mathcal{M}^{\mathrm{a}}_{Q\star}$ as big as possible. Since the computation of $R^{\mathrm{a}}_{Q\star}$-transitions requires disjunctions of abstract states for the source of transitions, there are 2^{2^n} possible states for n predicates. Cartesian abstraction [4] brings this upper bound down to 3^n, alas at a likely loss of precision.

6.2 Cartesian Abstraction

Continuing our discussion of Section 6.1, let $\Sigma_{Q'}$ be the set of tri-vectors over $\{0, 1, *\}$ of length n. We define $Q' \subseteq \Sigma \times \Sigma_{Q'}$ by $(s, t) \in Q'$ iff for all i, $t_i \neq *$ implies ($t_i = 1$ iff $s \models \psi_i$). The relation Q' is left-total and right-total. The relation Q' is also measurable, where $\mathcal{F}_Q = \mathcal{P}(\Sigma_Q)$. Given $B \subseteq \Sigma'_Q$ and $t \in B$, let ϕ'_t and ϕ_B be defined as in Section 6.1, where every empty conjunction ϕ'_t is understood to be tt. Then $Q.B = \llbracket \phi_B \rrbracket_\rho \in \mathcal{F}$. Again, since Q' represents a function Q^{ps} relates μ to μ_Q only.

Example 3. Continuing Example 2, let Σ_Q be $\{*\}$ with four R^{c}-transitions, one for each transition in the concurrent, labelled Markov chain. These transitions lead back to $*$ with probability 1. All state observables p and q are in $L^{\mathrm{c}}(*)$. There are no R^{a}-transitions and $L^{\mathrm{a}}(*)$ is empty. Every finite concurrent, labelled Markov chain with such observables is a possibilistic/probabilistic refinement of that system.

Theorem 8 (Cartesian Abstraction Has Measurable Navigation). *Theorem 7 extends to the use of Cartesian abstraction.*

Our results on possibilistic and probabilistic abstraction confirm existing work on sound abstraction of universal properties (e.g. reachability [15]) but extend such work to the scope of an unrestricted logic; unfortunately, the quality of abstraction-based probabilistic model checking for unrestricted properties depends crucially on the precision of the abstract model: whether we are able to compute few $R^{\mathrm{c}}_{Q\star}$ and many $R^{\mathrm{a}}_{Q\star}$ transitions. In any event, the presence of three-valued propositional state observables may improve the precision of reachability analyses.

7 Conclusions

We presented a quantitative notion of Kripke structures whose quantities stem from a partial order, whose properties have denotations in a σ-algebra, and whose possibilistic abstraction is sound for all verifications of properties from a quantitative logic with negation and quantification. State transitions in such models map states to quantitative measures. Modal probabilistic systems are a special instance of such models, have a probabilistic abstraction — for which we proved soundness for full probabilistic LTL — , and their possibilistic property semantics approximates the probabilistic one. Predicate-based abstractions specify abstract models in the quantitative and probabilistic case.

Acknowledgements

We wish to thank the anonymous referees for their most helpful comments.

References

1. S. Andova and J. C. M. Baeten. Abstraction in Probabilistic Process Algebras. In T. Margaria and W. Yi, editors, *Tools and Algorithms for the Construction and Analysis of Systems (TACAS 2001)*, volume 2031 of *Lecture Notes in Computer Science*, pages 204–219, Genova, Italy, April 2-6 2001. Springer Verlag.
2. C. Baier, E.M. Clarke, V. Hartonas-Garmhausen, M. Kwiatkowska, and M. Ryan. Symbolic Model Checking for Probabilistic Processes. In *Proc. ICALP'97*, volume 1256 of *Lecture Notes in Computer Science*, pages 430–440, 1997.
3. C. Baier and M. Kwiatkowska. Model checking for a probabilistic branching-time logic with fairness. *Journal of Distributed Computing*, 11:125–155, 1998.
4. T. Ball, A. Podelski, and S.K. Rajamani. Boolean and Cartesian Abstraction for Model Checking C Programs. In T. Margaria and W. Yi, editors, *Proceedings of TACAS 2001*, volume 2031 of *LNCS*, pages 268–283, Genova, Italy, April 2001. Springer Verlag.
5. G. Bruns and P. Godefroid. Model Checking Partial State Spaces with 3-Valued Temporal Logics. In *Proceedings of the 11th Conference on Computer Aided Verification*, volume 1633 of *Lecture Notes in Computer Science*, pages 274–287. Springer Verlag, July 1999.
6. G. Bruns and P. Godefroid. Generalized Model Checking: Reasoning about Partial State Spaces. In *Proceedings of CONCUR 2000 (11th International Conference on Concurrency Theory)*, volume 1877 of *Lecture Notes in Computer Science*, pages 168–182. Springer Verlag, August 2000.
7. R. R. Bryant. Symbolic Boolean Manipulation with Ordered Binary-Decision Diagrams. *ACM Computing Surveys*, 24(3):293–318, September 1992.
8. D. Clark, C. Hankin, S. Hunt, and R. Nagarajan. Possibilistic Information Flow is safe for Probabilistic Non-Interference. In *Workshop on Issues in the Theory of Security (WITS '00)*, Geneva, Switzerland, 7-8 July 2000.
9. E. Clarke, A. Biere, R. Raimi, and Y. Zhu. Bounded Model Checking Using Satisfiability Solving. *Formal Methods in System Design*, 19(1), July 2001.
10. E.M. Clarke, O. Grumberg, and D.E. Long. Model checking and abstraction. *ACM Transactions on Programming Languages and Systems*, 16(5):1512–1542, 1994.
11. R. Cleaveland, P. Iyer, and D. Yankelevich. Optimality in abstractions of model checking. In *SAS'95: Proc. Second Static Analysis Symposium*, Lecture Notes in Computer Science 983, pages 51–63. Springer, 1995.
12. C. Courcoubetis and M. Yannakakis. The Complexity of Probabilistic Verification. *Journal of the Association of Computing Machinery*, 42(4):857–907, July 1995.
13. P. Cousot and R. Cousot. Abstract interpretation: a unified lattice model for static analysis of programs. In *Proc. 4th ACM Symp. on Principles of Programming Languages*, pages 238–252. ACM Press, 1977.
14. P. Cousot and R. Cousot. Temporal abstract interpretation. In *Conference Record of the 27th Annual ACM SIGPLAN-SIGACT Symposium on Principles of Programming Languages*, pages 12–25, Boston, Mass., January 2000. ACM Press, New York, NY.

15. P.R. D'Argenio, B. Jeannet, H.E. Jensen, and K.G. Larsen. Reachability Analysis of Probabilistic Systems by Successive Refinements. In L. de Alfaro and S. Gilmore, editors, *Process Algebra and Probabilistic Methods: Performance Modelling and Verification*, volume 2165 of *Lecture Notes in Computer Science*, pages 39–56, Aachen, Germany, September 12-14 2001. Springer Verlag.
16. C. Derman. *Finite-State Markovian Decision Processes.* Academic Press, New York, 1970.
17. J. Desharnais, V. Gupta, R. Jagadeesan, and P. Panangaden. Approximating Labeled Markov Processes. In *15th Annual IEEE Symposium on Logic in Computer Science (LICS'00)*, Santa Barbara, California, 26-29 June 2000. IEEE Computer Society Press.
18. E.-E. Doberkat. The Converse of a Probabilistic Relation. Technical Report 113, Fachbereich Informatik, Universität Dortmund, June 2001.
19. E.-E. Doberkat. The Demonic Product of Probabilistic Relations. In *Foundations of Software Science and Computation Structures*, Lecture Notes in Computer Science, Grenoble, France, April 6-14 2002. Springer Verlag. To appear.
20. D. Dubois, J. Lang, and H. Pade. *Possibilistic logic*, volume 3 of *Handbook of Logic in Artificial Intelligence and Logic Programming*, pages 439–514. Oxford University Press, 1992.
21. M. Giry. A categorical approach to probability theory. In B. Banaschewski, editor, *Categorical Aspects of Topology and Analysis*, volume 915 of *Lecture Notes in Mathematics*, pages 68–85. Springer Verlag, 1981.
22. P. Godefroid, M. Huth, and R. Jagadeesan. Abstraction-based Model Checking using Modal Transition Systems. In *Proceedings of the International Conference on Theory and Practice of Concurrency*, volume 2154 of *Lecture Notes in Computer Science*, pages 426–440. Springer Verlag, August 2001.
23. P.R. Halmos. *Measure Theory.* Graduate Texts in Mathematics 18. Springer Verlag, 1950.
24. H. Hansson. *Time and Probability in Formal Design of Distributed Systems.* PhD thesis, Department of Computer Science, Uppsala University, Uppsala, Sweden, 1991.
25. A. Harding, M. Ryan, and P.-Y. Schobbens. Approximating ATL* in ATL. In *Third International Workshop on Verification, Model Checking and Abstract Interpretation*, volume 2294 of *Lecture Notes in Computer Science*, pages 289–301, Venice, Italy, January 21-22 2002. Springer Verlag.
26. J. Hillston. *A Compositional Approach to Performance Modelling.* Cambridge University Press, 1996.
27. C.A.R. Hoare. *Communicating Sequential Processes.* Prentice-Hall, 1985.
28. M. Huth. Model checking modal transition systems using Kripke structures. In *Third International Workshop on Verification, Model Checking and Abstract Interpretation*, volume 2294 of *Lecture Notes in Computer Science*, pages 302–316, Venice, Italy, January 21-22 2002. Springer Verlag.
29. M. Huth, R. Jagadeesan, and D. Schmidt. Modal transition systems: a foundation for three-valued program analysis. In Sands D., editor, *Proceedings of the European Symposium on Programming (ESOP 2001)*, pages 155–169. Springer Verlag, April 2001.
30. B. Jonsson and K.G. Larsen. Specification and Refinement of Probabilistic Processes. In *6th Annual IEEE Symposium on Logic in Computer Science*, pages 266–277, Amsterdam, The Netherlands, 15-18 July 1991. IEEE Computer Society Press.

31. P. Kelb. Model checking and abstraction: a framework preserving both truth and failure information. Technical Report OFFIS, University of Oldenburg, Germany, 1994.
32. K.G. Larsen and A. Skou. Bisimulation through probabilistic testing. *Information and Computation*, 94(1):1–28, September 1991.
33. K.G. Larsen and B. Thomsen. A Modal Process Logic. In *Third Annual Symposium on Logic in Computer Science*, pages 203–210. IEEE Computer Society Press, 1988.
34. A. McIver. A Generalization of Stationary Distributions, and Probabilistic Program Algebra. In *MFPS 2001: Seventeenth Conference on the Mathematical Foundations of Programming Semantics*, volume 45 of *Electronic Notes in Theoretical Computer Science*, Aarhus, Denmark, 23-26 May 2001. Elsevier.
35. R. Milner. *Communication and Concurrency*. Prentice-Hall, 1989.
36. D. Monniaux. Abstract interpretation of programs as Markov decision processes. Technical report, Départment d'Informatique, École Normale Supérieure, 45, rue d'Ulm, 75230 Paris cedex 5, France, 2001.
37. C. Morgan, A. McIver, K. Seidel, and J.W. Sanders. Refinement-oriented probability for CSP. *Formal Aspects of Computing*, 8(6):617–647, 1996.
38. Prakash Panangaden. The Category of Markov Kernels. In M. Kwiatkowska C. Baier, M. Huth and M. Ryan, editors, *Electronic Notes in Theoretical Computer Science*, volume 22. Elsevier Science Publishers, 2000.
39. D.M.R. Park. Concurrency and automata on infinite sequences. In P. Deussen, editor, *In Proc. of the 5th GI Conference*, volume 104 of *Lecture Notes in Computer Science*, pages 167–183. Springer Verlag, 1989.
40. A. Di Pierro, C. Hankin, and H. Wiklicky. Approximate non-interference. Submitted, February 2002.
41. A. Di Pierro and H. Wiklicky. Concurrent Constraint Programming: Towards Probabilistic Abstract Interpretation. In *Proc. of the 2nd Int'l ACM SIGPLAN conference on Principles and Practice of Declarative Programming (PPDP 2000)*, pages 127–138, Montreal, Canada, September 20-23 2000. ACM Press.
42. D.E. Rumelhart, J.L. McClelland, and the PDP Research Group. *Parallel Distributed Processing*, volume 1 of *Explotations in the Microstructure of Cognition*. The MIT Press, 1986.
43. A. Sabelfeld and D. Sands. A per model of secure information flow in sequential programs. In *Programming Languages and Systems, 8th European Symposium on Programming (ESOP'99)*, volume 1576 of *Lecture Notes in Computer Science*, pages 40–58. Springer Verlag, 1999.
44. D.A. Schmidt. From Trace Sets to Modal-Transition Systems by Stepwise Abstract Interpretation. *Electronic Notes in Theoretical Computer Science*, March 2001. Proc. Workshop on Structure Preserving Relations, Amagasaaki, Japan. To appear.
45. M. Vardi. Automatic verification of probabilistic concurrent finite-state programs. In *Proc. 26th IEEE Symp. on Foundations of Computer Science*, pages 327–338, Portland, Oregon, October 1985.
46. M. Vardi. Probabilistic Linear-Time Model Checking: an Overview of The Automata-Theoretic Approach. In J.-P. Katoen, editor, *Formal Methods for Real-Time and Probabilistic Systems, 5th Int'l AMAST Workshop (ARTS'99)*, volume 1601 of *Lecture Notes in Computer Science*, pages 265–276, Bamberg, Germany, 26-28 May 1999. Springer Verlag.
47. L. Zuck, A. Pnueli, and Y. Kesten. Automatic Verification of Probabilistic Free Choice. In A. Cortesi, editor, *Third International Workshop on Verification, Model Checking and Abstract Interpretation*, volume 2294, pages 208–224, Venice, Italy, 21-22 January 2002. Springer Verlag.

Out-of-Core Solution of Large Linear Systems of Equations Arising from Stochastic Modelling

Marta Kwiatkowska and Rashid Mehmood

School of Computer Science, University of Birmingham
Birmingham B15 2TT, United Kingdom
{mzk,rxm}@cs.bham.ac.uk

Abstract. Many physical or computer systems can be modelled as Markov chains. A range of solution techniques exist to address the state-space explosion problem, encountered while analysing such Markov models. However, numerical solution of these Markov chains is impeded by the need to store the probability vector(s) explicitly in the main memory. In this paper, we extend the earlier out-of-core methods for the numerical solution of large Markov chains and introduce an algorithm which uses a disk to hold the probability vector as well as the matrix. We give experimental results of the implementation of the algorithm for a Kanban manufacturing system and a flexible manufacturing system. Next, we describe how the algorithm can be modified to exploit sparsity structure of a model, leading to better performance. We discuss two models, a cyclic server polling system and a workstation cluster system, in this context and present results for the polling models. We also introduce a new sparse matrix storage format which can provide 30% or more saving over conventional schemes.

1 Introduction

Discrete-state Markovian models are widely employed for the analysis of communication networks and computer systems. It is often convenient to model such a system as a Continuous Time Markov Chain (CTMC), provided the probability distribution is restricted to be exponential. A CTMC may be represented by a set of states and a transition rate matrix Q containing state transition rates as coefficients. Hence, the problem of analysing Markov models is reducible to the classical problem of solving a system of linear equations of the form $Ax = b$ where $b = 0$. A range of solution techniques exist to combat the so-called state-explosion (also known as largeness) problem in this area. These, among others, include symbolic techniques [6,12,16,25], Kronecker methods [23], and *on-the-fly* methods [9]. However, the solution processes for these methods have high time complexity. While some new developments such as Matrix Diagrams (MDs) [2,5,22] and hybrid MTBDD methods [21] can allow a very compact and time efficient encoding of the CTMC matrices, they are obstructed by explicit storage of the probability vector(s) needed during the numerical solution phase.

H. Hermanns and R. Segala (Eds.): PAPM-PROBMIV 2002, LNCS 2399, pp. 135–151, 2002.

An established direction of research in the area of Markov modelling has concerned numerical solution techniques which store the CTMC matrix explicitly, using some sparse storage format. These are the most generally applicable techniques, fast though not as compact as the symbolic methods. Improvements have been obtained through using disks to store the CTMC matrices [7,8] and parallelising the disk-based numerical solutions [19,1]. However, the memory requirements for these methods are also dominated by the storage of a vector of size proportional to the number of states in the model.

This paper makes two novel contributions to the numerical solution of Markov chains. The first contribution is an extended out-of-core algorithm which relaxes the memory limitations for the explicit numerical solution methods caused by the need to store the probability vector(s) explicitly. This approach is also applicable to implicit methods, see §6. We use the term "implicit methods" in this paper for the numerical solution methods which use some kind of symbolic data structure (MTBDDs, MDs, etc) for the storage of a CTMC matrix. The term "explicit" will be used to denote the numerical solution methods which store the CTMC matrix using a sparse data structure (Modified Sparse Row etc). Algorithms that are designed to achieve high performance when their data structures are stored on disk are known as *out-of-core algorithms*; Toledo presents a survey of such algorithms in numerical linear algebra, see [27].

The algorithm introduced in this paper assumes no structure in the sparse matrix, and hence can be applied to any CTMC model. The algorithm divides the underlying CTMC matrix and the probability vector into a certain number of blocks and stores these blocks onto a disk. It reads into main memory a block of the matrix and of the probability vector, performs the required calculation and writes back updated elements of the probability vector onto the disk. Moreover, the algorithm provides fairly fast convergence due to the Gauss-Seidel iterative method being employed. To obtain high performance from the algorithm, the work is divided between two concurrent processes; one performs the computation while the other schedules the disk reading and writing. The memory requirement of the algorithm is dependent on the number of blocks; the higher the number of blocks that the matrix and the probability vector is divided into, the less memory is required by the algorithm. Virtually, it leaves no upper bound on the size of a numerically solvable CTMC matrix on a standard workstation provided sufficient disk space is available. However, there is an obvious tradeoff between the memory requirement of the algorithm and the resulting solution time. We demonstrate the behaviour of the algorithm by implementing it for an UltraSPARC-II 360MHz machine with 128MB physical memory (RAM) and a 6GB local disk. We test the implementation on a Kanban manufacturing system [3] and a Flexible Manufacturing System (FMS) [4] and present results.

The performance of the algorithm can be improved if a certain sparsity structure can be found in the underlying CTMC matrix. This fact is demonstrated by including results of a modified algorithm applied to a Cyclic Server Polling System [17].

The second contribution of this paper is a new sparse matrix storage format, called the *Compact Modified Sparse Row* (CMSR) or *compact MSR* format. We believe that this storage format, which can also be used with in-core iterative solutions of sparse linear systems, would be helpful in further pushing the upper limit on the size of solvable models in the area of Markov modelling. We have used the compact MSR format in the implementations of various serial and parallel algorithms, including the algorithm presented in this paper.

The Tool PRISM. All the CTMC matrices used in this paper are generated by the PRISM tool [21,20]. PRISM, a probabilistic model checker, is a tool for analysing probabilistic systems currently being developed at the University of Birmingham. It supports three models: discrete-time Markov chains, continuous-time Markov chains and Markov decision processes. For the numerical solution phase, it provides three choices: one using pure MTBDDs, one based on in-core sparse iterative methods, and a third using a hybrid approach. Several case studies have been performed, see `http://www.cs.bham.ac.uk/~dxp/prism/` for details. This paper is part of an effort to improve the range of solution methods provided by PRISM.

2 Numerical Methods

We focus in this paper on numerical solutions of continuous time Markov chains. The task of solving a CTMC to obtain the steady-state probabilities vector can be mathematically written as:

$$\pi Q = 0, \ \sum_{i=0}^{n-1} \pi_i = 1, \tag{1}$$

where $Q \in \mathbb{R}^{n\times n}$ is the state transition rate matrix, and $\pi \in \mathbb{R}^n$ is the steady state probability vector. The matrices Q are usually very sparse; further details about the properties of these matrices can be found in [26]. The equation (1) can be reformulated as $Q^T\pi^T = 0$, and well-known methods for the solution of systems of linear equations of the form $Ax = b$, provided $A = Q^T$, $x = \pi^T$, and $b = 0$, can be used. For a discrete-time Markov chain, the problem of model checking a probabilistic *until* formulae [13] reduces to solving $Ax = b$ where A is a sub-matrix of the transition probability $n \times n$ matrix P and b is a vector of the appropriate size.

The numerical solution methods for linear systems of the form $Ax = b$ are broadly classified into direct methods and the iterative methods. For large systems, direct methods become impractical due to the phenomenon of *fill-in*, caused by the generation of new entries during the factorisation phase. Iterative methods generate a sequence of approximations that only converges in the limit to the solution. Beginning with a given approximate solution, these methods modify the components of the approximation in each iteration, until a required convergence is achieved. See [11] for further information on direct methods and [24,26] for iterative methods.

We concentrate here on the Gauss-Seidel (GS) iterative method, which we employed for our out-of-core algorithm. In each iteration of the Gauss-Seidel algorithm we calculate

$$\pi_i^{(k)} = -\frac{1}{q_{ii}} \left(\sum_{j=0}^{i-1} \pi_j^{(k)} q_{ij} + \sum_{j=i+1}^{n-1} \pi_j^{(k-1)} q_{ij} \right), \tag{2}$$

for $0 \leq i < n$. Unlike Jacobi iteration, each component of the solution vector in the Gauss-Seidel method must be updated sequentially. However, Gauss-Seidel has advantages over Jacobi in that it has a faster convergence rate and that it requires only one vector of size n. We chose the relative measure

$$\max_i \left(\frac{\mid \pi_i^{(k)} - \pi_i^{(k-1)} \mid}{\mid \pi_i^{(k)} \mid} \right) < \epsilon, \tag{3}$$

for some ϵ $(0 < \epsilon \ll 1)$ to test for convergence in all the experiments presented in this paper.

3 Sparse Storage Considerations

Iterative solution of a system of linear equations of the form $\pi Q = 0$ involves explicit storage of the whole transition rate matrix Q. A usual choice for the data structure is the *modified sparse row* (MSR) format [24]. The MSR format stores all the off-diagonal non-zero entries in an array, say *Val*, of floats (4 bytes) or doubles (8 bytes). The corresponding column indices are stored in a separate array of integers (4 bytes), say *Col*. Diagonal entries may be stored in the first or the last n locations of the array *Val*. Another option is to store the n diagonal entries in a separate array. Further, an array of $n+1$ integers is used, the ith entry of which points to the first entry of the ith row in the arrays *Val* and *Col*. The total memory requirements, if non-zeros are stored as floats, is $8(a+n)$ bytes, and is $12(a+n)$ bytes when doubles are used to store the non-zero entries; a is the number of the off-diagonal non-zero entries in the matrix. The *modified sparse column* (MSC) storage format differs from MSR in that it stores the matrix entries and their positions by column.

Table 1. Storage requirements in Mbytes for the Polling models

Poll	*states* (n)	*off-diagonal nonzero* (a)	a/n	MB required for Q		MB
				MSR	*distinct-once*	per π
17	3,342,336	31,195,136	9.33	260	198	26
18	7,077,888	69,599,232	9.83	584	438	54
20	31,457,280	340,787,200	10.83	2839	2129	240
21	66,060,288	748,683,264	11.33	6323	4421	513

Table 1 lists some statistics about the matrices taken from a case study based on a Cyclic Server Polling system [17], obtained with the help of the tool PRISM. The table also lists the memory requirements to store the CTMC matrix and the solution vector for each model. Column 1 lists the parameter *Poll*, each value representing a separate CTMC model, and column 2 shows the resulting number of states n which equals the dimension of the matrix. Column 3 lists the number of the off-diagonal non-zero entries a. Column 4 gives the average off-diagonal non-zero entries per row, an indication of the sparsity of the matrices.

The number of distinct values in the CTMC matrices resulting from the models we considered is relatively small. This important characteristic can lead to appreciable memory savings. The distinct values may be stored in a small array, and the index to these values can be stored as a short (2 bytes). Column indices may be stored in a separate array of integers (4 bytes). The number representing the count of the off-diagonal non-zero entries in a row can be stored as an unsigned char (1 byte). The storage of the diagonal will require 8 bytes (doubles) per each entry. The total memory requirements for this scheme is $6a+9n$ bytes plus the storage for the actual distinct values of the matrix nonzero entries. Since the storage for the actual distinct values is relatively small for large models (in comparison with the total memory requirement), we do not consider it in future discussions.

Let us mention here a possible transformation of the system $\pi Q = 0$ into an equivalent system $R^T y^T = 0$ [1]. If we define $R = -QD^{-1}$, then the system $\pi Q = 0$ can be equivalently written as $\pi DQD^{-1} = -yR = 0$, with $y = -\pi D$. The matrix D is defined as the diagonal matrix with $d_{ii} = q_{ii}$ for $0 \leq i < n$. Effectively, we solve the equivalent system $R^T y^T = 0$, with all the diagonal entries of the matrix R being -1. Along the lines of the storage scheme $(6a+9n)$ described in the previous paragraph, the storage requirements for this system is $6a + n$ with an additional advantage of runtime savings of n divisions per each iteration. For future references (in this paper) we call this memory scheme the **distinct-once** memory scheme, implying that the distinct values in the matrix are stored only once. This storage format is currently being used in iterative solutions for the explicit storage of CTMC matrices [1,7,19].

3.1 A Compact MSR Sparse Storage Format

This section introduces and explains a new storage scheme that renders a 30% or more reduction over the distinct-once storage scheme. The scheme, used to store the sparse CTMC matrix, preserves the general applicability of the explicit iterative solutions of Markov chains, whether in-core, out-of-core or parallel. Moreover, it does not increase the runtime of the solution process; indeed, it reduces the runtime due to better caching effect. We call it the *Compact Modified Sparse Row* or *compact MSR* storage format. A similar *compact MSC* is possible.

In the distinct-once scheme, which is described earlier in the parent section (§3), the distinct values are stored once in an array of doubles and the indices to these values are stored in another array of shorts. In total, this scheme requires $6a + n$ bytes if matrix R is stored, and requires additional $8n$ bytes for diagonal elements if matrix Q is stored. The number of distinct values in the matrix R for various models are around ten thousand to twenty thousand [1,19], hence the indices are stored as shorts.

We observe that the operation QD^{-1} increases the number of distinct values in Q and the number of distinct values in D. The number of distinct values to be stored may be decreased by storing the matrix Q instead of storing the matrix $R = QD^{-1}$. Furthermore, diagonal elements also have relatively few distinct values; the same philosophy of distinct values for transition rate matrix can be applied to the diagonal vector. The distinct values in the diagonal elements may be stored as *1/value* instead of *value*. The storage of diagonal vector would require n or $2n$ bytes depending on the maximum possible number of distinct values.

Among the various models used in the literature, probably FMS models [4] have the highest number of distinct values [1,19]; they are a good example to demonstrate the effectiveness of the compact MSR storage scheme. Table 2 summarises a fact sheet for FMS models comparing storage requirements in MSR, distinct-once, and compact MSR formats. The model *fms10* has 25.4 million states, contains 125 (not given in the table) distinct values in the off-diagonal elements of matrix Q, and 1157 (not given in the table) distinct values in the diagonal elements of matrix Q. Counting up to the index "125" requires 7 binary bits of memory. A data type int uses 32 bits of storage. If we steal 7 bits from the allocated space for an unsigned int, we would still have 25 bits; these are sufficient to address up to 33.5 million states. The diagonal elements require additional $2n$ bytes for the storage of n short indices to an array of doubles, which stores the distinct values in the diagonal vector. The total memory requirements in the compact MSR format for FMS models is thus $4a + 3n$ bytes.

The proposed scheme of bitwise packing of the index of distinct values together with a state index within an int variable is space efficient but subject to a trade-off between the number of distinct elements and the model size. For example, the model *Poll20*, which has 31.45 million states (see Table 1), contains just 3 (not given in the table) distinct values in the off-diagonal elements of matrix Q, and 77 distinct values in the diagonal elements of matrix Q. Since only 2 bits of memory are required to index off-diagonal distinct values, this allows us to address models of up to one billion states. The compact MSR format for the Polling models requires $4a + 2n$ bytes. The limitation on the maximum number

Table 2. Storage requirements for the FMS models in various formats

fms	states	nonzero (*off-diagonal*)	memory required for Q (MB) *MSR format*	*distinct-once*	*compact MSR*
8	4,459,455	38,533,968	326	225	160
10	25,397,658	234,523,289	1975	1366	967
12	111,414,940	1,078,917,632	9042	6279	4433

of states is not a serious concern for parallel and disk-based solutions, where local indices within a matrix block are typically used. Out-of-core solution will also benefit through a reduction in disk access time.

We now turn our attention to performance issues of the scheme, which will also help explain the implementation issues. We know that modern microprocessors have several functional units which are pipelined to deliver performance in parallel. These independent units help in overlapping the execution of instructions when the instructions are independent of each other. This potential overlap among instructions is called **instruction-level parallelism** (ILP) since the instructions can be evaluated in parallel [15]. The availability of several functional units enables parallel execution of mutually independent instructions. Encoding (compact MSR format) the data (indices to distinct values and their positions in matrix) into a 4 byte structure (unsigned int) may be carried out using bitwise operations. Most modern processors have two simple integer units which can execute this type of operations in a single clock cycle. If the instructions in the solution process are arranged in a manner that allows instruction-level parallelism to be exploited by the microprocessor, the additional operations added due to the compact MSR format do not slow down the code.

We have described the compact MSR format here, which offers a 30% memory savings. The savings increase with the increase in the value of a/n. The average number of nonzero elements per row (a/n) in a matrix normally increases with the increase in the size of the model. It can be calculated that for $a/n = 11$, the new scheme provides 32.7% savings for FMS models and 34.4% savings for the Polling models over the distinct-once format. All performance related claims about the compact MSR format in this section are made after implementations and careful analysis of the storage scheme.

Serial Solutions. Serial out-of-core algorithms benefit from compact MSR in two ways: by reducing the main memory and file space requirements, and by reducing the amount of disk I/O. Moreover, dividing the CTMC matrix into blocks (see §4) with local state numbering relaxes the limits on the largest model solvable by serial iterative solutions. We use the term "*serial out-of-core*" to refer to a solution where one process is used to do major computations and another process is used to schedule disk I/O.

Parallel Solutions. In-core parallel and parallel out-of-core solutions would both benefit from the compact MSR format. Local state numbering for all blocks, as is the case in parallel solutions, would relax the limits on the largest models solvable using this format.

4 A Complete Out-of-Core Solution

The idea of using disk to store data structures is motivated by the problems associated with the very high memory requirements of the numerical solution process. Schemes such as MTBDDs [6], Kronecker methods [23,10], MDs [2,22,5] and the hybrid MTBDD approach of [21] provide a way of storing a CTMC matrix in a compact form. However, obtaining high performance from these methods is dependent on a certain structure in the model.

An alternative approach in stochastic modelling, introduced by Deavours and Sanders [7,8], is to use a *disk-based* (out-of-core) solution technique. They claimed, by comparing their tool with Kronecker [23] and "on-the-fly" [9] methods, that a disk-based approach is the method of choice for solving large Markov models provided sufficient disk space is available. Further work on the parallelisation of out-of-core methods in [18,19,1] has justified their claim.

A limitation common to the symbolic methods and to the serial and parallel out-of-core methods is the need to store one or more solution vectors of size n, the number of states in the CTMC model. Table 1, in its last column, lists the storage requirements for a probability vector. In this section we present an algorithm which relaxes these limits. The algorithm assumes no structure in the underlying CTMC matrix. The idea behind the algorithm is to store the CTMC matrix and the probability vector onto the disk, read a subset of the vector and of the matrix into RAM, perform computations, update the approximations to the probability vector, and write back the new approximation onto disk. Concurrency and synchrony play an important role in the performance of the algorithm.

4.1 The Algorithm

The out-of-core algorithm, given in Figure 1, comprises two concurrent processes: the *Disk-IO Process* and the *Compute Process*. Performance of the algorithm relies mainly on synchronised execution of the two processes; these issues are discussed in §4.2. For now, let us assume that the two processes in Figure 1 execute concurrently and synchronously; they synchronise by calling the functions *Wait()* and *Signal()*.

The algorithm assumes that the CTMC matrix to be solved is stored in a file on disk. The algorithm does not make a distinction between the storage format used. However, we have used the compact MSR format in our implementations. The CTMC matrix is divided into $B \times B$ square blocks of dimension $n/B \times n/B$, where some of the blocks might be empty. The algorithm assumes $N \ mod \ B = 0$. Figure 2 shows the sparsity pattern of a matrix resulting from a Flexible Manufacturing System (FMS) model [4]. The figure also explains the division of the matrix into 4×4 blocks. These blocks are numbered Q_{ij} for $0 \leq i, j < B$. The matrix blocks are stored in a file in such a way that all the row blocks – all $(i, 0)$ to $(i, 3)$ for row i in Figure 2 – are adjacent to each other. The aforementioned tool PRISM (§1) has been adapted to generate files according to these requirements.

The probability vector π is also divided into B blocks or subsets, each with n/B elements. In order to avoid confusion between the j-th element of the probability vector π and its j-th block, we use π_j to indicate the j-th element of the vector π and Π_j to indicate the j-th block of the vector π; Π_{ij} stands for the j-th element of the subset Π_i. The algorithm also assumes that the initial

Integer constant: B (*number of blocks*)
Semaphores: S_1, S_2: occupied
Shared variables: R_0, R_1 (*To read matrix Q blocks into RAM*)
Shared variables: Πbox_0, Πbox_1 (*To read solution vector π blocks into RAM*)

Disk-IO Process

```
1. Local variable: h, i, j, k
2. k ← B − 1
3. while not converged
4.    for i ← 0 to B − 1
5.       if i = 0 then j ← B − 1
6.       else j ← i − 1
7.       for h ← 0 to B − 1
8.          if not an empty block
9.             read Q_ij from disk
10.            if h ≠ 0
11.               read Π_j from disk
12.         Signal(S_1)
13.         Wait(S_2)
14.         if h = 0
15.            write Π_k to disk
16.         k ← k + 1 mod B
17.         if j = 0 then j ← B − 1
18.         else j ← j − 1
```

Compute Process

```
1. Local variable: i, j
2. while not converged
3.    for i ← 0 to B − 1
4.       if i = 0 then j ← B − 1
5.       else j ← i − 1
6.       for 0 to B − 1
7.          Wait(S_1)
8.          Signal(S_2)
9.          if j ≠ i
10.            if not an empty block
11.               Accumulate Q_ij Π_j
12.         else
13.            Update Π_i
14.            check for convergence
15.         if j = 0 then j ← B − 1
16.         else j ← j − 1
```

Fig. 1. An *out-of-core* Gauss-Seidel iterative algorithm

approximation for all the blocks of the probability vector π is stored on the disk except for the last block Π_{B-1}.

Compute Process. This process is responsible for all the numerical computations involved in the steady-state solution of a CTMC model. The numerical iterative method we employed in this algorithm is the Gauss-Seidel method, which was described in §2. We stress that we are using the point version of Gauss-Seidel iterative method, as opposed to the block version which applies the computations given by equation (2) at block level. The matrix Q and the vector π are read into RAM one block after another. Therefore, in order to calculate the next approximation for the subset Π_i, we need to accumulate the products $Q_{ij}\Pi_j$ in an array of *doubles* of size n/B for all $0 \leq j < B$ save $j \neq i$. These products are accumulated by line 11 of the Compute Process in a local array. The shared memory blocks R_x and Πbox_x, two each, are used to hold matrix and probability vector blocks respectively. All the elements of the subset Π_i are updated using equation (2), during the last iteration ($j = i$) of the inner loop (lines 6 – 16). The convergence test, as given by equation (3), is carried out before updating each element of the subset Π_i. Before proceeding to calculations in each iteration

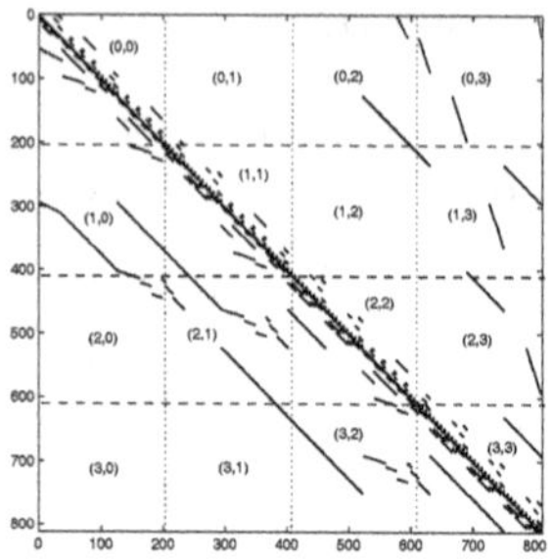

Fig. 2. Sparsity structure of FMS CTMC matrices

of the inner loop, the Compute Process waits for the signal from the Disk-IO Process – an indication that the Disk-IO Process has loaded the shared memory blocks with the new data.

Disk-IO Process. The Disk-IO Process is responsible for reading the matrix and the probability vector blocks from disk into RAM, and for writing the new approximations of the probability vector onto disk. The lines 5 – 6 along with lines 17 – 18 of the Disk-IO Process determine the indices of the matrix block and the probability vector block to be read during an inner loop (lines 7 – 18). The process reads a matrix block and a probability vector block in each iteration of inner loop and signals the Compute Process. It proceeds to the next iteration of the inner loop if the block is empty. To reduce the amount of I/O, only the range of the required elements of the probability vector are loaded from disk (except during the last iteration of the inner loop). For this purpose, we keep track of the lowest and the highest index of the required elements of a particular matrix block. During the first iteration of the inner loop, the newly calculated subset of the probability vector is written to disk, except otherwise when blocks of vector π are loaded from disk into the main memory.

4.2 Implementation

The design of an out-of-core numerical solver involves concurrency, and hence requires communication and synchronisation. We have two options available in UNIX: multithreaded code or multiprocess code. Threads in a process share address space and thus provide a convenient way of communication. Processes in UNIX are also allowed to do interprocess communication (IPC).

We have implemented our out-of-core algorithm using two separate processes communicating via shared memory and synchronising using semaphores. In UNIX, the shared memory facility normally provides the most efficient interprocess communication of all the IPC mechanisms. It allows more than one process at a time to attach a segment of physical memory to its virtual address space. The communication among the processes is made possible by allowing the

Table 3. Numerical solution times for the FMS and the Kanban models

Model	l	states	blocks	file size (*MB*)	RAM (*MB*)	*Time1* (sec/it)	*Time2* (sec/it)	Iter.	MB per π
Kanban	5	2,546,432	32	119	9	9	7	214	20
	6	11,261,376	64	575	20	90	69	289	86
	7	41,644,800	48	2164	105	561	308	374	317
FMS	8	4,459,455	27	195	16	34	21	893	34
	9	11,058,190	22	500	50	82	57	1018	84
	9	11,058,190	55	570	24	105	67	1018	84
	10	25,397,658	54	1266	54	280	178	1146	194

write access for more than one process. The semaphore mechanism can be used for synchronising the processes and to prevent inconsistencies. Our out-of-core algorithm achieves communication between the two processes by using the shared memory blocks R_x and Πbox_x, and synchronises through two semaphores, S_1 and S_2.

Generating Files. It has already been stated that the files for the out-of-core algorithm (in its required format) are generated by the tool PRISM. The times for the generation process are proportional to the times required to convert a model from BDD to a sparse format. Memory requirement for the file generation process can be either optimised for time or for memory. Optimising for memory can be achieved by keeping in main memory the data structures of the size of a small submatrix of the underlying CTMC matrix Q, and writing them onto file when the conversion is done; the process is repeated until the whole matrix is converted and written to the file. The generation process can be optimised for time by carrying out whole of the above process in one step, i.e, converting the whole model into sparse format in one step and then writing to file. Obviously, for efficiency reasons binary I/O is used. However, our concern in this paper is numerical solution of large linear systems, for the out-of-core numerical algorithm is independent of the modelling tool used to generate the data file.

4.3 Results

We tested the implemented out-of-core algorithm on four large models available in the literature. These are: a Cyclic Server Polling System [17], a Kanban System [3], a Workstation Cluster [14], and a Flexible Manufacturing System [4]. The Polling and the Cluster models have sparsity structures which can be exploited to obtain better performance; they are discussed in §5. In Table 3, we present solution times for FMS and Kanban CTMC matrices along with the memory requirements and the number of iterations the models required to converge. The convergence was tested according to equation (3) for $\epsilon = 10^{-6}$. The results were taken on an UltraSPARC-II 360MHz machine with 128MB RAM and a 6GB local

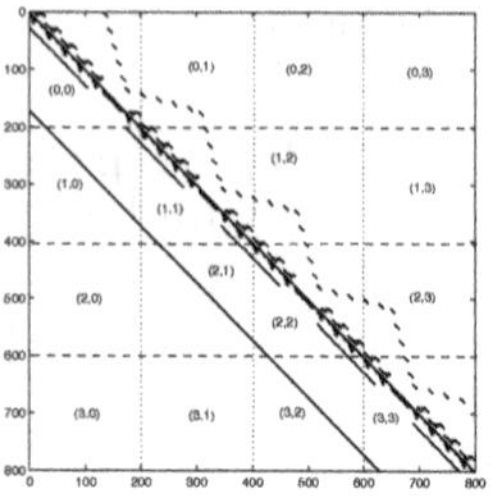

Fig. 3. Sparsity structure of matrices for the Cluster models

disk; *Time1* in column 7 reports these results. All the solution times reported are real times.

Column 4 in Table 3 lists the number of blocks the matrix of a particular model is partitioned into and column 6 lists the corresponding amount of memory used. Column 5 gives the sizes of the files that store the complete matrices on disk. The last column gives an idea of the amount of memory required if the probability vector had to be kept in main memory; for *kanban7* and *FMS10*, this is simply not possible (we do not consider virtual memory). It is clear that the memory requirement for a particular model can be further reduced by increasing the number of blocks. However, this may lead to an increase of the solution time. This fact is demonstrated by listing the results for the *FMS9* model with two different numbers of block. Finally, we stress that the results presented in the table are for the out-of-core method where both the matrix and the probability vector are stored on disk, though for models smaller than $14 - 15$ million states the vector can be stored in main memory and faster run time may be obtained.

Among the recent developments concerning implicit methods, we would like to compare the results of our algorithm with the work on MDs presented in [5,22] and with the work on the hybrid MTBDD method presented in [21]. For comparison purposes we have also executed our out-of-core algorithm on a workstation with an UltraSPARC-II 440MHz CPU; column 8 (*Time2*) in Table 3 lists these results. The algorithm in [5] requires more memory, for it needs to keep a vector of n doubles (the number of states) in the main memory. However, despite the higher memory requirements, the run times per iteration presented in [5] are considerably slower than our results.

The other approach we consider for comparison is the hybrid data structure [21] where MTBDDs are used to store the matrix, and the vector is stored explicitly. The results (run times per iteration) presented in [21] for the Kanban models are considerably better than ours. Their results presented for smaller FMS models[1] are also better than ours. The differences between their per iteration run time results and ours become less pronounced for larger FMS models. However, the critical factor is that the hybrid method employs JOR or Power

[1] `http://www.cs.bham.ac.uk/~dxp/prism/fms.html`.

methods which requires two solution vectors and have, for the models concerned, considerably slower convergence rate than Gauss-Seidel. Finally, we note that the hybrid method for the *FMS10* model requires at least 388MB ($2 \times n$ doubles) using Jacobi, while our Gauss-Seidel out-of-core algorithm with 54MB RAM delivers faster run times and better convergence rate.

Finally, we stress that the out-of-core algorithm introduced in this paper is an attempt to relax the memory limitations in the analysis of Markov models. It enables one to extend the range of solvable Markov models on a standard workstation. The runtime per iteration comparison between out-of-core and implicit methods would only be fair if the probability vector is kept in-core, and the matrix alone stored on disk; see [8] for such a comparison.

5 Exploiting Sparsity Structure

In parallel computing, it might be possible to exploit the *sparsity* structure of a matrix while decomposing and distributing the tasks to parallel processors. A few examples of the matrix structures that can be exploited are block-tridiagonal structure, symmetric structures, and *banded* matrices, in which the nonzero elements are confined within a band around the principal diagonal. The Cluster models are an example of banded matrices. The out-of-core solution may also borrow these ideas from parallel computing to obtain better performance. For example, if a large enough Cluster matrix is decomposed into a certain number of square blocks, the Gauss-Seidel calculation of a subset of the probability vector π requires at most two other subsets (blocks) of the vector π. These considerations may be exploited to modify the out-of-core algorithm presented in §4 leading to relatively balanced, regular and lesser amount of disk I/O. Figure 3 explains these facts.

Another sparsity structure that can be exploited by out-of-core and parallel algorithms is present in the matrices for the Polling models. Figure 4 shows the sparsity pattern of the Polling CTMC matrices. We observed that the Polling matrices require just another subset (in addition to the one being updated) of the probability vector π for the Gauss-Seidel computation of a subset of the vector π. We present the results of a modified algorithm which exploits the sparsity structure of the polling models in Table 4. However, the model in Figure 4 retains this structure as long as we keep an upper bound on the number of

Table 4. Numerical solution times for the Polling models

Poll	states	blocks	RAM (MB)	Time (sec/it)	Iter.	MB per π
18	7,077,888	16	13	44	34	54
19	14,942,208	16	27	98	35	114
20	31,457,280	16	62	220	36	240

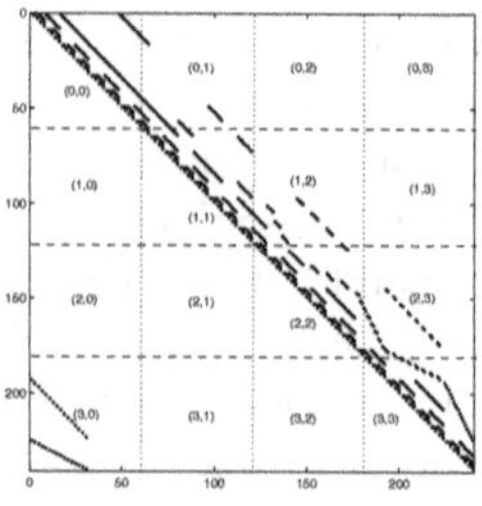

Fig. 4. Sparsity structure of matrices for the Polling models

blocks that the probability vector is decomposed into. For *Poll20* this limit is 16. This is the reason why the number of blocks for all the models in Table 4 are 16; column 3 shows the number of blocks for the vector π. Higher memory requirements for the lower number of blocks can be avoided by dividing the vector π and the matrix Q into a different number of blocks. All matrices in Table 4 are decomposed into 512 or more blocks. The convergence was tested according to equation (3) for $\epsilon = 10^{-7}$. We note in Table 4 that a model with 31 million states can be solved on an average speed workstation (360 MHz) in just over two hours using 62 MB RAM.

6 Future Directions for Research

We have introduced in this paper a new out-of-core algorithm for the numerical solution of large systems of linear equations. The algorithm relaxes the size limitation on implicit and explicit numerical solution methods, caused by the need to store the probability vector(s) explicitly. It can be applied in the numerical solution of any CTMC matrix, for it makes no assumptions about the structure of the matrix. The algorithm employs the compact MSR sparse storage format which itself was introduced in this paper. The scheme offers memory savings over current sparse storage schemes and it can be used with in-core, out-of-core or with parallel iterative numerical solutions of CTMC matrices. However, the correctness of the algorithm does not rely on the memory format employed.

We have described an implementation of the algorithm for an average-sized workstation. The physical memory of the workstation can at most hold one vector of 15 million doubles. By decomposing the matrix and the probability vector into a certain number of blocks, we were able to solve systems with up to 41 million states on a workstation with 128 MB RAM. We have also explained that even larger systems can be solved on a workstation of this average capacity. We further demonstrated how the sparsity structure of various models can be exploited to obtain better performance from the out-of-core algorithm. The paper opens new directions for future research which we outline below.

- The CPU usage for the implementations of the out-of-core algorithms described in this paper is around 60%. The CPU remains idle for 40% of the time due to disk I/O. Symbolic methods such as MTBDDs or MDs provide very compact encodings at a cost of CPU time. We envisage that keeping the matrices in main memory in aforementioned implicit encodings can improve the performance of the out-of-core algorithm. Moreover, this proposed work would also eliminate the hindrance of the implicit methods which is caused by the need to store the probability vector(s) explicitly in the main memory.
- Non-stationary iterative methods such as *Conjugate Gradient Squared* require more than two vectors of size proportional to the number of states in the system. Storing the vector π onto disk and keeping a certain number of vectors of size equal to a subset of the vector π means it is possible to use non-stationary methods even if the memory required to hold complete vectors is not available.
- Distributing a certain iterative computation such as Jacobi to a cluster of workstations means very high communication costs, especially if the link available on the cluster is not fast. Out-of-core methods which store the matrix onto disk and keep the probability vector in main memory have already been parallelised. We believe that the performance of out-of-core methods can be improved by parallelising the out-of-core algorithm introduced in this paper because it will enable very large models to be solved on clusters of modest sizes by reducing the main memory requirement and by providing work to overlap with the idle time during interprocessor communications.
- The out-of-core algorithm presented in this paper suffers from the problems associated with the sparsity structure of the CTMC models, for it causes either of the two Compute and Disk-IO processes to wait while one is busy doing its job. Further research is required to find out better out-of-core algorithms that adapt to the sparsity structure, and are capable of enjoying regular and balanced File I/O.

Acknowledgements

We would like to thank Stephen Gilmore for a discussion which motivated us to do this work. Gethin Norman and David Parker are acknowledged for the frequent useful conversations we had with them. We are also grateful to Daniel D. Deavours and William H. Sanders with whom we talked in Aachen.

References

1. Alexander Bell and Boudewijn R. Haverkort. Serial and Parallel Out-of-Core Solution of Linear Systems arising from Generalised Stochastic Petri Nets. In *Proc. High Performance Computing 2001*, Seattle, USA, April 2001.
2. G. Ciardo and A. Miner. A Data Structure for the Efficient Kronecker Solution of GSPNs. In *Proc. 8th International Workshop on Petri Nets and Performance Models (PNPM'99)*, Zaragoza, 1999.

3. G. Ciardo and M. Tilgner. On the use of Kronecker Operators for the Solution of Generalized Stocastic Petri Nets. ICASE Report 96-35, Institute for Computer Applications in Science and Engineering, 1996.
4. G. Ciardo and K. S. Trivedi. A Decomposition Approach for Stochastic Reward Net Models. *Performance Evaluation*, 18(1):37–59, 1993.
5. Gianfrance Ciardo. What a Structural World. In Reinhard German and Boudewijn Haverkort, editors, *Proceedings of the 9th International Workshop on Petri Nets and Performance Models*, pages 3–16, Aachen, Germany, September 11-14 2001.
6. E. Clarke, M. Fujita, P. McGeer, J. Yang, and X. Zhao. Multi-Terminal Binary Decision Diagrams: An Effificient Data Structure for Matrix Representation. In *Proc. International Workshop on Logic Synthesis (IWLS'93)*, pages 6a:1–15, Tahoe City, May 1993. Also available in *Formal Methods in System Design*, 10(2/3), 1997.
7. Daniel D. Deavours and William H. Sanders. An Efficient Disk-based Tool for Solving Very Large Markov Models. In *Lecture Notes in computer Science: Proceedings of the 9th International Conference on Modelling Techniques and Tools (TOOLS '97)*, pages 58–71, St. Malo, France, June 1997. Springer-Verlag.
8. Daniel D. Deavours and William H. Sanders. An Efficient Disk-based Tool for Solving Large Markov Models. *Performance Evaluation*, 33(1):67–84, 1998.
9. Daniel D. Deavours and William H. Sanders. "On-the-fly" Solution Techniques for Stochastic Petri Nets and Extensions. *IEEE Transactions on Software Engineering*, 24(10):889–902, 1998.
10. S. Donatelli. Superposed Stochastic Automata: A Class of Stochastic Petri Nets Amenable to Parallel Solution. *Performance Evaluation*, 18:21–36, 1993.
11. I. S. Duff, A. M. Erisman, and J. K. Reid. *Direct Methods for Sparse Matrices.* Oxford Science Publications. Clarendon Press Oxford, (with corrections)1997.
12. G. Hachtel, E. Macii, A. Pardo, and F. Somenzi. Markovian analysis of large finite state machines. *IEEE Transactions on CAD*, 15(12):1479–1493, December 1996.
13. H. Hansson and B. Jonsson. A Logic for Reasoning about Time and Probability. *Formal Aspects of Computing*, 6:512–535, 1994.
14. B. Haverkort, H. Hermanns, and J.-P. Katoen. On the Use of Model Checking Techniques for Dependability Evaluation. In *Proc. 19th IEEE Symposium on Reliable Distributed Systems (SRDS 2000)*, pages 228–237, Erlangen, Germany, October 2000.
15. John L. Hennessy and David A. Patterson. *Computer Architecture: A Quantitative Approach.* Morgan Kaufmann Publishers, Inc., 2nd edition, 1996.
16. H. Hermanns, J. Meyer-Kayser, and M. Siegle. Multi Terminal Binary Decision Diagrams to Represent and Analyse Continuous Time Markov Chains. In *Proc. Numerical Solutions of Markov Chains (NSMC'99)*, Zaragoza, 1999.
17. O. Ibe and K. Trivedi. Stochastic Petri Net Models of Polling Systems. *IEEE Journal on Selected Areas in Communications*, 8(9):1649–1657, 1990.
18. William J. Knottenbelt. *Parallel Performance Analysis of Large Markov Models.* PhD thesis, Imperial College of Science, Technology and Medicine, University of London, 1999.
19. William J. Knottenbelt and Peter G. Harrison. Distributed Disk-based Solution Techniques for Large Markov Models. In *Proc. Numerical Solution of Markov Chains (NSMC'99)*, Zaragoza, 1999.
20. M. Kwiatkowska, G. Norman, and D. Parker. PRISM: Probabilistic Symbolic Model Checker. In *Proc. TOOLS 2002*, April 2002.
21. M. Kwiatkowska, G. Norman, and D. Parker. Probabilistic Symbolic Model Checking with PRISM: A Hybrid Approach. In *Proc. TACAS 2002*, April 2002.

22. Andrew S. Miner. Efficient Solution of GSPNs using Canonical Matrix Diagrams. In Reinhard German and Boudewijn Haverkort, editors, *Proceedings of the 9th International Workshop on Petri Nets and Performance Models*, pages 101–110, Aachen, Germany, September 2001.
23. B. Plateau. On the Stochastic Structure of Parallelism and Synchronisation Models for Distributed Algorithms. In *Proc. 1985 ACM SIGMETRICS Conference on Measurement and Modeling of Computer Systems*, pages 147–153, Austin, TX, USA, May 1985.
24. Yousaf Saad. *Iterative Methods for Sparse Linear Systems.* PWS Publishing Company, 1996.
25. Markus Siegle. Advances in Model Representations. In Luca de Alfaro and Stephen Gilmore, editors, *Proc. PAPM/PROBMIV 2001, Available as Volume 2165 of* LNCS, pages 1–22, Aachen, Germany, September 2001. Springer Verlag.
26. W.J. Stewart. *Introduction to the Numerical Solution of Markov Chains.* Princeton University Press, 1994.
27. Sivan Toledo. A Survey of Out-of-Core Algorithms in Numerical Linear Algebra. In James Abello and Jeffrey Scott Vitter, editors, *External Memory Algorithms and Visualization*, DIMACS Series in Discrete Mathematics and Theoretical Computer Science. American Mathematical Society Press, Providence, RI, 1999.

Model Checking CSL until Formulae with Random Time Bounds*

Marta Kwiatkowska[1], Gethin Norman[1], and António Pacheco[2]

[1] School of Computer Science, University of Birmingham, Edgbaston
Birmingham B15 2TT, United Kingdom
[2] Department of Mathematics and CEMAT, Instituto Superior Técnico
Av. Rovisco Pais, 1049-001 Lisboa, Portugal

Abstract. Continuous Time Markov Chains (CTMCs) are widely used as the underlying stochastic process in performance and dependability analysis. Model checking of CTMCs against Continuous Stochastic Logic (CSL) has been investigated previously by a number of authors [2,4,13]. CSL contains a time-bounded until operator that allows one to express properties such as "the probability of 3 servers becoming faulty within 7.01 seconds is at most 0.1". In this paper we extend CSL with a random time-bounded until operator, where the time bound is given by a random variable instead of a fixed real-valued time (or interval). With the help of such an operator we can state that the probability of reaching a set of goal states within some generally distributed delay while passing only through states that satisfy a certain property is at most (at least) some probability threshold. In addition, certain transient properties of systems which contain general distributions can be expressed with the extended logic. We extend the efficient model checking of CTMCs against the logic CSL developed in [13] to cater for the new operator. Our method involves precomputing a family of coefficients for a range of random variables which includes Pareto, uniform and gamma distributions, but otherwise carries the same computational cost as that for ordinary time-bounded until in [13]. We implement the algorithms in MATLAB and evaluate them by means of a queueing system example.

1 Introduction

Continuous time Markov chains (CTMCs) are widely used as the underlying stochastic process in performance and dependability analysis. CTMCs are characterised by allowing only exponential distributions – the time that the system remains in a state is governed by a (negative) exponential distribution. The restriction to exponential distributions yields efficient analysis techniques for both transient and steady-state probabilities, and hence also for calculating standard performance measures such as throughput, mean waiting time and average cost. Recently extensions of temporal logic have been proposed which can express such

* Partly supported by EPSRC grant GR/M13046, FCT, grant SFRH/BSAB/ 251/01 and the projects POSI/34826/CPS/2000 and POSI/42069/CPS/2001.

H. Hermanns and R. Segala (Eds.): PAPM-PROBMIV 2002, LNCS 2399, pp. 152–168, 2002.

properties. The temporal logic CSL (Continuous Stochastic Logic) introduced by Aziz *et al.* [1,2] and since extended by Baier *et al.* [4] is based on the temporal logics CTL [5] and PCTL [11] and provides a powerful means to specify both path-based and traditional state-based performance measures on CTMCs in a concise, flexible and unambiguous way. CSL contains a time-bounded until operator that allows one to express properties such as "the probability of 3 servers becoming faulty within 7.01 seconds is at most 0.1" (more generally, one can additionally require that the executions step through states that satisfy a given property). Model checking of CTMCs against CSL has been improved in [3,13] through the use of uniformisation [10,12] and transient analysis, and the usefulness of this approach demonstrated by a number of case studies.

However, in practice it is often the case that exponential distributions are not an adequate modelling tool for capturing the behaviour of stochastic systems. Examples of such situations include modelling file transfer over the Internet, timeouts in communication protocols and the residence time in a wireless cell. For these cases the modelling framework must be capable of handling *general distributions*, such as Pareto, Erlang, gamma or phase-type. An unfortunate consequence of including general distributions within the modelling framework, as has been demonstrated recently, for example in [8] and [15], is a considerable increase in the complexity of performance analysis, or if using phase-type distributions a substantial increase in the size of the state space.

In this paper we make an alternative proposal, namely, to remain in the CTMC framework and instead extend the logic CSL with a variant of the time-bounded until operator which allows *random* (generally distributed) *time bound*, by replacing the constant time bound with a random variable. With the new operator one can specify properties such as the probability of reaching a set of goal states of a CTMC within some generally distributed delay while passing only through states that satisfy a certain property is at most (at least) some probability value. Although general distributions are not added explicitly to the model, using the random time-bounded until operator enables one to establish specific transient properties of systems which include generally distributed delays. As an example application consider a queue where the customers arrive with some generally distributed delay, then by letting the random time bound have *the same* distribution we can express (and verify) properties such as: "if the queue if full then, with probability at least p, when the next customer arrives there will be at most k customers in the queue".

The semantics of the new random time-bounded until operator involves a Riemann–Stieltjes integral, and so one would expect that the resulting complexity of its model checking is prohibitive. The central observation of this paper is that this integral reduces to an infinite summation involving mixed Poisson probabilities (equivalent to the α-factors of [8]). This result is similar to the case of the ordinary time-bounded until, except for an additional vector of coefficients that can be precomputed beforehand. Thanks to this observation we can propose an efficient model checking algorithm for the new operator, which is derived from that of [13] and which does not carry an increase in computational cost except the precomputation of coefficients. This somewhat surprising result yields a powerful and fast method for analysing certain properties of stochastic systems with generally distributed delays. Moreover, for a large class of general distributions which includes Pareto, gamma and Erlang that have been observed in stochastic systems, we also provide methods for pre-computing the mixed Poisson probabilities.

Finally, we model a queueing system and describe, with the help of a MATLAB implementation of the algorithms, the experimental results obtained when verifying the system against random time-bounded until formulae. We are able to demonstrate the inaccuracy of using exponential distributions where the actual arrivals warrant the use of heavy-tailed or other general distributions.

Outline of Paper: We begin by recalling the definition of CTMCs and the logic CSL. Next we introduce the new random time-bounded until operator, give its semantics and a model checking algorithm based on [13] which uses a family of coefficients. In the remainder of the paper we calculate the coefficients for a number of well-known distributions and describe the experimental results.

2 Continuous Time Markov Chains and the Logic CSL

In this section we briefly recall the basic concepts of CTMCs and the logic CSL, concentrating on the time bounded until operator. Let AP be a finite set of atomic propositions.

Definition 1. *A (labelled) CTMC $\mathcal{C}$ is a tuple $(S, \mathbf{R}, L)$ where S is a finite set of* states, $\mathbf{R} : S \times S \to \mathbb{R}_{\geq 0}$ *is the* rate matrix *and $L : S \to 2^{AP}$ is a* labelling *function which assigns to each state $s \in S$ the set $L(s)$ of atomic propositions that are valid in s.*

For any state $s \in S$, the probability of leaving state s within t time units is given by $1 - e^{-E(s) \cdot t}$ where $E(s) = \sum_{s' \in S} \mathbf{R}(s, s')$. If $\mathbf{R}(s, s') > 0$ for more than one $s' \in S$, then there is a *race* between the transitions leaving s, where the probability of moving to s' in a single step equals the probability that the delay corresponding to moving from s to s' "finishes before" the delays of any other transition leaving s. Note that, as in [3,4], we do allow self-loops.

A path through a CTMC is an alternating sequence $\sigma = s_0\, t_0\, s_1\, t_1\, s_2 \ldots$ such that $\mathbf{R}(s_i, s_{i+1}) > 0$ and $t_i \in \mathbb{R}_{>0}$ for all $i \geq 0$. The time stamps t_i denote the amount of time spent in state s_i. Let $Path^{\mathcal{C}}(s)$ denote the set of paths of $\mathcal{C}$ which start in state s (i.e. $s_0 = s$); $\sigma@t$ denote the state of σ occupied at time t, i.e. $\sigma@t = \sigma[i]$ where i is the largest index such that $\sum_{j=0}^{i-1} t_j \leq t$; and $\Pr_s$ denote the unique probability measure on sets of paths that start in s [4].

We now recall the logic CSL first introduced in [1,2] and extended in [4].

Definition 2 (Syntax of CSL). *The syntax of* CSL *is defined as follows:*

$$\Phi ::= \mathtt{true} \;\Big|\; a \;\Big|\; \Phi \wedge \Phi \;\Big|\; \neg\Phi \;\Big|\; \mathcal{S}_{\bowtie p}(\Phi) \;\Big|\; \mathcal{P}_{\bowtie p}(\Phi\, \mathcal{U}^{\leq t}\, \Phi)$$

where $a \in$ AP, $p \in [0, 1]$, $t \in \mathbb{R}_{>0}$ and $\bowtie \in \{\leq, \geq\}$.

The semantics of CSL for can be found in [4]. Here we concentrate on the time bounded until operator defined by:

$$s \models \mathcal{P}_{\bowtie p}(\Phi\,\mathcal{U}^{\leq t}\,\Psi) \quad\Leftrightarrow\quad \mathit{Prob}^{\mathcal{C}}(s, \Phi\,\mathcal{U}^{\leq t}\,\Psi) \bowtie p$$

where $\mathit{Prob}^{\mathcal{C}}(s, \Phi\,\mathcal{U}^{\leq t}\,\Psi)$ is given by:

$$\mathit{Prob}^{\mathcal{C}}(s, \Phi\,\mathcal{U}^{\leq t}\,\Psi) \stackrel{\text{def}}{=} \Pr_s\{\sigma \in \mathit{Path}^{\mathcal{C}}(s) \mid \sigma \models \Phi\,\mathcal{U}^{\leq t}\,\Psi\}$$

and $\Phi\,\mathcal{U}^{\leq t}\,\Psi$ asserts that Ψ will be satisfied at some time instant in the interval $[0, t]$ and that at all preceding time instants Φ holds:

$$\sigma \models \Phi\,\mathcal{U}^{\leq t}\,\Psi \quad\Leftrightarrow\quad \exists x \leq t.\ (\sigma@x \models \Psi \wedge \forall y < x.\, \sigma@y \models \Phi)\ .$$

In the remainder of this section we describe the model checking algorithm for time bounded until formulae originally presented in [13], based on a reduction to transient analysis and model checking PCTL. Further details on model checking the other CSL operators are available in, for example, [4,3,13]. To begin with we define uniformisation, a transformation of a CTMC into a DTMC (discrete-time Markov chain).

Definition 3. *For CTMC* $\mathcal{C} = (S, \mathbf{R}, L)$ *the* uniformised *DTMC is given by* $\mathit{unif}(\mathcal{C}) = (S, \mathbf{P}^{\mathit{unif}(\mathcal{C})}, L)$ *where* $\mathbf{P} = \mathbf{I} + \mathbf{Q}/q$ *for* $q \geq \max\{E(s) \mid s \in S\}$ *and* $\mathbf{Q} = \mathbf{R} - \mathit{diag}(\underline{E})$.

Using this transformation, in [13, Proposition 2] it is shown that for any $t \in \mathbb{R}$:

$$\mathit{Prob}^{\mathcal{C}}(s, \Phi\,\mathcal{U}^{\leq t}\,\Psi) = \sum_{k=0}^{\infty} \gamma(k, q{\cdot}t){\cdot}\mathit{Prob}^{\mathit{unif}(\mathcal{C})}(s, \Phi\,\mathcal{U}^{\leq k}\,\Psi) \qquad (1)$$

where $\gamma(k, q{\cdot}t)$ is the kth Poisson probability with parameter $q{\cdot}t$, i.e. $\gamma(k, q{\cdot}t) = e^{-q\cdot t}{\cdot}(q{\cdot}t)^k/k!$, and $\mathit{Prob}^{\mathit{unif}(\mathcal{C})}(s, \Phi\,\mathcal{U}^{\leq k}\,\Psi)$ is the probability that, in the DTMC $\mathit{unif}(\mathcal{C})$, from s a state satisfying Ψ is reached within k (discrete) steps while passing through only states that satisfy Φ.

Applying the model checking algorithm for PCTL presented in [11], calculating $\mathit{Prob}^{\mathit{unif}(\mathcal{C})}(s, \Phi\,\mathcal{U}^{\leq k}\,\Psi)$ for all $s \in S$ reduces to computing the vector of probabilities $(\mathbf{P}^{\mathit{unif}(\mathcal{C})[\neg\Phi\vee\Psi]})^k{\cdot}\underline{\iota}_\Psi$, where for any $s, s' \in S$:

$$\mathbf{P}^{\mathit{unif}(\mathcal{C})[\neg\Phi\vee\Psi]}(s, s') \stackrel{\text{def}}{=} \begin{cases} 1 & \text{if } s \models \neg\Phi \vee \Psi \text{ and } s = s' \\ 0 & \text{if } s \models \neg\Phi \vee \Psi \text{ and } s \neq s' \\ \mathbf{P}^{\mathit{unif}(\mathcal{C})}(s, s') & \text{otherwise} \end{cases}$$

and $\underline{\iota}_\Psi$ characterises $\mathit{Sat}(\Psi)$, i.e. $\underline{\iota}_\Psi(s) = 1$ if $s \models \Psi$, and 0 otherwise. It then follows that calculating $\mathit{Prob}(s, \Phi\,\mathcal{U}^{\leq t}\,\Psi)$ for *all* states amounts to computing the following vector of probabilities:

$$\underline{\mathit{Prob}}(\Phi\,\mathcal{U}^{\leq t}\,\Psi) = \sum_{k=0}^{\infty} \gamma(k, q{\cdot}t){\cdot}(\mathbf{P}^{\mathit{unif}(\mathcal{C})[\neg\Phi\vee\Psi]})^k{\cdot}\underline{\iota}_\Psi\ .$$

Note that, as iterative squaring is not attractive for stochastic matrices due to fill-in [20], the matrix product is typically computed in an iterative fashion: $\mathbf{P}^{\mathit{unif}(\mathcal{C})[\neg\Phi\vee\Psi]}{\cdot}(\cdots(\mathbf{P}^{\mathit{unif}(\mathcal{C})[\neg\Phi\vee\Psi]}{\cdot}\underline{\iota}_\Psi))$.

3 Random Time-Bounded until Formulae

In this section we extend the logic CSL to include *random* time-bounded until formulae and consider model checking algorithms for such formulae. Let T denote a *nonnegative random variable.* We let F_T denote the (cumulative) distribution function of T (i.e. $F_T(t) = \mathbf{P}[T \leq t]$), and assume that the support of T is contained in the interval $[L_T, R_T]$, where L_T may either be zero or positive and R_T may either be finite or infinite. Note that, in particular, we have $F_T(t)$ is zero for $t < L_T$ and one for $t \geq F_T$.

We now extend CSL by allowing formulae of the form $\mathcal{P}_{\bowtie p}(\Phi\, \mathcal{U}^{\leq T}\, \Psi)$ where T is a *nonnegative random variable*[1]. The formula asserts that, with probability $\bowtie p$, by the *random* time T a state satisfying Ψ will be reached such that all preceding states satisfy Φ. Formally, the semantics is given by:

$$s \models \mathcal{P}_{\bowtie p}(\Phi\, \mathcal{U}^{\leq T}\, \Psi) \quad \Leftrightarrow \quad Prob^{\mathcal{C}}(s, \Phi\, \mathcal{U}^{\leq T}\, \Psi) \bowtie p$$

where $Prob^{\mathcal{C}}(s, \Phi\, \mathcal{U}^{\leq T}\, \Psi)$ is defined as a Riemann–Stieltjes integral involving (deterministic) time-bounded until probabilities:

$$Prob^{\mathcal{C}}(s, \Phi\, \mathcal{U}^{\leq T}\, \Psi) \stackrel{\text{def}}{=} \int_{L_T}^{R_T} Prob^{\mathcal{C}}(s, \Phi\, \mathcal{U}^{\leq t}\, \Psi)\, dF_T(t)\,.$$

We now give our main observation concerning model checking random time-bounded until formulae, which allows us to replace the Riemann–Stieltjes integral with summation, at the cost of pre-computing coefficients.

Proposition 1. *For any $s \in S$:*

$$Prob^{\mathcal{C}}(s, \Phi\, \mathcal{U}^{\leq T}\, \Psi) = \sum_{k=0}^{\infty} \alpha_T(k, q) \cdot Prob^{\text{unif}(\mathcal{C})}(s, \Phi\, \mathcal{U}^{\leq k}\, \Psi)$$

where $\alpha_T(k, q) = \int_{L_T}^{R_T} \gamma(k, q{\cdot}t)\, dF_T(t)$.

Proof. By definition we have:

$$\begin{aligned}
Prob^{\mathcal{C}}(s, \Phi\, \mathcal{U}^{\leq T}\, \Psi) &= \int_{L_T}^{R_T} Prob^{\mathcal{C}}(s, \Phi\, \mathcal{U}^{\leq t}\, \Psi)\, dF_T(t) \\
&= \int_{L_T}^{R_T} \left(\sum_{k=0}^{\infty} \gamma(k, q{\cdot}t) \cdot Prob^{\text{unif}(\mathcal{C})}(s, \Phi\, \mathcal{U}^{\leq k}\, \Psi) \right) dF_T(t) && \text{by (1)} \\
&= \sum_{k=0}^{\infty} \left(\int_{L_T}^{R_T} \gamma(k, q{\cdot}t)\, dF_T(t) \right) \cdot Prob^{\text{unif}(\mathcal{C})}(s, \Phi\, \mathcal{U}^{\leq k}\, \Psi) && \text{rearranging} \\
&= \sum_{k=0}^{\infty} \alpha_T(k, q) \cdot Prob^{\text{unif}(\mathcal{C})}(s, \Phi\, \mathcal{U}^{\leq k}\, \Psi) && \text{by definition}
\end{aligned}$$

as required. □

[1] One condition we impose on the random variables is that they are *independent* of the CTMC under study.

It then follows from the algorithm of [13] for calculating $Prob(s, \Phi\, \mathcal{U}^{\leq t}\, \Psi)$ outlined in Section 2 that calculating $Prob(s, \Phi\, \mathcal{U}^{\leq T}\, \Psi)$ for all states of the CTMC reduces to computing the following vector of probabilities:

$$\underline{Prob}(\Phi\, \mathcal{U}^{\leq T}\, \Psi) = \sum_{k=0}^{\infty} \alpha_T(k, q) \cdot (P^{unif(\mathcal{C})[\neg\Phi\vee\Psi]})^k \cdot \underline{\iota}_\Psi\,. \tag{2}$$

The coefficients $\alpha_T(k, q)$ take values on the interval $[0, 1]$ since they are probabilities. Namely, $\alpha_T(k, q) = \mathbf{P}(N_q(T) = k)$ where N_q is a Poisson process with rate q, independent of T. Moreover,

$$\sum_{k=0}^{\infty} \alpha_T(k, q) = \sum_{k=0}^{\infty} P(N_q(T) = k) = 1\,.$$

Following Grandell [9], we call the coefficients $\{\alpha_T(k, q)\}_{k\in\mathbb{N}}$ *mixed Poisson probabilities*; $\alpha_T(k, q)$ is the k-th probability of a Poisson process with rate q at random time T. Note that the mixed Poisson probabilities are equivalent to the α-factors introduced in [8] which are used in the calculation of steady-state probabilities for non-Markovian stochastic Petri nets.

Figure 1 presents the pseudo-code for a generic algorithm for computing $Prob^{\mathcal{C}}(s, \Phi\mathcal{U}^{\leq T}\, \Psi)$ with an error of at most ε. Note that the DTMC $unif(\mathcal{C})[\neg\Phi \vee \Psi]$ may reach steady state before $R(\epsilon)$ and, in this case, the summation can be truncated at this earlier point [17]. In the next section we will consider methods for calculating these coefficients and the bound $R(\varepsilon)$.

input : $\alpha_T(0, q), \ldots, \alpha_T(R(\varepsilon), q)$ such that $\sum_{k=R(\varepsilon)+1}^{\infty} \alpha_T(k, q) < \varepsilon$
$\mathbf{P} := \mathbf{P}^{unif(\mathcal{C})[\neg\Phi\vee\Psi]}$
$\underline{b} := \underline{\iota}_\Psi$
$\underline{sol} := \underline{0}$
for $k = 0$ **to** $R(\varepsilon)$
  $\underline{sol} := \underline{sol} + \alpha_T(k, q) \cdot \underline{b}$
  $\underline{b} := \mathbf{P} \cdot \underline{b}$
endfor
// $\underline{Prob}(\Phi\, \mathcal{U}^{\leq T}\, \Psi) = \underline{sol}$

Fig. 1. Generic algorithm for computing $\underline{Prob}(\Phi\, \mathcal{U}^{\leq T}\, \Psi)$

4 Computation of the Mixed Poisson Probabilities

Under complete generality the computation of mixed Poisson probabilities relies on the evaluation of the integrals:

$$\alpha_T(k, q) = \int_{L_T}^{R_T} \gamma(k, q{\cdot}t)\, dF_T(t) \quad \text{for } k \in \mathbb{N}.$$

Next we will develop algorithms to compute the mixed Poisson probabilities $\alpha_T(k,q)$ for the case where the distribution of the random time T has a finite discrete, uniform, gamma or Pareto distribution, or is a finite mixture of distributions of these types. These algorithms can then be integrated with that presented in Figure 1 to compute $\underline{Prob}(\Phi\ \mathcal{U}^{\leq T}\ \Psi)$ when the distribution of T belongs to this class of distributions.

We start by showing how the mixed Poisson probabilities for a finite mixture can be computed in terms of the mixed Poisson probabilities of the random times involved in the mixture.

4.1 Finite Mixture of Random Times

Suppose that T is a mixture of n random variables, $T_1, T_2, \ldots, T_n$ with weights $a_1, a_2, \ldots, a_n$ ($a_i > 0$ for $1 \leq i \leq n$ and $\sum_{i=1}^{n} a_i = 1$). In this case, the distribution function of T is given by:

$$F_T(t) = \sum_{i=1}^{n} a_i \cdot F_{T_i}(t) \quad \text{for } t \in \mathbb{R}\,,$$

and hence the mixed Poisson probabilities are given by

$$\alpha_T(k,q) = \sum_{i=1}^{n} a_i \cdot \int_0^\infty \gamma(k, q \cdot t)\, dF_{T_i}(t) = \sum_{i=1}^{n} a_i \cdot \alpha_{T_i}(k,q)\,. \tag{3}$$

Therefore, the mixed Poisson probabilities for a mixture are a linear combination of the mixed Poisson probabilities of the random variables involved in the mixture, and the coefficients of the linear combination are the weights of the associated random times.

For a given precision ε, the following algorithm may be used to compute the coefficients $\alpha_T(j,q)$, $j = 0, 1, \ldots, R(\varepsilon)$, such that $\sum_{j=R(\varepsilon)+1}^{\infty} \alpha_T(j,q) < \varepsilon$.

```
sum := 0
k := -1
while sum ≤ 1 - ε do
    k := k + 1
    c[k] := 0
    for i = 1 to n do
        c[k] := c[k] + a_i·α_{T_i}(k, q)
    endfor
    sum := sum + c[k]
endwhile
Output : R(ε) = k and α_T(j, q) = c[j], j = 0, 1, ..., k
```

4.2 Random Times with Finite Discrete Distribution

Suppose that T is a (finite) discrete random variable taking nonnegative values $t_1, t_2, \ldots, t_n$ with probabilities $p_1, p_2, \ldots, p_n$, then its distribution function is given by:

$$F_T(t) = \sum \{p_i \mid 1 \leq i \leq n \ \wedge \ t_i \leq t\} \quad \text{for } t \in \mathbb{R}\,.$$

Hence, in this case the mixed Poisson probabilities are given by

$$\alpha_T(k,q) = \int_0^\infty \gamma(k, q{\cdot}t)\, dF_T(t) = \sum_{i=1}^{n} p_i{\cdot}\gamma(k, q{\cdot}t_i) \quad \text{for } k = 0, 1, 2, \ldots \tag{4}$$

Therefore, if T is a discrete random variable, the amount of time needed to compute a mixed Poisson probability is of the same order as the time needed to compute a single Poisson probability.

If T is almost surely constant and equal to t, $\mathbf{P}(T = t) = 1$, then we have

$$\alpha_T(k,q) = \int_0^\infty \gamma(k, qu)\, dF_T(u) = \gamma(k, q{\cdot}t) \quad \text{for } k = 0, 1, 2, \ldots$$

Thus, we obtain the Poisson probabilities for fixed time t. Notice also that (4) is a consequence of the previous equation and (3), as a discrete random variable is also a mixture of almost surely constant random variables (the values that these random variables assume with probability one are the values to which the given discrete random variable assigns positive probabilities).

Now to compute the mixed Poisson probabilities for discrete distributions we may use the algorithm presented in Section 4.1 by replacing $\alpha_{T_i}(k,q)$ and a_i by $\gamma(k, q.t_i)$ and p_i respectively. Note that the Fox-Glynn algorithm [7] can be used to avoid overflow when computing the Poisson probabilities $\gamma(k, q{\cdot}t_i)$ for large values of $q{\cdot}t_i$.

4.3 Random Times with Gamma Distribution

Suppose that T has a gamma distribution with (positive) parameters r and λ, $T \sim \text{Gamma}(r, \lambda)$, then T has probability density function:

$$f_T(t) = \frac{\lambda{\cdot}(\lambda{\cdot}t)^{r-1}{\cdot}e^{-\lambda{\cdot}t}}{\Gamma(r)} \quad \text{for } t > 0$$

where Γ is the gamma function, that is, for any $r > 0$:

$$\Gamma(r) = \int_0^\infty e^{-t}{\cdot}t^{r-1}\, dt\,.$$

Using the fact that $\Gamma(j+1) = j!$ for all $j \in \mathbb{N}$ we have:

$$\begin{aligned}
\alpha_T(k,q) &= \int_0^\infty e^{-q\cdot t}\frac{(q\cdot t)^k}{\Gamma(k+1)}\cdot\frac{\lambda\cdot(\lambda\cdot t)^{r-1}\cdot e^{-\lambda\cdot t}}{\Gamma(r)}\,dt \\
&= \frac{\lambda^r\cdot q^k}{\Gamma(k+1)\cdot\Gamma(r)}\int_0^\infty e^{-(q+\lambda)\cdot t}\cdot t^{(k+r)-1}\,dt && \text{rearranging} \\
&= \frac{\lambda^r\cdot q^k}{\Gamma(k+1)\cdot\Gamma(r)}\int_0^\infty e^{-u}\cdot u^{(k+r)-1}\cdot(q+\lambda)^{-(k+r)}\,du && \text{setting } u=\frac{t}{q+\lambda} \\
&= \frac{1}{\Gamma(k+1)\cdot\Gamma(r)}\left(\frac{q}{\lambda+q}\right)^k\left(\frac{\lambda}{\lambda+q}\right)^r\int_0^\infty e^{-u}\cdot u^{(k+r)-1}\,du && \text{rearranging} \\
&= \frac{\Gamma(r+k)}{\Gamma(k+1)\cdot\Gamma(r)}\left(\frac{q}{\lambda+q}\right)^k\left(\frac{\lambda}{\lambda+q}\right)^r && \text{by definition of } \Gamma \\
&= p_{\mathrm{NegBin}(r,\frac{\lambda}{\lambda+q})}(k)
\end{aligned}$$

where $p_{\mathrm{NegBin}(r,\lambda/(\lambda+q))}(k)$ is the k-th probability of the negative binomial distribution with parameters r and $\lambda/(\lambda+q)$. Thus, if T has a gamma distribution, the mixed Poisson probabilities are negative binomial probabilities. Therefore, the coefficients $\alpha_T(k,q)$ may be computed recursively by the following scheme, for $k \in \mathbb{N}$:

$$\alpha_T(0,q) = \left(\frac{\lambda}{\lambda+q}\right)^r \quad \text{and} \quad \alpha_T(k+1,q) = \left(\frac{k+r}{k+1}\right)\cdot\left(\frac{q}{\lambda+q}\right)\cdot\alpha_T(k,q)\,.$$

Note that this recursion is an instance of the Panjer recursion [18]. Furthermore, for sufficiently large k, the coefficients $\alpha_T(k,q)$ exhibit an exponential decay towards zero. For a given precision ε, the following algorithm may be used to compute the coefficients $\alpha_T(k,q)$ recursively.

```
// Compute α_T(j,q) for j = 0,1,...,R(ε) if T ~ Gamma(r,λ)
k := 0
c[k] := (λ/(λ+q))^r
sum := c[k]
while sum ≤ 1 − ε do
    k := k + 1
    c[k] := ((k+r−1)/k)·(q/(λ+q))·c[k − 1]
    sum := sum + c[k]
endwhile
Output : R(ε) := k and α_T(j,q) := c[j], j = 0,1,...,k
```

4.4 Random Times with Erlang or Exponential Distribution

In computer network systems the most used particular case of the gamma distribution is the Erlang distribution, which corresponds to the case where the parameter r is integer. In that case it is common to call r the number of phases of the Erlang distribution and λ the rate. The popular and intensively used exponential distribution with rate λ corresponds to the particular case of $r = 1$. We also note that the Erlang(r, λ) is distributed as the sum of r independent exponential random variables with rate λ, so that these exponential random variables may be seen as the r phases of the Erlang distribution.

Although the Erlang and exponential distributions are special cases of the gamma family of distributions, they deserve special treatment due to their importance for applications. Moreover, these distributions lead to a clearer interpretation of the results obtained for the mixed Poisson probabilities. In fact these results, which we will comment on briefly, are very well known results in probability and statistics that appear recurrently in applications.

If T has an exponential distribution with rate λ, the mixed Poisson probabilities are geometric probabilities with parameter (success probability) $\lambda/(\lambda + q)$:

$$\alpha_T(k, q) = \left(\frac{q}{\lambda + q}\right)^k \cdot \left(\frac{\lambda}{\lambda + q}\right) \quad \text{for } k = 0, 1, 2, \ldots$$

The success probability represents the probability that the random time T is smaller than the first arrival epoch in the uniformising Poisson process with rate q. Due to the memoryless property of the exponential, the number of arrivals in the uniformising Poisson process with rate q that occur before time T corresponds to the number of trials observed before the occurrence of the first success and has a geometric distribution with parameter $\lambda/(\lambda + q)$.

Note that $\sum_{k=R+1}^{\infty} \alpha_T(k, q) = (q/(\lambda + q))^{k+1}$ for all $R \in \mathbb{N}$, and hence as a consequence for all $\varepsilon \in (0, 1)$:

$$\sum_{k=R+1}^{\infty} \alpha_T(k, q) \leq \varepsilon \quad \text{if and only if} \quad R \geq \left\lceil \frac{\ln(\varepsilon)}{\ln(q/(\lambda + q))} \right\rceil - 1\,.$$

That is, to achieve precision ε in the computation of (2) we need to evaluate order $\ln(\varepsilon)$ mixed Poisson probabilities.

If T has Erlang distribution with parameters r and λ, then the coefficients $\alpha_T(k, q)$ are the probabilities associated to a negative binomial random variable with parameters r and $\lambda/(\lambda + q)$, and hence for $k \in \mathbb{N}$:

$$\alpha_T(k, q) = \binom{r + k - 1}{k} \cdot \left(\frac{q}{\lambda + q}\right)^k \cdot \left(\frac{\lambda}{\lambda + q}\right)^r .$$

This is a natural result as the sum of r independent random variables with geometric distribution with parameter $\lambda/(\lambda+q)$ has a negative binomial distribution with parameters r and $\lambda/(\lambda + q)$, which may be interpreted as the number of trials that are observed before the observation of the r-th success. If, again, a success is seen as an arrival in a Poisson process with rate λ and a failure as an arrival in the independent uniformising Poisson process with rate q, the value of the negative binomial random variable corresponds to the number of arrivals in the uniformising Poisson process with rate q that occur prior to the r-th arrival in the Poisson process with rate λ.

4.5 Random Times with Uniform Distribution

If T has uniform distribution on $[L_T, R_T]$, then the coefficients $\alpha_T(k,q)$ are given by

$$\alpha_T(k,q) = \int_{L_T}^{R_T} \frac{1}{R_T - L_T} \cdot e^{-q \cdot t} \frac{(q \cdot t)^k}{k!} \, dt \ \text{ for } k = 0, 1, \ldots$$

It follows that $\alpha_T(0,q) = \frac{e^{-q \cdot L_t} - e^{-q \cdot R_t}}{q \cdot (R_T - L_T)}$ and for $k = 0, 1, 2, \ldots$

$$\alpha_T(k+1,q) = \alpha_T(k,q) + \frac{1}{q \cdot (R_T - L_T)} \cdot \left[e^{-q \cdot L_t} \frac{(q \cdot L_T)^{k+1}}{(k+1)!} - e^{-q \cdot R_t} \frac{(q \cdot R_T)^{k+1}}{(k+1)!} \right]$$

which provides a recursive scheme of computing the coefficients $\alpha_T(k,q)$. Moreover, using induction, we conclude that for $k = 0, 1, 2, \ldots$:

$$\begin{aligned} \alpha_T(k,q) &= \frac{1}{q \cdot (R_t - L_T)} \cdot \left[\sum_{j=0}^{k} e^{-q \cdot L_t} . \frac{(q \cdot L_T)^j}{j!} - \sum_{j=0}^{k} e^{-q \cdot R_t} . \frac{(q \cdot R_T)^j}{j!} \right] \\ &= \frac{1}{q \cdot (R_t - L_T)} \cdot \left[\sum_{j=0}^{k} \gamma(k, q \cdot L_T) - \sum_{j=0}^{k} \gamma(j, q \cdot R_T) \right] \\ &= \frac{1}{q \cdot (R_T - L_T)} \cdot \left[F_{\mathrm{Poisson}(q \cdot L_T)}(k) - F_{\mathrm{Poisson}(q \cdot R_T)}(k) \right] \end{aligned}$$

where $F_{\mathrm{Poisson}(\lambda)}(\cdot)$ is the distribution function of a Poisson random variable with parameter λ. Note that if, in particular, $L_T = 0$, then

$$\alpha_T(k,q) = \frac{1 - F_{\mathrm{Poisson}(q \cdot R_T)}(k)}{q \cdot R_T} = \frac{e^{-q \cdot R_t}}{q \cdot R_t} \cdot \sum_{j=k+1}^{\infty} \frac{(q \cdot R_T)^j}{j!} .$$

For a given precision ε, the following algorithm may be used to compute the coefficients $\alpha_T(k,q)$ recursively.

```
// Compute α_T(j,q) for j = 0,1,...,R(ε) if T ~ Uniform([L_T, R_T])
k := 0
d := e^{-q·L_T} / (q·(R_T − L_T))
e := e^{-q·R_T} / (q·(R_T − L_T))
c[k] := d − e
sum := c[k]
while sum ≤ 1 − ε do
    k := k + 1
    d := d·(q·L_T)/k
    e := e·(q·R_T)/k
    c[k] := c[k − 1] + d − e
    sum := sum + c[k]
endwhile
Output : R(ε) := k and α_T(j,q) := c[j], j = 0,1,...,k
```

4.6 Random Times with Pareto Distribution

The Pareto distribution has recently gained high importance in telecommunications as it has been shown, e.g., that it fits the distribution of times in-between the start of Internet sessions [16] and the size of files available in the Web [6]. Suppose that T has a Pareto distribution with (positive) parameters κ and β, $T \sim \text{Pareto}(\kappa, \beta)$; i.e., T has probability density function:

$$f_T(t) = \begin{cases} 0 & \text{if } x \leq \kappa \\ \frac{\beta\cdot\kappa^\beta}{t^{\beta+1}} & \text{if } x > \kappa\,. \end{cases}$$

This is a heavy-tailed distribution that has infinite variance for $\beta \leq 2$ and infinite expected value for $\beta \leq 1$. The mixed Poisson probabilities in this case are:

$$\begin{aligned}
\alpha_T(k,q) &= \int_\kappa^\infty e^{-q\cdot t}.\frac{(q\cdot t)^k}{k!}\cdot\frac{\beta\cdot\kappa^\beta}{t^{\beta+1}}\,dt \\
&= \frac{\beta\cdot\kappa^\beta\cdot q^{\beta+1}}{k!}\cdot\int_\kappa^\infty e^{-q\cdot t}\cdot(q\cdot t)^{k-\beta-1}\,dt && \text{rearranging} \\
&= \frac{\beta\cdot\kappa^\beta\cdot q^\beta}{k!}\cdot\int_{q\kappa}^\infty e^{-y}\cdot y^{k-\beta-1}\,dy && \text{letting } y = q\cdot t \\
&= \beta\cdot(q\kappa)^\beta\cdot\frac{\Gamma(k-\beta, q\kappa)}{k!}
\end{aligned}$$

for $k = 0, 1, 2, \ldots$, where $\Gamma(\cdot,\cdot)$ is the incomplete gamma function. Now, for any $x > 0$ integrating by parts we have

$$\Gamma(a+1, x) = \int_x^\infty e^{-y}\cdot y^a\,dy = e^{-x}\cdot x^a + a\cdot\Gamma(a,x) \tag{5}$$

and, as a consequence for $a \neq 0$:

$$\Gamma(a,x) = \frac{1}{a}\left[\Gamma(a+1,x) - e^{-x}\cdot x^a\right]\,. \tag{6}$$

Applying these results we get that the following upward and backward recursions for $\alpha_T(k,q)$, for $k = 0, 1, 2, \ldots$:

$$\begin{aligned}
\alpha_T(k+1,q) &= \beta\cdot(q\kappa)^\beta\cdot\frac{\Gamma(k+1-\beta, q\kappa)}{(k+1)!} \\
&= \beta\cdot(q\kappa)^\beta\cdot\left[\frac{k-\beta}{k+1}\cdot\frac{\Gamma(k-\beta,q\kappa)}{k!} + e^{-q\kappa}.\frac{(q\kappa)^{k-\beta}}{(k+1)!}\right] && \text{by (5)} \\
&= \frac{1}{k+1}\cdot\left[(k-\beta)\cdot\alpha_T(k,q) + \beta\cdot e^{-q\kappa}.\frac{(q\kappa)^k}{k!}\right] && \text{rearranging.}
\end{aligned}$$

Similarly, using (6) for any $k \neq \beta$:

$$\alpha_T(k,q) = \frac{1}{k-\beta}\cdot\left[(k+1)\cdot\alpha_T(k+1,q) - \beta\cdot e^{-q\kappa}.\frac{(q\kappa)^k}{k!}\right]\,.$$

For a given precision ε, the following algorithm may be used to compute the coefficients $\alpha_T(k,q)$ recursively resorting to a single evaluation of the incomplete gamma function.

```
// Compute α_T(j,q) for j = 0,1,...,R(ε) if T ~ Pareto(κ,β)
k_aux := ⌈β⌉
c[k_aux] := β·(qκ)^β · Γ(k_aux − β, qκ) / k_aux!
d_aux := β·e^(−qκ) · (qκ)^k_aux / k_aux!
sum := c[k_aux]
// Backward loop
k := k_aux
d := d_aux
while k > 0 do
    k := k − 1
    d := d·(k+1)/(qκ)
    c[k] := [(k + 1)·c[k + 1] − d]/(k − β)
    sum := sum + c[k]
endwhile
// Forward loop
k := k_aux
d := d_aux
while sum ≤ 1 − ε do
    k := k + 1
    c[k] := [(k − 1 − β)·c[k − 1] + d]/k
    d := d·(qκ)/k
    sum := sum + c[k]
endwhile
Output : R(ε) := k and α_T(j,q) := c[j], j = 0,1,...,k
```

This algorithm is stable for any values of the parameters, including the situations where the variance, and even the expected value, is infinite. The computing time increases linearly with the number of mixed Poisson probabilities computed which, for a given precision, grows considerably as the shape parameter β decreases. As reported in [8], $R(\varepsilon)$ exhibits a fast increase when ε decreases, contrarily to the other distributions considered. Moreover, for a given precision ε, $R(\varepsilon)$ takes values several orders of magnitude larger than for non heavy-tailed distributions.

5 Example

We consider the $GI/M/a/a+c$ queueing system. In this system the arrival process is a renewal process with a fixed (but otherwise general) inter-arrival time distribution, the service times are exponential (with rate λ), there are a identical servers, and there are c positions for waiting. We can model-check transient properties of this queueing system involving the time at which a new customer arrives by:

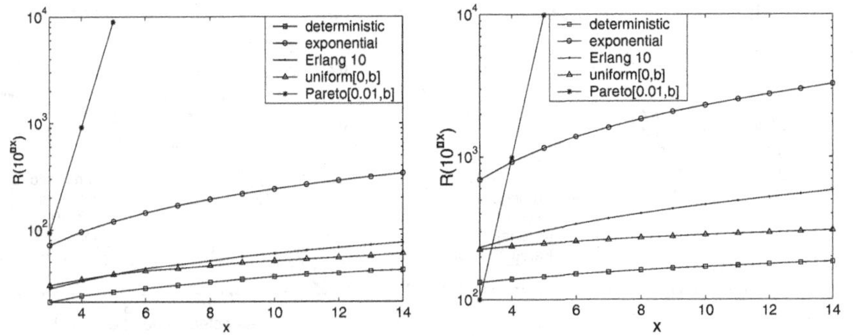

Fig. 2. Change in $R(\varepsilon)$ as the accuracy ε changes

- constructing a model of the queueing system in which transitions corresponding to new arrivals are removed;
- verifying this restricted model against formulae which use the inter-arrival time distribution as their random time-bound.

Note that, in this restricted model, all transitions have an exponential delay, that is, it is a CTMC, and since the service time distribution is independent of the arrival time distribution, the inter-arrival time distribution is independent of this model.

Formally, the CTMC is given by $\mathcal{C} = (S, \mathbf{R}, L)$ with $S = \{s_0, s_1, \ldots, s_{a+c}\}$ where s_k denotes the state in which there are k customers in the queue, and the rate matrix is given by:

$$\mathbf{R}(s_i, s_j) = \begin{cases} a{\cdot}\lambda & \text{if } a \leq i \leq a+c \text{ and } j = i-1 \\ i{\cdot}\lambda & \text{if } 0 \leq i < a \text{ and } j = i-1 \\ 0 & \text{otherwise} \end{cases}$$

where λ is the rate of service of each server. Now, if T represents the inter-arrival time distribution, then the satisfaction of $\mathcal{P}_{\bowtie p}(\Phi\, \mathcal{U}^{\leq T}\, \Psi)$ in $\mathcal{C}$ corresponds to the following property holding in the $GI/M/a/a+c$ queueing system: with probability $\bowtie p$, when the next arrival occurs, a state satisfying Ψ will be reached such that all preceding states satisfy Φ.

We fix $a = 10$ and $c = 5$ and calculate, for each state s of the CTMC $\mathcal{C}$, the value of $Prob^{\mathcal{C}}(s, \texttt{true}\, \mathcal{U}^{\leq T}\, \Psi)$ where Ψ is true in the states in which there are at most $k(= 3)$ customers in the queue (including those being served). In other words, we calculate for each state of the $GI/M/a/a+c$ queueing system the probability of there being at most k customers in the queue when the next arrival occurs. We consider five different distributions for T: deterministic (which represents the standard time-bounded until formula), exponential, Erlang (with 10 phases), uniform and Pareto.

The results were obtained with a prototype implementation in MATLAB with an accuracy $\varepsilon = 10^{-8}$. We note that the mixed Poisson probabilities for the Pareto distribution were slow to converge; however, in this case, we use the fact

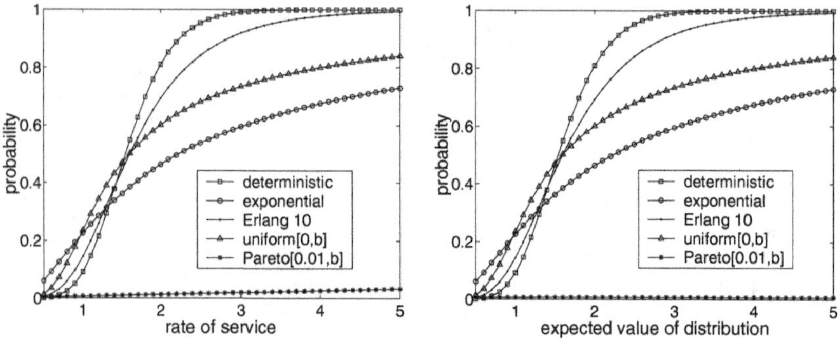

Fig. 3. Value of $Prob^{\mathcal{C}}(s, \mathtt{true}\ \mathcal{U}^{\leq T}\ \Psi)$ when the queue is full

that we can stop iterating when the DTMC $unif(\mathcal{C})[\neg\Phi \vee \Psi]$ reaches steady state. To illustrate this point, in Figure 2 we have plotted the values of $R(\varepsilon)$ for different values of ε. Note that we have used a logarithmic scale on the y-axis to allow the results for the Pareto to be plotted in the same graph as the remaining distributions. The graph on the left corresponds to the case when T has an expected value of 1, while the one on the right to the case when the expected value of T is 10.

As can be seen from Figure 2, the case of the Pareto distribution quickly becomes unmanageable as we increase the accuracy. This result follows from the fact that the Pareto distribution is *heavy tailed*. In fact, the differences in the values of $R(\varepsilon)$ for the different distributions correspond to the difference in their "tails": the heavier the tail the larger the value of $R(\varepsilon)$. For example, the deterministic distribution is zero for all values greater than the expected value, whereas the uniform distribution is zero for any value greater than two times the expected value, and the exponential distribution has a heavier tail than an Erlang (with more than one phase). This correspondence can also be seen in the difference between the case when T has expected value 1 and 10 (the graphs on the left and right respectively): as the expected time increases then so does the probability of the distribution taking larger values.

In Figure 3 we plot the value of $Prob^{\mathcal{C}}(s, \mathtt{true}\ \mathcal{U}^{\leq T}\ \Psi)$ when s is the state in which the queue is full. The left graph corresponds to the case when we vary the service rate and fix the expected value of T at 1, while in the graph on the right the service rate is fixed at 1 and we vary the expected value of T. Similarly, in Figure 4 we plot the values of $Prob^{\mathcal{C}}(s, \mathtt{true}\ \mathcal{U}^{\leq T}\ \Psi)$ when s is the state in which there are $k+1$ customers in the queue.

The graphs in Figure 3 and Figure 4 demonstrate the expected result: as either the expected time of T increases (the expected time between consecutive arrivals increases) or the rate of service increases (the expected duration of a service decreases) the probability of there being at most k customers in the queue at time T increases. Furthermore, the probability is much lower for the state where the queue is full (Figure 3) than for the state where there are only $k+1$ customers in the queue (Figure 4). This follows from the fact that, from

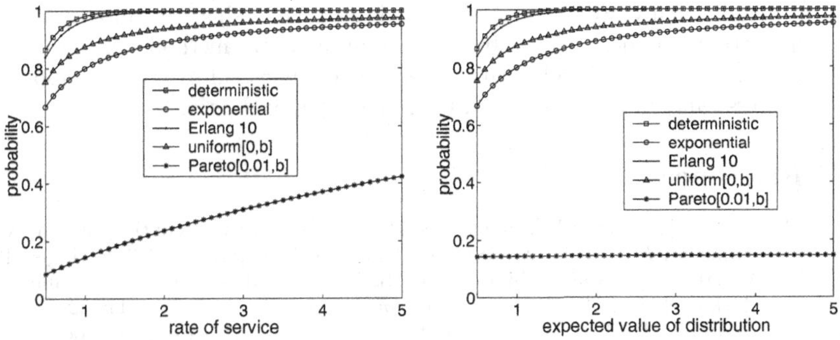

Fig. 4. Value of $Prob^{\mathcal{C}}(s, \mathtt{true}\ \mathcal{U}^{\leq T}\ \Psi)$ when there are $k+1$ customer in the queue

the state in which the queue is full, to reach a Ψ state $s+c-k$ customers need to be served, as opposed to only 1 customer from the state where there are $k+1$ customers in the queue.

The first point to notice about the results is that approximating general distributions with exponential distributions leads to inaccurate results. For example, compare the results when T is exponential (customers arrive with an exponential distribution) to when T has a Pareto distribution (customers arrive with a Pareto distribution). It is also apparent that the computed probabilities are much smaller for the Pareto distribution than for the remaining distributions, whose survival functions exhibit exponential decay to zero. This holds because, although the Pareto distribution is heavy-tailed, it assumes very small values with much higher probability than the other distributions. Significant differences may also arise for light-tailed distributions such as the exponential and Erlang distributions, as illustrated in Figure 3. Finally, note the close similarity between the cases when T has a deterministic and Erlang distribution; this is also to be expected as the Erlang distribution is often used as a continuous approximation of a (discrete) deterministic distribution.

6 Conclusions

This paper presents an extension of CSL with until formulae where the time bound is given as a general nonnegative random variable. This extension allows one, in certain cases, to calculate transient measures of systems which include general distributions. We demonstrate that model checking for such formulae can be efficiently carried out by first precomputing a vector of mixed Poisson probabilities and then using a straightforward adaptation of the algorithm for ordinary time-bounded until.

Currently, we have only considered a prototype implementation of these algorithms in MATLAB. In future we plan to implement these algorithms in the probabilistic symbolic model checker PRISM [14,19] in order to tackle the verification of more complex models.

Additionally, we would like to work on generalising this approach to other important families of distributions; apply analytic methods to finding upper bounds for $R(\varepsilon)$; consider random expected time; and extend our approach to express random time intervals rather than simply the time bound T.

References

1. A. Aziz, K. Sanwal, V. Singhal, and R. Brayton. Verifying continuous time Markov chains. In *Proc. CAV'96*, volume 1102 of *LNCS*, pages 269–276. Springer, 1996.
2. A. Aziz, K. Sanwal, V. Singhal, and R. Brayton. Model checking continuous time Markov chains. *ACM Transactions on Computational Logic*, 1(1):162–170, 2000.
3. C. Baier, B. Haverkort, H. Hermanns, and J.-P. Katoen. Model checking continuous-time Markov chains by transient analysis. In *Proc. CAV 2000*, volume 1855 of *LNCS*, pages 358–372, 2000.
4. C. Baier, J.-P. Katoen, and H. Hermanns. Approximative symbolic model checking of continuous-time Markov chains. In *Proc. CONCUR'99*, volume 1664 of *LNCS*, pages 146–162. Springer, 1999.
5. E. Clarke, E. Emerson, and A. Sistla. Automatic verification of finite-state concurrent systems using temporal logic specifications. *ACM Transactions on Programming Languages and Systems*, 8(2):244–263, 1986.
6. M. Crovella and A. Bestavros. Self-similarity in world wide Web traffic: evidence and possible causes. *IEEE/ACM Transactions on Networking*, 5(6):835–846, 1997.
7. B. Fox and P. Glynn. Computing Poisson probabilities. *Communications of the ACM*, 31(4):440–445, 1988.
8. R. German. *Performance Analysis of Communication Systems: Modeling with Non-Markovian Stochastic Petri Nets.* John Wiley and Sons, 2000.
9. J. Grandell. *Mixed Poisson Processes.* Chapman & Hall, 1997.
10. D. Gross and D. Miller. The randomization technique as a modeling tool and solution procedure for transient Markov processes. *Operations Research*, 32(2):343–361, 1984.
11. H. Hansson and B. Jonsson. A logic for reasoning about time and probability. *Formal Aspects of Computing*, 6:512–535, 1994.
12. A. Jensen. Markov chains as an aid in the study of Markov processes. *Skandinavisk Aktuarietidsskrift, Marts*, pages 87–91, 1953.
13. J.-P. Katoen, M. Kwiatkowska, G. Norman, and D. Parker. Faster and symbolic CTMC model checking. In *Proc. PAPM-PROBMIV 2001*, volume 2165 of *LNCS*, pages 23–38. Springer, 2001.
14. M. Kwiatkowska, G. Norman, and D. Parker. Probabilistic symbolic model checking with PRISM: A hybrid approach. In *Proc. TACAS 2002*, volume 2280 of *LNCS*, pages 52–66. Springer, 2002.
15. G.I. Lópes, H. Hermanns, and J.-P. Katoen. Beyond memoryless distributions. In *Proc PAPM-PROBMIV 2001*, volume 2165 of *LNCS*, pages 57–70. Springer, 2001.
16. S. Molnár and I. Maricza, editors. *Source characterization in broadband networks.* COST 257 Mid-term seminar interim report on source characterization, 2000.
17. J. Muppala and K. Trivedi. *Queueing Systems, Queueing and Related Models*, chapter Numerical Transient Solution of Finite Markovian Queueing Systems, pages 262–284. Oxford University Press, 1992.
18. H. Panjer. Recursive evaluation of a family of compound distributions. *Astin Bulletin*, 12(1):22–26, 1982.
19. PRISM web page. `http://www.cs.bham.ac.uk/~dxp/prism/`.
20. W. J. Stewart. *Introduction to the Numerical Solution of Markov Chains.* Princeton, 1994.

Probabilistic Model Checking of the IEEE 802.11 Wireless Local Area Network Protocol*

Marta Kwiatkowska[1], Gethin Norman[1], and Jeremy Sproston[2]

[1] School of Computer Science, University of Birmingham
Birmingham B15 2TT, United Kingdom
[2] Dipartimento di Informatica, Università di Torino, 10149 Torino, Italy

Abstract. The international standard IEEE 802.11 was developed recently in recognition of the increased demand for wireless local area networks. Its medium access control mechanism is described according to a variant of the Carrier Sense Multiple Access with Collision Avoidance (CSMA/CA) scheme. Although collisions cannot always be prevented, randomised exponential backoff rules are used in the retransmission scheme to minimise the likelihood of repeated collisions. More precisely, the backoff procedure involves a uniform probabilistic choice of an integer-valued delay from an interval, where the size of the interval grows exponentially with regard to the number of retransmissions of the current data packet. We model the two-way handshake mechanism of the IEEE 802.11 standard with a fixed network topology using probabilistic timed automata, a formal description mechanism in which both nondeterministic choice and probabilistic choice can be represented. From our probabilistic timed automaton model, we obtain a finite-state Markov decision process via a property-preserving discrete-time semantics. The Markov decision process is then verified using PRISM, a probabilistic model checking tool, against probabilistic, timed properties such as "at most 5,000 microseconds pass before a station sends its packet correctly."

1 Introduction

Wireless communication devices are increasingly becoming part of our daily lives. In particular, *Wireless Local Area Networks* (WLANs) are often used in cases when data communication over a small area is required, but a wired network is not practical or economic. The international standard IEEE 802.11 was developed recently to cater for the burgeoning use of WLANs, and has enabled the use of heterogeneous communication devices from different vendors within the same network. In contrast to wired devices, stations of a wireless network cannot listen to their own transmission, and are therefore unable to employ medium access control schemes such as Carrier Sense Multiple Access with Collision Detection (CSMA/CD) in order to prevent simultaneous transmission on the channel. Instead, the IEEE 802.11 standard describes a Carrier Sense Multiple Access with

* Supported in part by the EPSRC grant GR/N22960, the CNR grant No.99.01716.CTO1, and the EU within the DepAuDE project IST-2001-25434.

H. Hermanns and R. Segala (Eds.): PAPM-PROBMIV 2002, LNCS 2399, pp. 169–187, 2002.

Collision Avoidance (CSMA/CA) mechanism, using a randomised exponential backoff rule to minimise the likelihood of transmission collision. The backoff procedure is implemented by first choosing an integer valued delay from a bounded interval, where the choice is made according to the uniform probability distribution over the interval. Then the station is required to wait for a length of time dependent on this integer-valued delay. An important characteristic of the backoff procedure, which results in the probability of repeated transmission collisions decreasing as the number of transmission collisions of a particular data packet increases, is that the size of the interval grows exponentially in the number of retransmissions.

Previous studies of the IEEE 802.11 standard have either concerned simulation [15] or analytic approaches from the field of performance evaluation [4,11]. In this paper, we consider automatic verification of a medium access control subprotocol of the IEEE 802.11 WLAN standard using *probabilistic model checking* [10,5]. Given a probabilistic model, expressed as a stochastic process such as a Markov decision process [9], and a (probabilistic) specification, such as "a data packet is delivered with probability 1", the probabilistic model checking algorithm determines which states of the model satisfy the specification.

We model a two-way handshake mechanism of the IEEE 802.11 medium access control scheme, operating in a fixed network topology consisting of two sending stations and two destination stations. Our modelling formalism is that of *probabilistic timed automata* [17], which, like Markov decision processes, allows both nondeterministic choice (used for example, to model asynchrony between sub-components of the system) and probabilistic choice (which, for example, is present in the randomised backoff procedure) to coexist in the same model. Probabilistic timed automata are an extension of timed automata [1]; that is, they are timed automata for which discrete probability distributions range over the edges of the control graph. Equivalently, probabilistic timed automata can be thought of as an extension of Markov decision processes for which the values of a set of real-valued clocks can influence the transitions from each state.

The initial stages of the modelling process employ the timed automata model checking tool UPPAAL [20] to automatically verify the soundness of several abstractions applied to our probabilistic timed automaton model, in order to reduce its complexity in anticipation of probabilistic model checking. The correctness of this process relies on equipping the non-probabilistic UPPAAL model with additional event labels to represent probabilistic choice. From the resulting, smaller probabilistic timed automaton, which nevertheless has an infinite number of states due to the presence of real-valued clock variables, we use a property preserving discrete time semantics to obtain a finite state Markov decision process. We then use the probabilistic model checking tool PRISM [16,21] to verify properties referring both to the likelihood of repeated transmission collision, and to the probability that a station sends a packet correctly within a certain deadline. In contrast to previous numerical analyses of the IEEE 802.11 medium access control scheme, such as [11], we use nondeterminism to model the interleaving which results from asynchronous parallel composition of system components, to model unspecified time delays, and as a conservative over-approximation mechanism when constructing abstractions. Following this methodology, the results we compute through probabilistic model checking give upper and lower bounds on the probability of satisfying the properties of interest.

The paper proceeds by first giving an informal description of a two-way handshake sub-protocol of the IEEE 802.11 standard in the next section. In Section 3, we introduce probabilistic timed automata, defining both their continuous and discrete-time semantics. Section 4 then explains how probabilistic timed automata can be used to model the sub-protocol, describes the construction of the abstract models, and presents the probabilistic model checking results. Section 5 concludes the paper.

2 The Basic Access Mechanism of the IEEE 802.11 DCF

Our focus is on a contention resolution protocol of a basic class of IEEE 802.11 WLAN. The class, referred to as the Independent Basic Service Set or "ad hoc networks", comprises a number of stations communicating over a shared channel in a peer-to-peer manner, without a centralised medium access control (MAC) protocol arbitrating requests to transmit on the channel. Instead, the aim of MAC schemes for such networks is to keep the number of *collisions* (simultaneous transmissions) on the channel to a minimum. For this purpose, the IEEE 802.11 standard defines a Distributed Coordination Function (DCF) based on a CSMA/CA protocol. An important feature of the DCF is that of a randomised, slotted exponential backoff, which is designed to break the symmetry between stations that are repeating previously failed transmissions i.e. transmissions which collided.

As the IEEE 802.11 standard specifies that a sending station monitors the channel prior to transmitting, collisions can occur if multiple stations are simultaneously in their *vulnerable period.* If a station is in this period, which occurs when it starts to send its data, then this transmission can only be detected by other stations after some delay (equal to the length of the vulnerable period); hence, another station may also decide to begin transmitting, resulting in a collision. The duration of the vulnerable period is given by the sum of (1) the time taken for a station to assess the channel and deliver its state to the MAC layer, (2) the time taken for the destination station to change from a receiving to a transmitting state, and (3) the air propagation time.

The standard defines two transmission mechanisms, of which we focus on Basic Access (BA). In this scheme, when a station in a WLAN is ready to transmit a new data packet, it must sense that the channel is free for a duration given by the DCF Interfame Space (DIFS), the length of which depends on the physical layer, and which should be at least as long as the vulnerable period. If the channel is free for this period, then the station can commence transmission of its frame to another station. Upon termination of the transmission, the sending station listens immediately to the channel, in order to detect whether another station is currently transmitting. If so, the sending station decides that a collision

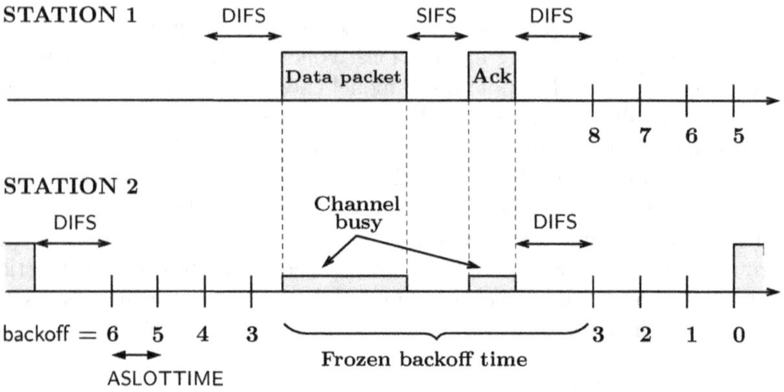

Fig. 1. An example of the backoff procedure

has occurred; if not, it then waits for an acknowledgement to be sent from the destination station. The importance of the acknowledgement in the context of wireless devices can be seen on consideration that a sending station cannot listen to its own transmission; if this was instead possible, then the station could detect that a transmission was successful.

The sending station enters the *backoff procedure* if either:

- the channel is not sensed idle for a DIFS;
- the channel is sensed busy after the station finishes a data transmission;
- a positive acknowledgement of successful transmission is not received from the destination station before a timeout; or
- the station receives an acknowledgement and wishes to send another packet.

The backoff procedure first consists of the station monitoring the channel; if it is busy, the station waits until it is free, after which it must continue to be sensed free for a DIFS. Next, there is a random choice of backoff value, which indicates the number of time periods called "slots" which must be passed through before the station can start transmitting. The duration of the slot is given by ASLOTTIME, and must be at least as large as the vulnerable period. If the channel is detected idle for an ASLOTTIME, the backoff value is decremented by 1. This decrementing procedure is temporarily suspended if a transmission is detected, and is resumed *only* after the channel is sensed free for DIFS time units. When the backoff value reaches 0, the station can commence its transmission.

An example of a backoff procedure is illustrated in Figure 1, which is adapted from [4]. We consider the case in which station 2 has just finished sending a data packet, whereas station 1 has yet to send a packet. After station 2 has finished transmission, its waits for DIFS time units before selecting randomly the backoff value of 6. Station 2 then proceeds to decrement the value of its backoff value by 1 for each duration of length ASLOTTIME passed through. However, after detecting that the channel is free for DIFS time units, station 1 decides to send its data packet before station 2 has finished its backoff countdown. The figure shows how the backoff value of station 2 is frozen while the channel is occupied, both when the data packet and the accompanying acknowledgement are sent, and also during the Short Interframe Space[1] (SIFS) which separates these transmissions. That the backoff countdown is frozen during SIFS follows from the fact that the timing parameters of the IEEE 802.11 standard specify that SIFS < DIFS, and that the channel must be detected free for DIFS time units before the backoff countdown can be resumed.

The random selection of backoff value is implemented as a uniform distribution over integers in the range $[0, CW]$ (the contention window), where $CW = (aCWmin + 1) \cdot 2^{bc} - 1$; the value $aCWmin$ is a constant given by the physical layer, whereas bc is a variable called the *backoff counter*, which represents the number of unsuccessful retransmissions of the pending data packet that have been made (therefore bc is initially 0). The backoff counter can increase to a ceiling imposed by MAX_BACKOFF, which again is a constant given by the physical layer (more precisely, it is calculated through the constant $aCWmax$, given by the physical layer, which defines the maximal contention window).

We impose a number of restrictions and assumptions when modelling the IEEE 802.11 Basic Access DCF mechanism. A fixed network topology, consisting of two sending stations and two destination stations is assumed, meaning that the Extended Interfame Space is not modelled. We also do not consider the Timing Synchronisation Function, which stipulates that short frames are periodically broadcast by a designated station in order to synchronise the local clocks of all other stations. Finally, for simplicity we assume that retry limits, which bound the number of retransmissions of a data packet, are set to infinity. However, we anticipate that these features could be included in our model in the future.

3 Probabilistic Timed Automata

3.1 Syntax of Probabilistic Timed Automata

Time, Clocks, Zones, and Distributions. Let $\mathbb{T} \in \{\mathbb{R}, \mathbb{N}\}$ be the *time domain* of either the non-negative reals or naturals. Let $\mathcal{X}$ be a finite set of variables called *clocks* which take values from the time domain $\mathbb{T}$. A point $v \in \mathbb{T}^{|\mathcal{X}|}$ is referred to as a *clock valuation*. Let $\mathbf{0} \in \mathbb{T}^{|\mathcal{X}|}$ be the clock valuation which assigns 0 to all clocks in $\mathcal{X}$. For any $v \in \mathbb{T}^{|\mathcal{X}|}$ and $t \in \mathbb{T}$, the clock valuation $v \oplus t$ denotes the *time increment* for v and t (we present two alternatives for $\oplus$ in Section 3.2; for the time domain $\mathbb{R}$ it is standard addition $+$). We use $v[X := 0]$ to denote the clock valuation obtained from v by resetting all of the clocks in $X \subseteq \mathcal{X}$ to 0, and leaving the values of all other clocks unchanged.

[1] The Short Interframe Space is the time the IEEE 802.11 standard specifies that a destination station should wait for after successfully receiving data.

Let $Zones(\mathcal{X})$ be the set of *zones* over $\mathcal{X}$, which are conjunctions of atomic constraints[2] of the form $x \sim c$ for $x \in \mathcal{X}$, $\sim \in \{\leq, =, \geq\}$, and $c \in \mathbb{N}$. The clock valuation v *satisfies* the zone ζ, written $v \triangleleft \zeta$, if and only if ζ resolves to true after substituting each clock $x \in \mathcal{X}$ with the corresponding clock value from v.

A discrete probability *distribution* over a countable set Q is a function $\mu : Q \to [0,1]$ such that $\sum_{q \in Q} \mu(q) = 1$. For a possibly uncountable set Q', let $\mathsf{Dist}(Q')$ be the set of distributions over countable subsets of Q'. For some element $q \in Q$, let $\mu_q \in \mathsf{Dist}(Q)$ be the distribution which assigns probability 1 to q.

Syntax of Probabilistic Timed Automata. We review the definition of probabilistic timed automata [17]. An added feature is that of urgent events, which are a well-established concept for classical timed automata [12,7].

Definition 1. *A* probabilistic timed automaton *is a tuple* $(L, \bar{l}, \mathcal{X}, \Sigma, inv, prob)$ *where:* L *is a finite set of* locations *including the* initial location $\bar{l} \in L$; Σ *is a finite set of* events, *of which* $\Sigma_u \subseteq \Sigma$ *are declared as being* urgent; *the function* $inv : L \to Zones(\mathcal{X})$ *is the* invariant condition; *the finite set* $prob \subseteq L \times Zones(\mathcal{X}) \times \Sigma \times \mathsf{Dist}(2^{\mathcal{X}} \times L)$ *is the* probabilistic edge relation.

A *state* of a probabilistic timed automaton is a pair (l, v) where $l \in L$ and $v \in \mathbb{T}^{|\mathcal{X}|}$ are such that $v \triangleleft inv(l)$. Informally, the behaviour of a probabilistic timed automaton can be understood as follows. The model starts in the initial location $\bar{l}$ with all clocks set to 0, and hence the initial state is $(\bar{l}, \mathbf{0})$. In this, and any other state (l, v), there is a nondeterministic choice of either (1) making a *discrete transition* or (2) letting *time pass*. In case (1), a discrete transition can be made according to any $(l, g, \sigma, p) \in prob$ with source location l which is *enabled*; that is, g is satisfied by the current clock valuation v. Then the probability of moving to the location l' and resetting all of the clocks in X to 0 is given by $p(X, l')$. In case (2), the option of letting time pass is available only if the invariant condition $inv(l)$ is satisfied while time elapses and there does not exist an enabled probabilistic edge with an urgent event.

Note that a *timed automaton* [1] is a probabilistic timed automaton for which every probabilistic edge (l, g, σ, p) is such that $p = \mu_{(X,l')}$ for some $(X, l') \in 2^{\mathcal{X}} \times L$.

Higher-Level Modelling. To aid higher-level modelling, a notion of urgency can be associated with locations, in addition to events. Once an urgent location is entered, it must be left immediately, without time passing. Urgent locations can be represented syntactically using an additional clock [7,26].

Integer variables with bounded ranges, which can be tested within enabling conditions and reset by edge distributions, can also be represented syntactically within the probabilistic timed automaton framework above by encoding the values of such variables within locations [25]. Indeed, the probabilistic choice of the

[2] Readers familiar with timed automata will note that we consider the syntax of *closed* zones, which do not feature atomic constraints of the form $x > c$ or $x < c$.

backoff procedure in the IEEE 802.11 WLAN sub-procedure that we study takes the form of a random assignment to an integer variable; hence notation such as backoff := $RANDOM()$, for an integer variable backoff, should be interpreted as a probabilistic choice between locations, as is standard for probabilistic timed automata.

It is often useful to define complex systems as the *parallel composition* of a number of interacting sub-components. The definition of the parallel composition operator $\|$ uses ideas from the theory of (untimed) probabilistic systems [23] and classical timed automata [1]. Let $\mathsf{PTA}_i = (L_i, \bar{l}_i, \mathcal{X}_i, \Sigma_i, inv_i, prob_i)$ for $i \in \{1, 2\}$.

Definition 2. *The* parallel composition *of two probabilistic timed automata* PTA_1 *and* PTA_2 *is the probabilistic timed automaton* $\mathsf{PTA}_1 \| \mathsf{PTA}_2 = (L_1 \times L_2, (\bar{l}_1, \bar{l}_2), \mathcal{X}_1 \cup \mathcal{X}_2, \Sigma_1 \cup \Sigma_2, inv, prob)$ *where* $inv(l, l') = inv_1(l) \wedge inv_2(l')$ *for all* $(l, l') \in L_1 \times L_2$ *and* $((l_1, l_2), g, \sigma, p) \in prob$ *if and only if one of the following conditions holds:*

- $\sigma \in \Sigma_1 \setminus \Sigma_2$ *and there exists* $(l_1, g, \sigma, p_1) \in prob_1$ *such that* $p = p_1 \otimes \mu_{(\emptyset, l_2)}$*;*
- $\sigma \in \Sigma_2 \setminus \Sigma_1$ *and there exists* $(l_2, g, \sigma, p_2) \in prob_2$ *such that* $p = \mu_{(\emptyset, l_1)} \otimes p_2$*;*
- $\sigma \in \Sigma_1 \cap \Sigma_2$ *and there exists* $(l_1, g_1, \sigma, p_1) \in prob_1$ *and* $(l_2, g_2, \sigma, p_2) \in prob_2$ *such that* $g = g_1 \wedge g_2$ *and* $p = p_1 \otimes p_2$

where for any $l_1 \in L_1$, $l_2 \in L_2$, $X_1 \subseteq \mathcal{X}_1$ *and* $X_2 \subseteq \mathcal{X}_2$: $p_1 \otimes p_2(X_1 \cup X_2, (l_1, l_2)) = p_1(X_1, l_1) \cdot p_2(X_2, l_2)$.

3.2 Semantics of Probabilistic Timed Automata

Probabilistic Systems. The semantics of probabilistic timed automata is defined in terms of transition systems exhibiting both nondeterministic and probabilistic choice. We call such models *probabilistic systems*; they are essentially equivalent to Markov decision processes [9], simple probabilistic automata [23], and probabilistic-nondeterministic systems [5].

Definition 3. *A* probabilistic system $\mathsf{PS} = (S, \bar{s}, Act, Steps)$ *consists of a set* S *of* states, *an* initial state $\bar{s} \in S$, *a set* Act *of* actions, *and a* probabilistic transition relation $Steps \subseteq S \times Act \times \mathsf{Dist}(S)$.

A *probabilistic transition* $s \xrightarrow{a,\mu} s'$ is made from a state $s \in S$ by first nondeterministically selecting an action-distribution pair (a, μ) such that $(s, a, \mu) \in Steps$, and second by making a probabilistic choice of target state s' according to the distribution μ, such that $\mu(s') > 0$. We refer to probabilistic transitions of the form $s \xrightarrow{a,\mu_{s'}} s'$ (recall $\mu_{s'}(s') = 1$) as *transitions*. A *transition system* $\mathsf{TS} = (S, \bar{s}, Act, Steps)$ is a probabilistic system for which every probabilistic transition is a transition.

We consider two ways in which a probabilistic system's computation may be represented. A *path* represents a particular resolution of both nondeterminism *and* probability. Formally, a path of a probabilistic system is a non-empty finite or infinite sequence of probabilistic transitions $\omega = s_0 \xrightarrow{a_0,\mu_0} s_1 \xrightarrow{a_1,\mu_1} \cdots$ such that

$s_0 = \bar{s}$. We denote by $\omega(i)$ the $(i+1)$th state of ω and $last(\omega)$ the last state of ω if ω is finite. On the other hand, an *adversary* represents a particular resolution of nondeterminism *only*. Formally, an adversary of a probabilistic system is a function A mapping every finite path ω to a pair (a, μ) such that $(last(\omega), a, \mu) \in Steps$ [27]. Let Adv_{PS} be the set of adversaries of PS. For any $A \in Adv_{\mathsf{PS}}$, let $Path^A_{ful}$ denote the set of infinite paths associated with A. Then, we define the probability measure $Prob^A$ over $Path^A_{ful}$ according to classical techniques [14].

The *maximal (minimal) reachability probability* is the maximum (minimum) probability with which a given set of states can be reached from the initial state. Formally, for a probabilistic system $\mathsf{PS} = (S, \bar{s}, Act, Steps)$, set $F \subseteq S$ of target states, and adversary $A \in Adv_{\mathsf{PS}}$, let:

$$ProbReach^A(F) \stackrel{\text{def}}{=} Prob^A\{\omega \in Path^A_{ful} \mid \exists i \in \mathbb{N} . \omega(i) \in F\} .$$

Then the maximal and minimal reachability probabilities $MaxProbReach_{\mathsf{PS}}(F)$ and $MinProbReach_{\mathsf{PS}}(F)$, respectively, are defined as follows:

$$MaxProbReach_{\mathsf{PS}}(F) \stackrel{\text{def}}{=} \sup_{A \in Adv_{\mathsf{PS}}} ProbReach^A(F)$$
$$MinProbReach_{\mathsf{PS}}(F) \stackrel{\text{def}}{=} \inf_{A \in Adv_{\mathsf{PS}}} ProbReach^A(F) .$$

Semantics of Probabilistic Timed Automata. We now give the semantics of probabilistic timed automata defined in terms of probabilistic systems. Observe that the definition is parameterised both by the time domain $\mathbb{T}$ and the time increment $\oplus$.

Definition 4. *Let* $\mathsf{PTA} = (L, \bar{l}, \mathcal{X}, \Sigma, inv, prob)$ *be a probabilistic timed automaton. The* semantics of PTA with respect to the time domain $\mathbb{T}$ and the time increment $\oplus$ *is the probabilistic system* $[\![\mathsf{PTA}]\!]^{\oplus}_{\mathbb{T}} = (S, \bar{s}, Act, Steps)$ *where:* $S \subseteq L \times \mathbb{T}^{|\mathcal{X}|}$ *and* $(l, v) \in S$ *if and only if* $v \triangleleft inv(l)$; $\bar{s} = (\bar{l}, \mathbf{0})$; $Act = \mathbb{T} \cup \Sigma$; *and* $((l, v), a, \mu) \in Steps$ *if and only if one of the following conditions holds:*

Time transitions. $a \in \mathbb{T}$ *and* $\mu = \mu_{(l, v \oplus t)}$ *such that:*
1. $v \oplus t' \triangleleft inv(l)$ *for all* $0 \leq t' \leq t$*, and,*
2. *for all probabilistic edges of the form* $(l, g, \sigma, -) \in prob$*, if* $v \triangleleft g$*, then* $\sigma \notin \Sigma_u$*;*

Discrete transitions. $a \in \Sigma$ *and there exists* $(l, g, \sigma, p) \in prob$ *such that* $v \triangleleft g$ *and for any* $(l', v') \in S$*:*

$$\mu(l', v') = \sum_{\substack{X \subseteq \mathcal{X}\ \& \\ v' = v[X:=0]}} p(X, l') .$$

The summation in the definition of discrete transitions is required for the cases in which multiple clock resets result in the same target state (l', v'). Note that the semantics of timed automata is given in terms of transition systems.

In our setting, the semantics falls into two classes, depending on whether the underlying model of time is the positive reals or the naturals. If $\mathbb{T} = \mathbb{R}$ we let

$\oplus$ equal $+$ and refer to $[\![\mathsf{PTA}]\!]_{\mathbb{R}}^{+}$ as the *continuous semantics* of the probabilistic timed automaton PTA. In contrast, if $\mathbb{T} = \mathbb{N}$, we let $\oplus$ equal $\oplus_{\mathbb{N}}$ which is defined below and refer to $[\![\mathsf{PTA}]\!]_{\mathbb{N}}^{\oplus_{\mathbb{N}}}$ as the *integer semantics* of PTA. Let PTA be a probabilistic automaton; for any $x \in \mathcal{X}$, let $\mathbf{k}_x$ denote the greatest constant the clock x is compared to in the zones of PTA. Then, for any clock valuation $v \in \mathbb{N}^{|\mathcal{X}|}$ and time duration $t \in \mathbb{N}$, let $v \oplus_{\mathbb{N}} t$ be the clock valuation of $\mathcal{X}$ which assigns the value $\min\{v_x + t, \mathbf{k}_x + 1\}$ to all clocks $x \in \mathcal{X}$ (although the operator $\oplus_{\mathbb{N}}$ is dependent on PTA, we elide a sub- or superscript indicating this for clarity).

Note that the definition of integer semantics for probabilistic timed automata is a generalisation of the analogous definition for the classical model in [3]. As we henceforth use the same type of time increment for a particular choice of time domain, we omit the $+$ and $\oplus_{\mathbb{N}}$ superscripts from the notation. The fact that the integer semantics of a probabilistic timed automaton is finite, and the continuous semantics of probabilistic timed automaton is generally uncountable, can be derived from the definitions. As noted by [18], the semantics of the parallel composition of two probabilistic timed automata corresponds to the semantics of the parallel composition of their individual semantic probabilistic systems, for both the continuous and integer semantics.

The following theorem is key to establishing the correctness of the integer semantics with regard to the probability of reaching a certain set of target locations of a probabilistic timed automaton. The theorem[3] states that both the maximal and minimal probabilities of reaching a target location are equal in the continuous and integer semantics, and is a probabilistic extension of a similar result established in [3]. The proof of the theorem appears in [19]. Let $L' \subseteq L$ be a set of target locations of a probabilistic timed automaton PTA, and let the set of all states in $[\![\mathsf{PTA}]\!]_{\mathbb{T}}$ corresponding to locations in L' be denoted by $F_{\mathbb{T}}^{L'} = \bigcup\{(l, v) \mid l \in L',\ v \in \mathbb{T}^{|\mathcal{X}|} \ \wedge \ v \triangleleft inv(l)\}$.

Theorem 1. *For every probabilistic timed automata* PTA *and target set* $L' \subseteq L$ *of locations:*

$$MaxProbReach_{[\![\mathsf{PTA}]\!]_{\mathbb{R}}}(F_{\mathbb{R}}^{L'}) = MaxProbReach_{[\![\mathsf{PTA}]\!]_{\mathbb{N}}}(F_{\mathbb{N}}^{L'})$$
$$MinProbReach_{[\![\mathsf{PTA}]\!]_{\mathbb{R}}}(F_{\mathbb{R}}^{L'}) = MinProbReach_{[\![\mathsf{PTA}]\!]_{\mathbb{N}}}(F_{\mathbb{N}}^{L'}).$$

Traces and Trace Distributions. A *trace* of a transition system is a sequence of actions which is obtained from a path by projecting all information except the actions. The notion of (finite or infinite) traces can be lifted to paths of probabilistic systems in a natural manner; for example, the trace of the infinite path $s_0 \xrightarrow{a_0,\mu_0} s_1 \xrightarrow{a_1,\mu_1} \cdots$ is the infinite sequence $a_0 a_1 \cdots$. Let A be an adversary of the probabilistic system $\mathsf{PS} = (S, \bar{s}, Act, Steps)$, and let $f : Path_{ful}^{A} \to Act^{\omega}$ be a function assigning the trace to each infinite path of A. The *trace distribution* [22] of A is a probability measure over traces characterised by $f(Prob^A)$. The set of trace distributions of PS, denoted by $\mathsf{tdist}(\mathsf{PS}) \subseteq Act^{\omega} \to [0, 1]$, comprises

[3] As in the non-probabilistic case [3], the theorem relies on the fact that only closed zones are used within the description of the probabilistic timed automaton.

the trace distributions corresponding to all of the adversaries of PS. Given two probabilistic systems PS_1, PS_2, we say that PS_1 *trace distribution refines* PS_2, denoted by $\mathsf{PS}_1 \preceq_D \mathsf{PS}_2$, if and only if $\mathsf{tdist}(\mathsf{PS}_1) \subseteq \mathsf{tdist}(\mathsf{PS}_2)$.

We next establish a result that is used to bridge the divide between reasoning about sets of states, as done in reachability analysis, and reasoning about actions, as done in trace-theoretic approaches.

Lemma 1. *Let* $\mathsf{PS}_1 = (S_1, \bar{s}_1, Act_1, Steps_1)$ *and* $\mathsf{PS}_2 = (S_2, \bar{s}_2, Act_2, Steps_2)$ *be two probabilistic systems such that* $\mathsf{PS}_1 \preceq_D \mathsf{PS}_2$. *If both of the following conditions hold:*

1. $F_1 \subseteq S_1 \setminus \{\bar{s}_1\}$, $F_2 \subseteq S_2 \setminus \{\bar{s}_2\}$, *and*
2. *there exists* $ReachActs \subseteq Act_1 \cap Act_2$ *such that for any* $i \in \{1,2\}$, $s_i \in S_i \setminus F_i$ *and* $s_i' \in F_i$, *we have* $s_i \xrightarrow{a,\mu} s_i'$ *if and only if* $a \in ReachActs$,

then we have:

$$MaxProbReach_{\mathsf{PS}_1}(F_1) \leq MaxProbReach_{\mathsf{PS}_2}(F_2)$$
$$MinProbReach_{\mathsf{PS}_1}(F_1) \geq MinProbReach_{\mathsf{PS}_2}(F_2)\,.$$

We lift the notion of trace distributions, sets of trace distributions and trace distribution refinement from probabilistic systems to probabilistic timed automata. For example, for two probabilistic timed automata PTA_1 and PTA_2, we say that PTA_1 trace distribution refines PTA_2, written $\mathsf{PTA}_1 \preceq_D \mathsf{PTA}_2$, if and only if $[\![\mathsf{PTA}_1]\!]_{\mathbb{R}} \preceq_D [\![\mathsf{PTA}_2]\!]_{\mathbb{R}}$.

The analogue of trace distribution refinement for transition systems and timed automata is *trace refinement.* Given two transition systems TS_1, TS_2, we say that TS_1 *trace refines* TS_2, denoted by $\mathsf{TS}_1 \preceq_T \mathsf{TS}_2$, if the set of traces of TS_1 is included in the set of traces of TS_2. Furthermore, for two timed automata TA_1 and TA_2, we say that TA_1 trace refines TA_2, written $\mathsf{TA}_1 \preceq_T \mathsf{TA}_2$, if and only if $[\![\mathsf{TA}_1]\!]_{\mathbb{R}} \preceq_T [\![\mathsf{TA}_2]\!]_{\mathbb{R}}$.

In the presence of urgent events, trace refinement is not a precongruence [13]. As the timed ready simulation preorder proposed by Jensen et al. is too fine a notion to imply a refinement relation for our models of the IEEE 802.11 BA protocol, we instead drop the requirement of urgency on events; then the set of traces of a timed automaton with urgency is contained within the set of traces of the timed automaton obtained by *changing urgent actions to non-urgent actions.* Formally, for a timed automaton $\mathsf{TA} = (L, \bar{l}, \mathcal{X}, \Sigma, inv, prob)$, let its *lazy timed automaton* $\mathsf{LTA} = (L, \bar{l}, \mathcal{X}, \Sigma', inv, prob)$ be such that $\Sigma' = \Sigma$ but $\Sigma'_u = \emptyset$. This gives us the following result; observe that the converse of the lemma does not hold.

Lemma 2. *Let* TA_1 *and* TA_2 *be two timed automata such that* $\Sigma_{u,2} \subseteq \Sigma_{u,1}$, *and let* LTA_1 *and* LTA_2 *be their lazy timed automata counterparts, respectively. Then:*

$$\mathsf{LTA}_1 \preceq_T \mathsf{LTA}_2 \;\Rightarrow\; \mathsf{TA}_1 \preceq_T \mathsf{TA}_2\,.$$

Formerly Probabilistic Timed Automata. For a probabilistic timed automaton $\mathsf{PTA} = (L, \bar{l}, \mathcal{X}, \Sigma, inv, prob)$, its *formerly probabilistic timed automaton*, denoted by $\mathsf{FPTA} = (L, \bar{l}, \mathcal{X}, \Sigma', inv, prob')$, is the timed automaton which agrees with PTA on all of its elements apart from the event set Σ' and the edge relation $prob'$, which are defined in the following manner. For each probabilistic edge $e = (l, g, \sigma, p) \in prob$, we define the set Σ'_e of events comprising of tuples $\langle\langle l, g, \sigma, p, X, l' \rangle\rangle$ for each (X, l') such that $\mu(X, l') > 0$, and let $\Sigma' = \bigcup_{e \in prob} \Sigma'_e$. The edge relation $prob'$ is defined to be the smallest set such that, for each event $\langle\langle l, g, \sigma, p, X, l' \rangle\rangle \in \Sigma'$, there exists an edge $(l, g, \langle\langle l, g, \sigma, p, X, l' \rangle\rangle, p_{(X,l')}) \in prob'$. The following lemma states formally a property first introduced in [18].

Lemma 3. *Let* PTA_1 *and* PTA_2 *to be two probabilistic timed automata, and let* FPTA_1 *and* FPTA_2 *be their formerly probabilistic timed automata. If* $\mathsf{FPTA}_1 \preceq_T \mathsf{FPTA}_2$ *then* $\mathsf{PTA}_1 \preceq_D \mathsf{PTA}_2$.

The intuition underlying the proof is that traces of FPTA_1 can be assembled into sets which define an adversary A of PTA_1; from $\mathsf{FPTA}_1 \preceq_T \mathsf{FPTA}_2$, an adversary of PTA_2 with the *same* trace distribution of A can also be constructed.

In practice, we would not construct a formerly probabilistic timed automaton as outlined above, because even for relatively small models the construction process would be laborious when done by hand. First, note that it suffices to include information of the form $\langle\langle l, g, \sigma, p, X, l' \rangle\rangle$ only for probabilistic edges not made with probability 1; for all other probabilistic edges of the form $(l, g, \sigma, p_{(X,l')}) \in prob$, we include the edge $(l, g, \langle\langle \sigma \rangle\rangle, p_{(X,l')}) \in prob'$. Lemma 3 continues to hold in this case, because we can nevertheless assemble sets of paths of FPTA_1 and FPTA_2 that result in the same trace distributions.

Let $\mathsf{PTA}_{1\|2} = \mathsf{PTA}_1 \| \mathsf{PTA}_2$. Observe that, in general, $\mathsf{FPTA}_1 \| \mathsf{FPTA}_2 \neq \mathsf{FPTA}_{1\|2}$, because probabilistic edges which do *not* assign probability 1 to a single outcome may synchronise in $\mathsf{PTA}_{1\|2}$, creating new product distributions of the form $p \otimes p'$ which do not have an analogue in $\mathsf{FPTA}_1 \| \mathsf{FPTA}_2$. For our case study, this does not cause a problem, since all transitions which do not occur with probability 1 (the transitions for setting the values of backoff) do not synchronise.

4 Modelling and Verification

4.1 Modelling Using UPPAAL

Detailed Model. Our initial step in modelling the IEEE 802.11 BA mechanism was to design a detailed model intended to represent the behaviour of the protocol when there is a collision; that is, when two stations send data packets at the same time. The model **WLAN** consists of five components operating in parallel, namely **Send**$_1$, **Send**$_2$ (sending stations), **Dest**$_1$, **Dest**$_2$ (destination stations), and **Chan** (the channel). In the following, we assume familiarity with the conventions for the graphical representation of timed automata. Note that we

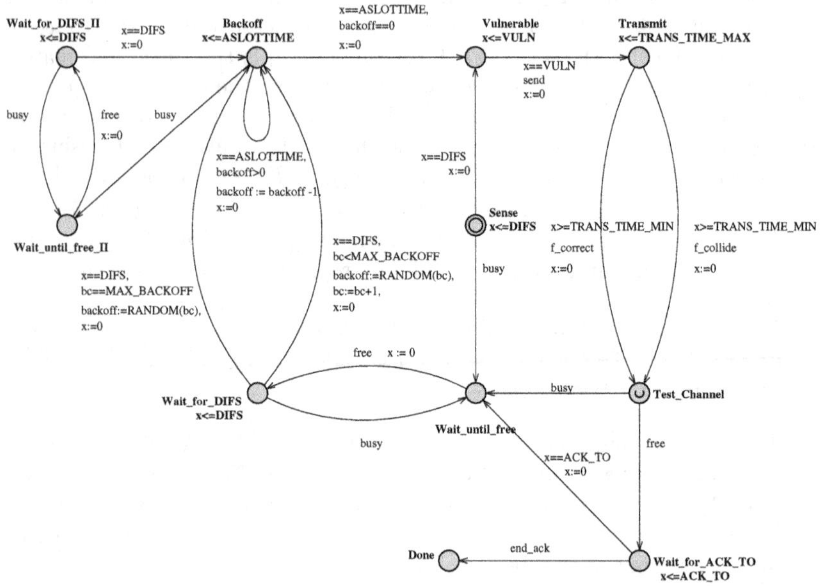

Fig. 2. Template for the sender stations

use the parameters of the Frequency Hopping Spread Spectrum (FHSS) physical layer, with a transmission bit rate of 2Mbps for the data payload.

The template for the senders is shown in Figure 2. Unless indicated otherwise, all transitions are made with probability 1. Note that the events busy and free are the urgent events of the sender. The initial location is indicated by the double circle. The sender begins with a data packet to send, and senses the channel. If the channel remains free for DIFS $= 128\mu s$, then the sender enters its vulnerable period (explained in Section 2) and starts sending a packet, otherwise the station enters backoff. The time taken to send a packet is nondeterministic (within TRANS_TIME_MIN $= 224\mu s$ and TRANS_TIME_MAX $= 15,717\mu s$) and the success of the transmission – whether the event f_correct (successful) or f_collide (unsuccessful) is performed – depends on whether a collision has occurred and is recorded by the channel. The sender then immediately tests the channel (represented by the urgent location Test_Channel). If the channel is busy, the sender enters the backoff procedure, otherwise it waits for an acknowledgement. If the acknowledgement arrives within ACK_TO $= 300\mu s$, then the packet has been sent correctly and the sender finishes; otherwise it times-out and enters the backoff procedure. In the backoff procedure the sender first waits for the channel to be free for DIFS and then sets its backoff value according to the

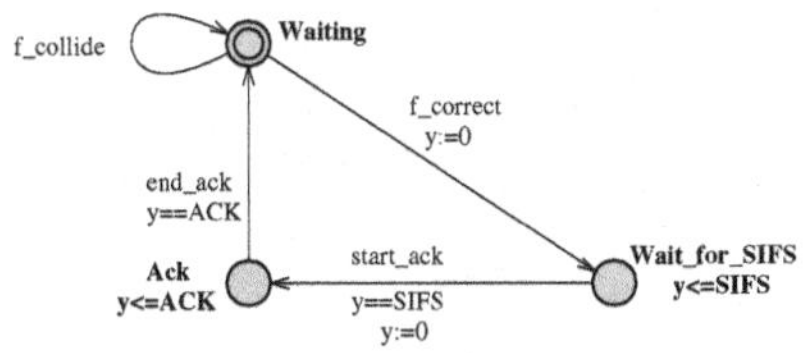

Fig. 3. Template for the destination stations

random assignment[4] backoff $:= RANDOM(bc)$, where bc, the backoff counter, is updated if its current value is less than its maximal value (MAX_BACKOFF $= 6$ since $aCWmax = 1023$). The sender then decrements backoff by 1 if the channel remains free for ASLOTTIME $= 50\mu s$. However, if the channel is sensed busy within this slot, it waits until the channel becomes free and then waits for DIFS before resuming its backoff procedure. When the value of backoff reaches 0 the sender starts re-sending its data packet.

The template for the destinations is shown in Figure 3. Each destination waits for an incoming packet. If a packet arrives correctly (event f_correct), then the destination waits for SIFS $= 28\mu s$ and subsequently sends the acknowledgement, which takes ACK $= 205\mu s$ time units to send. On the other hand, if the message arrives garbled (event f_collide), the destination does nothing.

The probabilistic timed automaton **Chan**, which represents the channel, is shown in Figure 4. The location FREE corresponds to the case in which the channel is available. From this location, receipt of a data packet from station 1 (event send_1, sent by $\mathbf{Send}_1$) triggers the transition to location RCV1; then this packet can either finish successfully (event $\mathsf{f\_correct}_1$, sent by $\mathbf{Send}_1$ again) and return the channel to the location FREE, or collide with a transmission by station 2 (event send_2 sent by $\mathbf{Send}_2$) and make the channel proceed to RCV1RCV2. From the latter location only $\mathsf{f\_collide}_i$ events can remove the data packets from the channel. The left-hand side of the figure shows the part of the model used to represent the receipt of the acknowledgement on the channel. Note that the situations in which an acknowledgement is sent at the same time as a data packet, and in which two acknowledgements collide, are not modelled in this automaton. Although our original channel model did include locations to cater for this possibility, they were removed after reachability analysis of our model using the timed automata model checker UPPAAL [20] (with the FHSS timing parameters) established that such collisions are not possible.

Abstract Model. Before constructing discrete-time models of our probabilistic timed automata, we first apply a number of abstractions. In particular, the

[4] A uniform choice over integers in the range $[0, (aCWmin + 1) \cdot 2^{bc} - 1]$ where $aCWmin = 15$.

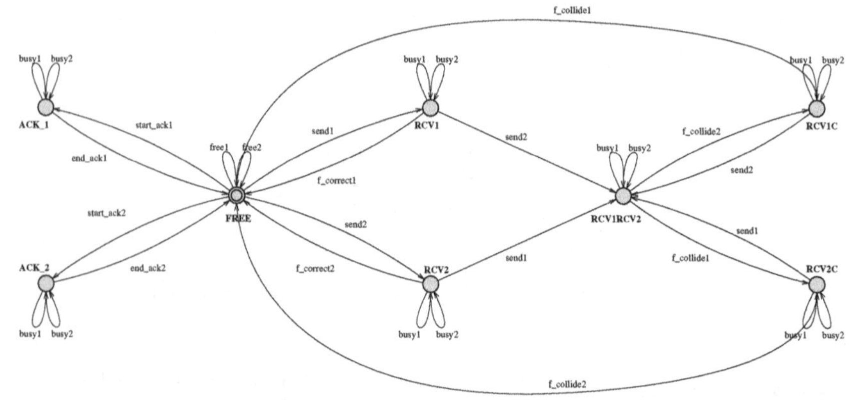

Fig. 4. Template for the channel

destination, the behaviour of which is deterministic, is incorporated into the sender stations to obtain the probabilistic timed automata $\mathbf{AbsSend}_i$, for $i \in \{1,2\}$. Furthermore, using the result that acknowledgements can never collide, the same locations of the channel are used for the receipt of data packets and acknowledgements, resulting in the probabilistic timed automaton **AbsChan**. Our aim is to verify the soundness of the above abstraction with respect to trace distribution refinement. That is, for the concrete and abstracted systems, denoted by

$$\mathbf{WLAN} = \mathbf{Send}_1 \| \mathbf{Dest}_1 \| \mathbf{Chan} \| \mathbf{Dest}_2 \| \mathbf{Send}_2$$
$$\mathbf{AbsWLAN} = \mathbf{AbsSend}_1 \| \mathbf{AbsChan} \| \mathbf{AbsSend}_2$$

respectively, we show that $\mathbf{WLAN} \preceq_D \mathbf{AbsWLAN}$.

Our abstraction methodology is illustrated in Figure 5. We first construct lazy formerly probabilistic timed automata models for each sub-component of the detailed and abstract models using the methodology in Section 3, which are denoted by a prime. Then, using the timed automata model checker UPPAAL, together with the methodology for testing trace refinement of timed automata presented in [24], we establish that $\mathbf{Send}'_i \| \mathbf{Dest}'_i \| \mathbf{Chan}' \preceq_T \mathbf{AbsSend}'_i$ for $i \in \{1,2\}$, and $\mathbf{Chan}' \preceq_T \mathbf{AbsChan}'$. This is illustrated in Figure 5 by the dashed lines from the sub-components of $\mathbf{WLAN}'$ to the sub-components of $\mathbf{AbsWLAN}'$ shown in the right-most box. As trace refinement is compositional, and as the set of traces of $(\mathsf{TA}_1 \| \mathsf{TA}_2) \| \mathsf{TA}_2$ equals that of $\mathsf{TA}_1 \| \mathsf{TA}_2$ for any timed automata TA_1, TA_2, we conclude that:

$$\mathbf{Send}'_1 \| \mathbf{Dest}'_1 \| \mathbf{Chan}' \| \mathbf{Dest}'_2 \| \mathbf{Send}'_2 \preceq_T \mathbf{AbsSend}'_1 \| \mathbf{AbsChan}' \| \mathbf{AbsSend}'_2 ,$$

as denoted by the dashed line from $\mathbf{WLAN}'$ to $\mathbf{AbsWLAN}'$ in the central box of Figure 5. Hence, using Lemma 2 and Lemma 3, it follows that:

$$\mathbf{Send}_1 \| \mathbf{Dest}_1 \| \mathbf{Chan} \| \mathbf{Dest}_2 \| \mathbf{Send}_2 \preceq_D \mathbf{AbsSend}_1 \| \mathbf{AbsChan} \| \mathbf{AbsSend}_2 .$$

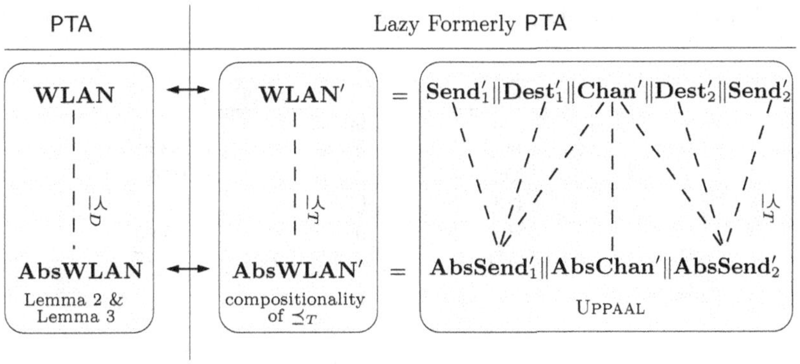

Fig. 5. A diagram illustrating the abstraction procedure

Thus, **WLAN** $\preceq_D$ **AbsWLAN** as required, as shown in Figure 5 by the dashed line from **WLAN** to **AbsWLAN** in the left-most box.

Note that, as the probabilistic choice involved in setting the backoff value involves 1024 possibilities, instead of constructing up to 1024 edges in the formerly probabilistic timed automata $\mathbf{Send}_i'$ and $\mathbf{AbsSend}_i'$, we model the choice of $\mathsf{backoff}_i$ using a "widget": a sub-automaton that determines the range of the choice of $\mathsf{backoff}_i$, sets $\mathsf{backoff}_i$ to the maximal value of the range, and then permits a nondeterministic choice between decrementing the value of $\mathsf{backoff}_i$ or proceeding to the location Backoff. The transitions of the widget are labelled appropriately to ensure correspondence between the values of $\mathsf{backoff}_i$ chosen by the formerly probabilistic timed automata $\mathbf{Send}_i'$ and $\mathbf{AbsSend}_i'$. Naturally, no time is permitted to elapse while in the widget.

A final abstraction step comprises an adjustment to the *time scale* of the model, in anticipation of analysis using probabilistic model checking in PRISM. Thus far, we have assumed that a time unit is equal to $1\mu s$; however, this leads to the model having a largest constant of 15,717, which results in a prohibitively large model size when using the discrete-time semantics introduced in Section 3.2. Therefore, we choose a new time unit of ASLOTTIME $= 50\mu s$, rounding upper bounds on the values of the clocks up, lower bounds down, and let $\mathbf{AbsWLAN_t}$ be the resulting probabilistic timed automaton, the largest constant of which is now 315. For a timed automaton TA, it is established in [2] that the trace set of a timed automaton $\mathsf{TA_t}$ after such a transformation includes that of the original model, denoted by $\mathsf{TA} \preceq_T \mathsf{TA_t}$. Therefore, we have $\mathbf{AbsWLAN}' \preceq_T \mathbf{AbsWLAN_t}'$, and, by Lemma 3, also that **AbsWLAN** $\preceq_D$ $\mathbf{AbsWLAN_t}$. By the transitivity of $\preceq_D$, we know that **WLAN** is a trace distribution refinement of our final abstract model $\mathbf{AbsWLAN_t}$.

Table 1. Maximum probability of either station's backoff counter reaching k

k	Iterations	Time per iteration (s)	Probability
1	18	0.052	1
2	89	0.206	0.183593
3	208	0.366	0.017032
4	423	0.549	7.942e-4
5	816	0.819	1.85e-5
6	1,571	1.32	2.17e-7

4.2 Verification Using PRISM

In this section, we use the probabilistic model checking tool PRISM to automatically verify the satisfaction of probabilistic reachability properties. The model that we build with PRISM is the discrete-time semantic model of the final abstraction given in the previous section ($[\![\mathbf{AbsWLAN_t}]\!]_{\mathbb{N}}$). Note that, for all the properties considered, the correspondence between **WLAN** and **AbsWLAN** required by Lemma 1 holds, and hence the results obtained for the abstract model are upper (lower) bounds for maximal (minimal) reachability probabilities of the full model.

Due to the size of the models, all experiments were performed with PRISM's most space efficient model checking engine, which uses Multi-Terminal Binary Decision Diagrams (MTBDDs) [6]. We ran all experiments on a 440 MHz SUN Ultra 10 workstation with 512 MB memory under the Solaris 2.7 operating system. All properties were checked with an accuracy of $\varepsilon = 10^{-6}$. Further details on the PRISM model and the model checking results are available from the PRISM web page [21].

The model took 72.1 seconds to construct and has 5,958,233 states. We then calculated the minimal probability of both stations eventually sending their packet correctly. As expected, this has probability 1, and hence the probability is also 1 for the detailed model. This calculation took 7,137 iterations and the time per iteration was 14.1 seconds. Furthermore, we calculated the maximum probability of either station's backoff counter reaching k, the results of which are presented in Table 1. Observe that the greater the value of the backoff counters, the greater the number of collisions, and hence the longer it takes for a data packet to be sent correctly. We observe that the probability of reaching k falls rapidly as k increases. Note that for $k = 1$ the probability is 1, which corresponds to the fact that we model the case where stations initially collide.

We then investigated soft deadline properties, namely calculating the *minimum probability of a station delivering a packet within some deadline*. In particular, we considered this property for a variety of deadlines and different values of TRANS_TIME_MAX. For example, for the case when TRANS_TIME_MAX $=$ $2,500$ and the deadline equals $10,000\mu s$, the model has 583,661,380 states, took 658 seconds to construct and the minimum probability of a station sending a

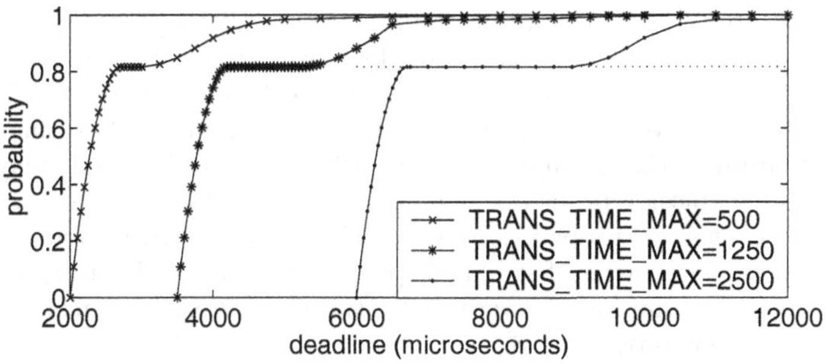

Fig. 6. A graph showing the results for the deadline properties

packet within the deadline is 0.918914, where the verification procedure required 354 iterations and the time per iteration was 3.22 seconds.

In Figure 6 we have plotted the minimum probability of a packet being sent for different deadlines and values of TRANS_TIME_MAX. Note that since stations initially collide with probability 1, the probability will be zero for any deadline which does not allow the stations to collide, enter the backoff procedure and then resend their data packets. This can be seen in the graph by noting that the probability is zero for all deadlines less than or equal to $2 \cdot \mathsf{TRANS\_TIME\_MAX} + 1{,}000$, i.e. the time for a station to send its packet twice plus a constant which includes the time to wait for an acknowledgement and enter backoff.

The dotted line in the graph corresponds to the minimum probability of a station sending a packet correctly while not entering backoff more than once; the results below this line correspond to deadlines where only the first backoff procedure can influence the outcome (that is, for these deadlines there is insufficient time for a station to enter the backoff procedure more than once and send its data correctly). Furthermore, the portions of the graph where the probability does not increase correspond to deadlines which are not large enough for a station to enter backoff more than once and successfully send its data packet, but are sufficient for all cases when backoff is entered at most once. We note that, for each deadline and value of TRANS_TIME_MAX, the minimum probability corresponds to the case when the adversary chooses TRANS_TIME_MAX as the time it takes to send each data packet.

5 Conclusions

We have presented an application of probabilistic model checking to a subprotocol the IEEE 802.11 standard for WLANs, also using non-probabilistic analysis in a "proof assistant" role to verify the soundness of several abstractions of our original protocol model. The use of nondeterminism allows us to model asynchronous behaviour of stations, in addition to providing a conservative approximation mechanism when constructing smaller abstract models.

Future work could lift several simplifying assumptions that were made in this work, such as the fixed network topology in which sending stations cannot also be destination stations, and the absence of the Timing Synchronisation Function. It is straightforward to increase the number of sending and destination stations in our framework, although naturally such verification attempts may suffer from the state-explosion problem commonly encountered in model checking.

We considered probabilistic reachability properties, which are counter-parts of *transient* properties of stochastic processes. It is intended to extend the range of properties that can be verified by the tool PRISM, to include for example expected-time properties [8].

References

1. R. Alur and D. L. Dill. A theory of timed automata. *Theoretical Computer Science*, 126(2):183–235, 1994.
2. R. Alur, A. Itai, R.P. Kurshan, and M. Yannakakis. Timing verification by successive approximation. *Information and Computation*, 18(1):142–157, 1995.
3. D. Beyer. Improvements in BDD-based reachability analysis of timed automata. In *Proc. FME'01*, volume 2021 of *LNCS*, pages 318–343. Springer, 2001.
4. G. Bianchi. Performance analysis of the IEEE 802.11 distributed coordination function. *IEEE Journal on Selected Areas in Communication*, 18:535–547, 2000.
5. A. Bianco and L. de Alfaro. Model checking of probabilistic and nondeterministic systems. In *Proc. FSTTCS'95*, volume 1026 of *LNCS*, pages 499–513. Springer, 1995.
6. E.M. Clarke, M. Fujita, P. McGeer, K. McMillan, J. Yang, and X. Zhao. Multi-Terminal Binary Decision Diagrams: An efficient data structure for matrix representation. In *Proc. IWLS'93*, pages 6a:1–15, 1993. Also available in *Formal Methods in System Design*, 10(2/3), 1997.
7. C. Daws and S. Yovine. Two examples of verification of multirate timed automata with KRONOS. In *Proc. RTSS'95*, pages 66–75. IEEE Computer Society Press, 1995.
8. L. de Alfaro. Computing minimum and maximum reachability times in probabilistic systems. In *Proc. CONCUR'99*, volume 1664 of *LNCS*, pages 66–81. Springer, 1999.
9. C. Derman. *Finite-State Markovian Decision Processes*. Academic Press, 1970.
10. H. Hansson and B. Jonsson. A logic for reasoning about time and reliability. *Formal Aspects of Computing*, 6(5):512–535, 1994.
11. A. Heindl and R. German. Performance modeling of IEEE 802.11 wireless LANs with stochastic Petri nets. *Performance Evaluation*, 44:139–164, 2001.
12. T. Henzinger, P.-H. Ho, and H. Wong-Toi. A user guide to HYTECH. In *Proc. TACAS'95*, volume 1019 of *LNCS*, pages 41–71. Springer, 1995.
13. H. Jensen, K. Larsen, and A. Skou. Scaling up UPPAAL: Automatic verification of real-time systems using compositionality and abstraction. In *Proc. FTRTFT 2000*, volume 1926 of *LNCS*, pages 19–30. Springer, 2000.
14. J. G. Kemeny, J. L. Snell, and A. W. Knapp. *Denumerable Markov Chains*. Graduate Texts in Mathematics. Springer, 2nd edition, 1976.
15. A. Koepsel, J.-P. Ebert, and A. Wolisz. A performance comparison of point and distributed coordination function of an IEEE 802.11 WLAN in the presence of real-time requirements. In *Proc. MoMuC 2000*, 2000.

16. M. Kwiatkowska, G. Norman, and D. Parker. Probabilistic symbolic model checking with PRISM: A hybrid approach. In *Proc. TACAS 2002*, volume 2280 of *LNCS*, pages 52–66. Springer, 2002.
17. M. Kwiatkowska, G. Norman, R. Segala, and J. Sproston. Automatic verification of real-time systems with discrete probability distributions. *Theoretical Computer Science*, 286, 2002. To appear.
18. M. Kwiatkowska, G. Norman, and J. Sproston. Probabilistic model checking of deadline properties in the IEEE 1394 FireWire root contention protocol. Submitted. Extended abstract appears in *Proc. Int. Workshop on Application of Formal Methods to IEEE 1394 Standard*, 2001.
19. M. Kwiatkowska, G. Norman, and J. Sproston. Probabilistic model checking of the IEEE 802.11 wireless local area network protocol. Technical Report CSR-02-05, School of Computer Science, University of Birmingham, 2002.
20. K. G. Larsen, P. Pettersson, and W. Yi. UPPAAL in a nutshell. *Software Tools for Technology Transfer*, 1(1+2):134–152, 1997.
21. PRISM web page. `http://www.cs.bham.ac.uk/~dxp/prism/`.
22. R. Segala. *Modeling and Verification of Randomized Distributed Real-Time Systems.* PhD thesis, Massachusetts Institute of Technology, 1995.
23. R. Segala and N. A. Lynch. Probabilistic simulations for probabilistic processes. *Nordic Journal of Computing*, 2(2):250–273, 1995.
24. D. Simons and M. Stoelinga. Mechanical verification of the IEEE 1394a root contention protocol using UPPAAL2k. *Software Tools for Technology Transfer*, 3(4):469–485, 2001.
25. S. Tripakis. *The formal analysis of timed systems in practice.* PhD thesis, Université Joseph Fourier, 1998.
26. S. Tripakis. Timed diagnostics for reachability properties. In *Proc. TACAS'99*, volume 1579 of *LNCS*, pages 59–73. Springer, 1999.
27. M.Y. Vardi. Automatic verification of probabilistic concurrent finite-state programs. In *Proc. FOCS'85*, pages 327–338. IEEE Computer Society Press, 1985.

Deriving Symbolic Representations from Stochastic Process Algebras*

Matthias Kuntz and Markus Siegle

Friedrich-Alexander-Universität Erlangen-Nürnberg, Institut für Informatik
{mskuntz,siegle}@informatik.uni-erlangen.de

Abstract. A new denotational semantics for a variant of the stochastic process algebra TIPP is presented, which maps process terms to Multi-terminal binary decision diagrams. It is shown that the new semantics is Markovian bisimulation equivalent to the standard SOS semantics. The paper also addresses the difficult question of keeping the underlying state space minimal at every construction step.

1 Introduction

Motivation: Binary decision diagrams (BDD) enjoy great success for the compact representation, manipulation and analysis of very large state spaces. They have proved to be an efficient vehicle for alleviating the notorious problem of state space explosion. Recently, stochastic models have been represented symbolically with the help of Multi-terminal BDDs (MTBDD), and it has been shown that in addition to functional analysis, performance analysis and the verification of performability properties can also be carried out on such symbolic representations [7,14,19,21,25,27].

We employ stochastic process algebras (SPA) for model specification and wish to generate symbolic representations directly from the high-level model, instead of generating transition systems as an intermediate representation. For this purpose, we develop a denotational semantics which maps a given SPA specification directly to its underlying MTBDD. The semantics proceeds in a compositional fashion, according to the structure of the process term at hand, as it has been observed before [7,14,27] that structure exploitation is the key to achieving compact representations. The process algebra which we use is a restricted version of TIPP [15] which guarantees finiteness of the underlying state space. To our knowledge, this is the first complete BDD-based semantics for SPA which completely avoids the construction of transition systems.

The transition system encoded by the MTBDD resulting from our compositional semantics is not necessarily minimal with respect to Markovian bisimulation. Since minimality is a desirable feature, we address the difficult problem of keeping the underlying state space minimal at every step of the MTBDD construction. While it turns out that certain reductions can be performed with the help of some relatively simple heuristic algorithms, we cannot, in general, achieve the maximal reduction without applying a standard bisimulation algorithm.

* This work is supported by the DFG-funded project BDDANA (HE 1408/8) and by the DFG/NWO-funded project VOSS (SI 710/2).

H. Hermanns and R. Segala (Eds.): PAPM-PROBMIV 2002, LNCS 2399, pp. 188–206, 2002.

Related Work: The following is a short summary of related research. We are aware of three approaches to a BDD-based semantics for process algebras, but all of these only cover the functional case, whereas our approach covers stochastic process algebras (note, however, that with minor changes our method can also be applied to purely functional process algebras).

The earliest approach is that of [9], where the authors describe BDD-based procedures for parallel composition, relabelling and restriction of BDDs generated from CCS terms. These procedures assume that the operands are already available as BDDs, and it is shown that under certain conditions the size of the symbolic representation of a given term is proportional to the sum of the sizes of its components. The paper [28] considers the process algebra LOTOS and proposes to exploit the compositional nature of process algebras for building BDD representations of the underlying state space. The overall system specification is decomposed into its basic building blocks. For these, small transition systems are generated and converted into their corresponding BDD representations. The BDD representation of the overall system is then obtained by combining the BDDs for the components in an appropriate way. The approach presented in [8] is the most general, since this method is applicable to a large class of process algebras whose rule system is of GSOS format. A given process algebra term is interpreted as a Boolean formula, i.e. a minterm, and each operator that appears in the term is encoded by one or more Boolean variables, depending on its arity. The minterm is associated with a Boolean vector, and the application of the leading operator causes the changing of some values of the Boolean vector according to specific rules. In terms of transition systems, the original minterm corresponds to the source state and the modified minterm corresponds to the target state of a transition. The length of the encoding is kept constant, although the encoded term can become shorter if its leading operator is of dynamic nature. Since each operator is represented by at least one Boolean variable, the depth of the BDD can become larger than necessary. For the parallel composition operator, the BDD is not constructed from the BDDs representing the two operands, but in a monolithic way, which is not in accordance with the findings of [9,7,25,26,27].

In [24,3] a denotational semantics is presented which maps terms of a restricted stochastic process algebra to real-valued matrices (which could be represented by MTBDDs). This work is interesting for us since it also aims at constructing representations which are minimal with respect to Markovian bisimulation.

Outline: This paper is organised as follows: In Sect. 2 we define the SPA language and study some properties of Markovian bisimulation. Sect. 3 briefly reviews MTBDDs and provides definitions related to the encoding of finite sets. In Sect. 4 the MTBDD semantics is presented together with an example. The correctness of the new semantics is shown in Sect. 5, the problem of minimality is addressed in Sect. 6, and Sect. 7 concludes the paper.

2 Stochastic Process Algebras and Bisimulation

2.1 Definition of the Language

In this paper, a restricted version of the stochastic process algebra TIPP [15] is considered which will be referred to as restricted TIPP or R-TIPP for short.

Definition 1. (Language R-TIPP) *For a set of actions Act (including the internal action τ), let $a \in Act$ and $b \in Act \backslash \{\tau\}$. Let $L \subseteq Act \backslash \{\tau\}$ be a set of visible actions, let $\lambda \in I\!R^{>0}$ be a rate, and let $X \in Var$ be a process variable. R-TIPP is the language whose terms are given by the following grammar:*

$$P \;:=\; P|[L]|P \;\big|\; hide\; b\; in\; P \;\big|\; Q$$
$$Q \;:=\; stop \;\big|\; X \;\big|\; (a,\lambda);Q \;\big|\; Q+Q \;\big|\; recX : Q$$

All occurrences of process variables must be guarded[1].

We have restricted the language in order to keep the state space finite, since we will only be able to represent finite state spaces symbolically. Therefore, recursion is not allowed over static operators.

As underlying semantics we assume the standard SOS semantics given by the semantic rules in Fig. 1[2]. The semantic model is a multi-transition system, i.e. a transition system where the number of instances of a transition is recognised. This multi-transition system is defined as the tuple $(S, Act, \Longrightarrow, s^0)$, where S is the set of derivable process terms, Act is the set of actions, $s^0 \in S$ is the initial process and $\Longrightarrow = \{\!|(s,a,\lambda,t) \mid s,t \in S,\; a \in Act,\; \lambda \in I\!R^{>0}|\!\}$ is a multi-relation ($\{\!|$ and $|\!\}$ denote multiset brackets). The multiplicity of a certain transition is defined as the number of its distinct derivations according to the semantic rules in Fig. 1. For more details see for instance [12,17]. It is possible to flatten the multi-relation to an ordinary transition relation as follows: Transitions with multiplicity greater than one can be amalgamated into a single transition whose rate is the sum of the individual rates, and such a cumulation (which is actually a special case of Lemma 2 below) preserves the behaviour with respect to Markovian bisimulation[3]. Therefore, from now on, we can safely assume that multiple transitions are already cumulated and that the semantic model is a stochastic labelled transition system (with an ordinary transition relation), defined as follows:

Definition 2. (Stochastic Labelled Transition System (SLTS)) *Let S be a finite set of states, let $s^0 \in S$ be the initial state and let Act be a finite set of action labels. Let $\longrightarrow \subseteq S \times Act \times I\!R^{>0} \times S$. We call $\mathcal{T} = (S, Act, \longrightarrow, s^0)$ a stochastic labelled transition system. If $(s,a,\lambda,t) \in \longrightarrow$ we write $s \xrightarrow{a,\lambda} t$.*

---

1 For a definition of guardedness see e.g. [22].

2 In the rule for synchronisation, we adopt TIPP's concept of multiplying the rates of the two partner transitions. However, the MTBDD-based semantics described in this paper could also realise other synchronisation schemes, as long as the resulting rate is a function of the two partner rates.

3 In Sect. 4 we shall see that our MTBDD-based semantics automatically cumulates multiple transitions into a single one.

$$\frac{}{(a,\lambda);P \xrightarrow{a,\lambda} P}$$

$$\frac{P \xrightarrow{a,\lambda} P'}{P+Q \xrightarrow{a,\lambda} P'} \qquad \frac{Q \xrightarrow{a,\lambda} Q'}{P+Q \xrightarrow{a,\lambda} Q'}$$

$$\frac{P \xrightarrow{a,\lambda} P'}{P|[L]|Q \xrightarrow{a,\lambda} P'|[L]|Q}\ a \notin L \qquad \frac{Q \xrightarrow{a,\lambda} Q'}{P|[L]|Q \xrightarrow{a,\lambda} P|[L]|Q'}\ a \notin L \qquad \frac{P \xrightarrow{a,\lambda} P' \quad Q \xrightarrow{a,\mu} Q'}{P|[L]|Q \xrightarrow{a,\lambda * \mu} P'|[L]|Q'}\ a \in L$$

$$\frac{P \xrightarrow{b,\lambda} P'}{\text{hide } b \text{ in } P \xrightarrow{\tau,\lambda} \text{hide } b \text{ in } P'} \qquad \frac{P \xrightarrow{a,\lambda} P'}{\text{hide } b \text{ in } P \xrightarrow{a,\lambda} \text{hide } b \text{ in } P'}\ a \neq b$$

$$\frac{Q[(\text{rec}X : Q)/X] \xrightarrow{a,\lambda} Q'}{\text{rec}X : Q \xrightarrow{a,\lambda} Q'}$$

Fig. 1. Semantic rules for the language R-TIPP

2.2 Markovian Bisimulation

Bisimulation relations characterise equivalent behaviour at the level of the labelled transition system. Examples are strong and weak bisimulation [22], strong and weak Markovian bisimulation [13] and extended Markovian bisimulation [2] (strong Markovian bisimulation agrees with Hillston's strong equivalence [17]). In this paper, we do not consider weak bisimulations since in R-TIPP all internal transitions have an exponentially distributed delay, which can neither be ignored nor "merged" with another such delay.

Informally, two states are Markovian bisimilar if from both states the same equivalence classes can be reached in one step by the same actions and with the same "cumulative rate". Markovian bisimulation can be seen as a refinement of Markov chain lumpability [6], by distinguishing between different action names. Formally:

Definition 3. (Cumulative Rate γ) *The cumulative rate from a state $s \in S$ by action $a \in Act$ to a set of states $C \subseteq S$ is defined by the function $\gamma(s,a,C) = \sum_{\lambda \in E(s,a,C)} \lambda$, where $E(s,a,C) = \{\!|\, \lambda \mid s \xrightarrow{a,\lambda} s' \wedge s' \in C \,|\!\}$. ($\{\!|$ and $|\!\}$ denote multiset brackets.)*

Definition 4. (Markovian Bisimulation) *An equivalence relation $\mathcal{B}$ on the set of states S of an SLTS is a (strong) Markovian bisimulation, if $(s_1,s_2) \in \mathcal{B}$ implies that for all equivalence classes C of $\mathcal{B}$ and all actions a it holds that*

$$\gamma(s_1,a,C) = \gamma(s_2,a,C)$$

Two states s_1 and s_2 are Markovian bisimilar (written $s_1 \sim_M s_2$) if they are contained in a Markovian bisimulation.

If $C_1 = \{s_1, s_2, \ldots\}$ and C_2 are equivalence classes of a Markovian bisimulation $\mathcal{B}$ we sometimes write $\gamma(C_1,a,C_2)$ instead of $\gamma(s_i,a,C_2)$, knowing that the cumulative rate is the same for all $s_i \in C_1$. Given all Markovian bisimulations $\mathcal{B}_1, \mathcal{B}_2, \ldots$ on the same SLTS, one is typically interested in the largest (i.e. coarsest) one, namely $\mathcal{B} = \bigcup_i \mathcal{B}_i$.

Bisimulations are useful for reducing the state space of a given transition system, by replacing each class of equivalent states by a single macro state. In that case, the carrier set of the bisimulation relation is the state space S of the transition system to be reduced. From a slightly different perspective, in order to show that two transition systems are equivalent (written $\mathcal{T}_1 \sim_M \mathcal{T}_2$), we may show that their initial states s_1^0 and s_2^0 are bisimulation equivalent. In this case, the union of the two state spaces $S_1 \cup S_2$ can be used as the carrier set of the bisimulation relation. A third viewpoint, which arises when we consider parallel composition, is discussed in the sequel.

Equivalence Class Structure under Parallel Composition: We now discuss how equivalence classes of process $P = Q|[L]|R$ can be constructed from the equivalence classes of Q and R (we will use Lemma 1 in Sect. 5).

Lemma 1. *Let $Part^Q = \{C_1^Q, \ldots, C_{m_Q}^Q\}$ ($Part^R = \{C_1^R, \ldots, C_{m_R}^R\}$) be a partition of the state space of process Q (R) which corresponds to the equivalence classes of a Markovian bisimulation. The Cartesian product $C_{i,j}^P = C_i^Q \times C_j^R$ yields a partition $Part^P$ of the state space of process $P = Q|[L]|R$ (with equivalence classes $\{C_{i,j}^P | i = 1, \ldots, m_Q, j = 1, \ldots, m_R\}$) which again corresponds to a Markovian bisimulation.*

Lemma 1 can be proven by showing the equivalence of the cumulative rates [20], details are omitted here (Lemma 1 can also be seen as a consequence of the fact that Markovian bisimulation is a congruence [15,13]). Note that some equivalence classes of the combined process P may not be reachable due to synchronisation conditions. Note further that Lemma 1 does not assume that $Part^Q$ and $Part^R$ correspond to the largest bisimulation relations, nor does it claim that the equivalence classes $C_{i,j}^P$ are maximal.

Parallel Transitions: Two transitions $s_1 \xrightarrow{a_1,\lambda_1} t_1$ and $s_2 \xrightarrow{a_2,\lambda_2} t_2$ are called parallel if $s_1 = s_2$ and $a_1 = a_2$ and $t_1 = t_2$ (note that, in principle, both $\lambda_1 \neq \lambda_2$ and $\lambda_1 = \lambda_2$ is possible, although the latter case is ruled out if we only consider ordinary transition systems, as opposed to multi-transition systems). Parallel transitions can be created by applying the choice or hiding operators, or by applying the recursion operator in combination with choice. As we will see in Sect. 4, our MTBDD semantics does not represent parallel transitions separately, but cumulates their rates, which is correct by the following lemma:

Lemma 2. *Let $\mathcal{T}$ be an SLTS and let transition system $\mathcal{T}'$ be constructed from $\mathcal{T}$ by cumulating parallel transitions, i.e. by replacing each set of parallel transitions $\{s \xrightarrow{a,\lambda_i} t \mid i = 1, \ldots, n\}$ by a single transition $s \xrightarrow{a,\lambda} t$, where $\lambda = \sum_{i=1}^n \lambda_i$. Then $\mathcal{T} \sim_M \mathcal{T}'$.*

Lemma 2 can be shown by comparing cumulative rates [20].

3 Basis for Symbolic Representation

3.1 Multi-terminal Binary Decision Diagrams

MTBDDs [11] (also called ADDs [1]) are an extension of BDDs [5] for the graph-based representation of pseudo-Boolean functions, i.e. functions of type $\mathbb{B}^n \mapsto \mathbb{R}$. An MTBDD is a collapsed binary decision tree whose isomorphic subtrees have been merged and whose don't care vertices are skipped. We consider ordered MTBDDs where on every path from the root to a terminal vertex the variable labelling of the nonterminal vertices obeys a fixed ordering.

In the sequel we assume that the MTBDD variables have the following ordering, denoted by $\prec$. At the first $n_a \geq \lceil \log_2 |Act| \rceil$ levels from the root are the variables a_i encoding the action. On the remaining levels we have $2{*}n_s \geq 2{*}\lceil \log_2 |S| \rceil$ variables encoding the source and target state of a transition. The source state variables (s_i) and the target states variables (t_i) are ordered in an interleaved fashion, which yields the following overall variable ordering[4]:

$$\mathsf{a}_{n_a-1} \prec \ldots \prec \mathsf{a}_0 \prec \mathsf{s}_{n_s-1} \prec \mathsf{t}_{n_s-1} \prec \ldots \prec \mathsf{s}_0 \prec \mathsf{t}_0$$

The function represented by MTBDD M is denoted f_{M}. Given an MTBDD M, we use $\mathsf{M}\big|_{\mathsf{s}=0}$ or $\mathsf{M}\big|_{\mathsf{s}=1}$ to denote its restriction to the case $\mathsf{s} = 0$ or $\mathsf{s} = 1$. Note that the MTBDD resulting from such a restriction does no longer depend on Boolean variable s. Given two MTBDDs M_1 and M_2 and an arithmetic operator $\star \in \{+, -, *, \ldots\}$, we simply write $\mathsf{M} := \mathsf{M}_1 \star \mathsf{M}_2$ to obtain the MTBDD which represents $f_{\mathsf{M}_1} \star f_{\mathsf{M}_2}$. These standard arithmetic (and Boolean) operators can be implemented efficiently on the MTBDD data structure with the help of the so-called APPLY algorithm [11].

3.2 Encodings

Definition 5. (Encoding Function) *Let M be an arbitrary finite set. $Enc_M(m)$ denotes the injective encoding function that maps $m \in M$ to its binary encoding (a Boolean vector) of length n, i.e. $Enc_M : M \mapsto \mathbb{B}^n$, $n \geq \lceil \log_2 |M| \rceil$. If M is obvious from the context, the index of the encoding function can be omitted. We write $Enc_M(m) = \vec{m} = (m_{n-1}, \ldots, m_0)$.*

Definition 6. (Encoding Sets) *Let the length n of an encoding be given. PC is the set of all possible binary encodings, i.e. $PC := \mathbb{B}^n$. The set of used encodings UC contains those elements of PC that were already used to encode elements of a given set M, i.e. $UC := \{\vec{c} \mid \vec{c} \in PC \wedge \exists m \in M : (Enc_M(m) = \vec{c})\}$. The set of free encodings FC contains those elements of PC that are not in UC, i.e. $FC := PC \setminus UC$.*

[4] This interleaved ordering is the commonly accepted heuristics for obtaining small MTBDD sizes, see for instance [9,11,27].

Definition 7. (Extension of a Set of Encodings by a Leading Binary Digit) *Let C be a set of Boolean vectors of length n. $Ext_0(C)$ is obtained by adding a leading zero to the elements of C, i.e.:*

$$Ext_0(C) = \{\vec{c'} \mid \vec{c'} = 0 \circ \vec{c} \wedge \vec{c} \in C\}$$

Analogously we obtain $Ext_1(C)$ from C by adding a leading one. The function $Ext(C)$ adds an arbitrary leading digit to the vectors in C, i.e. $Ext(C) = Ext_0(C) \cup Ext_1(C)$.

Definition 8. (Choice of Encoding) *An element $\vec{c}$ of a given set of encodings C is chosen with respect to a total ordering relation $\bowtie$ by the function $Ch(C, \bowtie) := \vec{c} \in C$ such that $\forall \vec{c'} \in C : (\vec{c} \bowtie \vec{c'})$.*

Definition 9. (Literal, Normal Term, Minterm) *A literal is a Boolean variable (a) or its complement ($1 - \mathsf{a}$). A normal term is a term in which no variable occurs more than once. A minterm in n variables is a normal multiplication term in n literals.*

Note that since we are working with MTBDDs we use $(1 - \mathsf{a})$ instead of $\bar{\mathsf{a}}$ and multiplication $*$ instead of conjunction $\wedge$.

Definition 10. (Minterm Function) *Given n distinct Boolean variables $\mathsf{a}_1, \ldots, \mathsf{a}_n$ and a Boolean vector $(b_1, \ldots, b_n)$ of length n, $MT(\mathsf{a}_1, \ldots, \mathsf{a}_n, b_1 \ldots, b_n)$ denotes the minterm consisting of n literals, i.e.*

$$MT(\mathsf{a}_1, \ldots, \mathsf{a}_n, b_1 \ldots, b_n) \ := \ \mathsf{a}_1^* * \ldots * \mathsf{a}_n^*$$

where $\mathsf{a}_i^* = \mathsf{a}_i$ if $b_i = 1$ and $\mathsf{a}_i^* = (1 - \mathsf{a}_i)$ if $b_i = 0$.

Definition 11. (Transition encoding function) *A transition $x \xrightarrow{a,\lambda} y$ of an SLTS can be encoded using the minterm function:*

$$TR(x \xrightarrow{a,\lambda} y) \ := \ MT(\vec{\mathsf{s}}, Enc_S(x)) * MT(\vec{\mathsf{a}}, Enc_{\mathsf{Act}}(a)) * MT(\vec{\mathsf{t}}, Enc_S(y)) * \lambda$$

where $\vec{\mathsf{a}}$ denotes the vector of Boolean variables encoding the action, and $\vec{\mathsf{s}}$ and $\vec{\mathsf{t}}$ denote the vectors of Boolean variables encoding the source and target state of the transition. In the sequel $TR(x \xrightarrow{a,\lambda} y)$ will be written as $TR(x, a, \lambda, y)$.

4 MTBDD Semantics for R-TIPP

4.1 General Idea

The general idea behind our MTBDD semantics is to encode the transitions of a given process algebraic term P symbolically by an MTBDD. The symbolic representation $[\![P]\!]$ is constructed from the parse tree of P which is processed in a depth first manner, thereby constructing $[\![P]\!]$ inductively from simpler terms. Finally we have the pure MTBDD-based representation of the transitional behaviour of P.

Definition 12. (Symbolic Representation of Process Algebra Terms) *The symbolic representation of a process algebra term P is denoted $[\![P]\!]$. It consists of the following parts:*

- *an MTBDD $B(P)$ which encodes the transition relation,*
- *a list of encodings of process variables X that appear in P, denoted $Enc_S(X)$*[5]*,*
- *the encoding of the initial state $Enc_S(s_P^{DS})$.*

The list of action encodings $Enc_{Act}(a)$ is globally valid for all processes and therefore not included in $[\![P]\!]$.

In the following sections we describe how to obtain $[\![P]\!]$ from the symbolic representations of its constituents.

4.2 Process Variables and Stop Process

Verbal Description: A (guarded) process variable X specifies a reference state within a surrounding rec$X$ operator[6]. Therefore, process variables are encoded in a similar fashion as states, i.e. their encodings are taken from PC (the set of possible encodings). Within each sequential component[7] (within the scope of the same rec$X$ operator) process variables having the same name get the same encoding. Upon first appearance of a process variable X, the MTBDD associated with it is the 0-MTBDD (the MTBDD consisting of only the terminal vertex 0).

Formal Description:

if not first appearance of X within present sequential component **then**
    skip /* do nothing */
else if $FC = \emptyset$ **then** /* need to extend the set of possible encodings */
        $PC := Ext(PC);\ UC := Ext_0(UC);\ FC := PC \setminus UC$
    endif
    $Enc_S(X) := Ch(FC, <)$
    $B(X) := 0$
endif

The stop process is a special case of a process variable (a process constant). It has no emanating behaviour, i.e. it remains inactive forever. Therefore, the stop process is associated with the 0-MTBDD.

[5] As explained in Sect. 4.2, process variables correspond to states, therefore we use the encoding function Enc_S for both states and process variables.

[6] Note that R-TIPP does not allow defining equations where the behaviour originating in a process variable is specified in another equation.

[7] A sequential component is a process term which does not include the parallel composition operator.

4.3 Prefix $P = (a, \lambda); Q$

Verbal Description: To generate $[\![P]\!] = [\![(a, \lambda); Q]\!]$ from $[\![Q]\!]$ an additional transition has to be inserted into $B(Q)$, leading from the encoding of a new initial state to the encoding of the initial state of Q. A free encoding is chosen and used as the new initial state of the overall process. The path that encodes the new transition is added to the existing MTBDD. (If the set FC of free encodings is empty the set of possible encodings PC has to be extended first, and in the existing MTBDD $B(Q)$ a new source- and target-variable (s_n and t_n) have to be introduced, whose values remain constant.)

Formal Description:

if $FC = \emptyset$ **then**
$PC := Ext(PC);\ UC := Ext_0(UC);\ FC := PC \backslash UC$
$B(Q) := B(Q) * (1 - \mathsf{s}_n) * (1 - \mathsf{t}_n)$
endif
$Encs(s_P^{DS}) := Ch(FC, <)$
$B(P) := B(Q) + TR(s_P^{DS}, a, \lambda, s_Q^{DS})$

4.4 Choice $P = Q + R$

Verbal Description: When deriving the symbolic representation $[\![P]\!] = [\![Q + R]\!]$ from $[\![Q]\!]$ and $[\![R]\!]$, a new initial state is introduced for $Q + R$. All transitions emanating from the initial states of the subprocesses Q and R have to be copied, as they may also take place in the initial state of the overall process. (If the set FC of free encodings is empty the set of possible encodings PC has to be extended first, and the existing MTBDDs $B(Q)$ and $B(R)$ have to be adjusted accordingly.)

Formal Description:

if $FC = \emptyset$ **then**
$PC := Ext(PC);\ UC := Ext_0(UC);\ FC := PC \backslash UC$
$B(Q) := B(Q) * (1 - \mathsf{s}_n) * (1 - \mathsf{t}_n)$
$B(R) := B(R) * (1 - \mathsf{s}_n) * (1 - \mathsf{t}_n)$
endif
$Encs(s_P^{DS}) := Ch(FC, <)$
/* copy initial transitions from s_Q^{DS} and s_R^{DS}: */
$B(Q') := B(Q)\Big|_{\vec{\mathsf{s}} = Encs(s_Q^{DS})} * MT(\vec{\mathsf{s}}, Encs(s_P^{DS}))$
$B(R') := B(R)\Big|_{\vec{\mathsf{s}} = Encs(s_R^{DS})} * MT(\vec{\mathsf{s}}, Encs(s_P^{DS}))$
$B(P) := B(Q) + B(R) + B(Q') + B(R')$ /* put it all together */

At this point we observe that this procedure will cumulate parallel or multiple transitions correctly: In case Q contains a transition $s_Q^{DS} \xrightarrow{a,\lambda_1} t$ and R contains a transition $s_R^{DS} \xrightarrow{a,\lambda_2} t$ (for any common target state t, and for either $\lambda_1 \neq \lambda_2$ or $\lambda_1 = \lambda_2$), these two transitions will be cumulated, since they are represented in $B(Q')$ and $B(R')$ as parallel transitions emanating from s_P^{DS} and leading to t, and since the MTBDD addition on the last line realises the addition of rates.

As an optimisation of the above procedure, if Q does not possess a cyclic transition sequence of the form $s_0 \xrightarrow{a_1,\lambda_1} s_1 \xrightarrow{a_2,\lambda_2} s_2 \dots s_{l-1} \xrightarrow{a_l,\lambda_l} s_0$, the initial state of Q can be re-used as initial state of P (and similar for R).

4.5 Parallel Composition $P = Q|[L]|R$

Verbal Description: For symbolic parallel composition we follow the same basic strategy as described e.g. in [9,7,25,26,27], where it had been found that this scheme ensures that the size of the symbolic representation of the composed process is linear in the size of its components. $[\![P]\!] = [\![Q|[L]|R]\!]$ can be constructed from $[\![Q]\!]$ and $[\![R]\!]$ as follows[8]: The MTBDD which represents the transitions in which both processes participate is constructed by combining those parts of $B(Q)$ and $B(R)$ which correspond to transitions labelled by actions from L (we use L to denote the BDD which encodes the actions in L). The MTBDD which represents the transitions which Q (R) performs independently of R (Q) is constructed by multiplying the part of $B(Q)$ ($B(R)$) with a BDD Id_R (Id_Q) which denotes stability[9] of proces R (Q).

Formal Description:

$$Enc_S(s_P^{DS}) := Enc_S(s_Q^{DS}) \circ Enc_S(s_R^{DS})$$
$$B(P) := (B(Q) * \mathsf{L}) * (B(R) * \mathsf{L}) + B(Q) * (1 - \mathsf{L}) * \mathsf{Id}_R + B(R) * (1 - \mathsf{L}) * \mathsf{Id}_Q$$

4.6 Recursion $P = \mathbf{rec}X : Q$

Verbal Description: When constructing $[\![P]\!] = [\![\mathrm{rec}X : Q]\!]$ from $[\![Q]\!]$ we can distinguish two cases:

1. X does not appear (unbound) in Q: In this case we simply identify the symbolic representation of rec$X : Q$ with that of Q.
2. X appears in Q: In this case the process variable X is identified with the encoding of the initial state of Q.

[8] It is assumed that $B(Q)$ depends on the vectors of Boolean variables $\vec{\mathsf{a}}$, $\vec{\mathsf{s}}^Q$, $\vec{\mathsf{t}}^Q$ and $B(R)$ depends on $\vec{\mathsf{a}}$, $\vec{\mathsf{s}}^R$, $\vec{\mathsf{t}}^R$, i.e. their sets of state variables are disjoint.

[9] BDD Id_Q, depending on the vectors of Boolean variables $\vec{\mathsf{s}}^Q$ and $\vec{\mathsf{t}}^Q$, encodes the identity matrix of appropriate size, and has a very compact representation under the interleaved variable ordering. Similar for Id_R.

Formal Description (case 2 only):

$$Enc_S(s_P^{DS}) := Enc_S(s_Q^{DS})$$
$$B(P) := B(Q)*(1-MT(\vec{t}, Enc_S(X)))+B(Q)\Big|_{\vec{t}=Enc_S(X)} *MT(\vec{t}, Enc_S(s_Q^{DS}))$$

Note that recursion (in combination with the choice operator appearing within the scope of the recursion) may lead to parallel transitions which are cumulated correctly by the above procedure: In case process Q contains two transitions $s \xrightarrow{b,\lambda_1} s_Q^{DS}$ and $s \xrightarrow{b,\lambda_2} X$ (for any source state s and any action b), the latter of them will be redirected to the target state s_Q^{DS} and the two transitions will be cumulated into the single transition $s \xrightarrow{b,\lambda_1+\lambda_2} s_Q^{DS}$ by the addition of the two MTBDDs.

4.7 Hiding P = hide b in Q

Verbal Description: For constructing $[\![P]\!] = [\![\text{hide } b \text{ in } Q]\!]$ from $[\![Q]\!]$, the MTBDD $B(Q)$ is first cofactorised with respect to the encoding of action b. The result is multiplied with the minterm encoding the internal action τ. Finally, the part of the original MTBDD $B(Q)$ that does not correspond to action b is added.

Formal Description:

$$Enc_S(s_P^{DS}) := Enc_S(s_Q^{DS})$$
$$B(P) := B(Q)\Big|_{\vec{a}=Enc_{Act}(b)} *MT(\vec{a}, Enc_{Act}(\tau))+B(Q)*(1-MT(\vec{a}, Enc_{Act}(b)))$$

Again, this procedure cumulates parallel transitions correctly. For any pair of states s and t, a transition $s \xrightarrow{b,\lambda_1} t$ (which will be turned into an internal τ-transition) and an existing τ-transition $s \xrightarrow{\tau,\lambda_2} t$ will be cumulated by the addition of the two MTBDDs, leading to a single transition $s \xrightarrow{\tau,\lambda_1+\lambda_2} t$.

4.8 Example

In this section we show how to build the symbolic representation for the process term $P := \text{rec}X : ((a, \lambda); (b, \mu); X \; + \; (c, \gamma); \text{stop})$.

1. First, we generate the parse tree of P which is shown in Fig. 2 (a)
2. We set the action encoding as $Enc_{Act}(\tau) = 00$, $Enc_{Act}(a) = 01$, $Enc_{Act}(b) = 10$ and $Enc_{Act}(c) = 11$.
3. First, we generate the symbolic representation of the left hand side of P.
   (a) $[\![X]\!]$: For process variable X we allocate one state bit, s_0, and encode X as $\mathsf{s}_0 = 0$.
   (b) $[\![(b, \mu); X]\!]$: This yields a new state which we encode as $\mathsf{s}_0 = 1$. The MTBDD representation is shown in Fig. 2 (b). (In the graphical depiction of an MTBDD, dashed lines indicate zero-edges and solid lines indicate one-edges.)

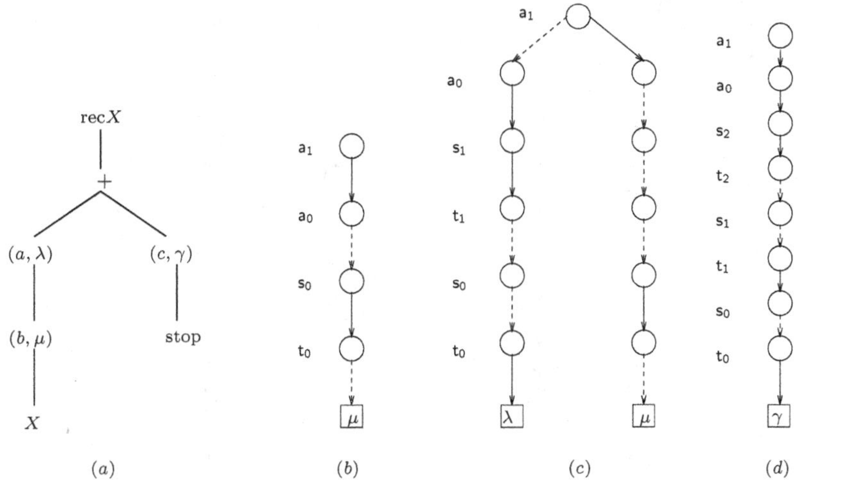

Fig. 2. (a) Parse tree of P, (b) MTBDD representing $(b, \mu); X$, (c)(d) MTBDD representations of the left hand side and right hand side of P

   (c) $[\![(a, \lambda); (b, \mu); X]\!]$: This yields a new state. We need to introduce an additional state variable, s_1, and we encode the new state as $\mathsf{s}_1\mathsf{s}_0 = \mathbf{10}$. Now the encoding of the left hand side of P is complete, we store its initial state and the encoding of the process variable X: $(Enc_S(s^{DS}_{(a,\lambda);(b,\mu);X}) = 10, Enc_S(X) = 00)$. The MTBDD representation is depicted in Fig. 2 (c).
4. Next we generate the symbolic representation of the right hand side of P:
   (a) $[\![stop]\!]$: We use the fresh encoding $\mathsf{s}_1\mathsf{s}_0 = \mathbf{11}$ to encode the process stop.
   (b) $[\![(c, \gamma); stop]\!]$: Since the set of free encodings is now empty, we have to add a third variable to encode this new state: $\mathsf{s}_2\mathsf{s}_1\mathsf{s}_0 = \mathbf{100}$. We store the following information: $(Enc_S(s^{DS}_{(c,\gamma);stop}) = 100, Enc_S(\text{stop}) = 011)$. The MTBDD generated for the right hand side of P is depicted in Fig. 2 (d).
5. The third state variable (s_2 and t_2) must also be introduced into the MTBDD corresponding to the left hand side of P (not shown in the figure).
6. The left hand side and the right hand side of P are now composed according to the semantics of the choice operator. As no subprocess has transitions leading back to its initial state, we can choose one of the subprocesses' initial states as the initial state of the overall process. In this case we choose the initial state of the left hand side: $Enc(s^{DS}_{(a,\lambda);(b,\mu);X+(c,\gamma);\text{stop}}) = 010$, yielding the MTBDD representation as shown in Fig. 3 (a).
7. In the last step we add recursion. The encoding of X is now identified with the encoding of the initial state, leading to the final MTBDD depicted in Fig. 3 (b).

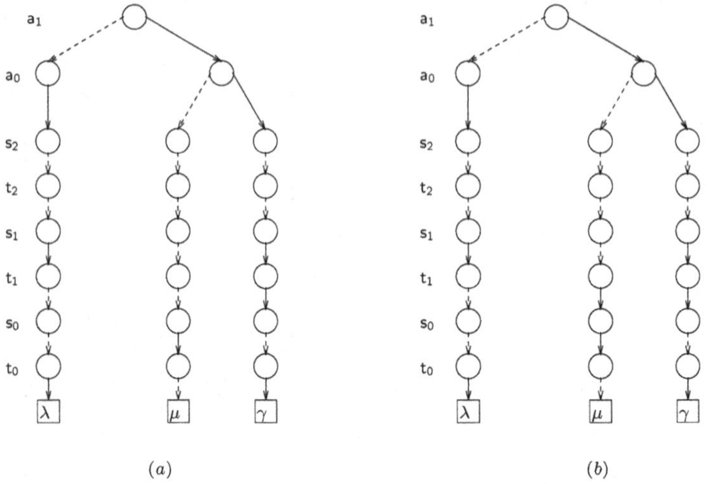

Fig. 3. (a): Left hand side and right hand side joined together using choice, (b) Recursion operator added

5 Correctness of the Semantics

In this section we show that our MTBDD semantics is bisimulation equivalent to the standard SOS-semantics. Roughly speaking, this is shown along the following steps:

- From the MTBDD-representation $[\![P]\!]$ of a given term P we derive an SLTS which we denote by $Tr([\![P]\!])$ and whose initial state we denote by s_P^{DS}. The SLTS $Tr([\![P]\!]) = (S_P^{DS}, Act, \longrightarrow, s_P^{DS})$ is obtained from $[\![P]\!]$ by a straightforward algorithm [20] which extracts the encoded transitions one by one from the MTBDD. The details are omitted here.
- Using induction on the term's structure and exploiting the congruence property of Markovian bisimulation we show that for an arbitrary term P the SLTS obtained by applying the SOS-rules, in the following denoted by $SOS(P)$ (with state space S_P^{SOS} and initial state s_P^{SOS}), is bisimulation equivalent to $Tr([\![P]\!])$[10]. (Induction on the term's structure means that for each operator we show that its addition preserves the bisimulation equivalence relation established for the shorter term.)

Theorem 1. *For any process term P from the language R-TIPP it holds that $Tr([\![P]\!]) \sim_M SOS(P)$.*

[10] Note that $SOS(P)$ and $Tr([\![P]\!])$ are bisimulation equivalent but not necessarily isomorphic, since $SOS(P)$ may contain parallel transitions, while $Tr([\![P]\!])$ cannot represent these separately.

Proof. The complete proof can be found in [20], here we only present part of it. We distinguish the following cases:

1. $P = \text{stop}$: This case constitutes the start of the induction. For the stop process, our MTBDD semantics generates a state encoding and the 0-MTBDD. The SLTS $Tr([\![\text{stop}]\!])$ derived from this symbolic representation consists of a single state with no outgoing transitions, which is isomorphic (and therefore Markovian bisimulation equivalent) with the SLTS $SOS(\text{stop})$. The case $P = \text{X}$ is similar.
2. $P = (a, \lambda); Q$: By construction, in $Tr([\![(a, \lambda); Q]\!])$ there is only a single transition emanating from the initial state, namely $s_P^{DS} \xrightarrow{a,\lambda} s_Q^{DS}$. Similarly, in $SOS((a, \lambda); Q)$ there is only a single transition emanating from the initial state, namely $s_P^{SOS} \xrightarrow{a,\lambda} s_Q^{SOS}$. Since by the induction hypothesis we have $s_Q^{DS} \sim_M s_Q^{SOS}$ it follows that $s_P^{DS} \sim_M s_P^{SOS}$ and thus $Tr([\![P]\!]) \sim_M SOS(P)$.

3. $P = Q + R$: We need to show that $s_P^{DS} \sim_M s_P^{SOS}$ which means that we have to show that for all actions $a \in \mathit{Act}$ and for all equivalence classes $C \subseteq S_P^{DS} \cup S_P^{SOS}$ we have $\gamma(s_P^{DS}, a, C) = \gamma(s_P^{SOS}, a, C)$. (Note that the bisimulation relation $\mathcal{B}$ is now defined on the union of the state spaces of $Tr([\![Q + R]\!])$ and $SOS(Q + R)$, i.e. each equivalence class C is a subset of that union.)

$$\begin{aligned}
\gamma(s_P^{DS}, a, C) &\overset{\text{(by construction)}}{=} \gamma(s_P^{DS}, a^Q, C) + \gamma(s_P^{DS}, a^R, C) \\
&\overset{\text{(by construction)}}{=} \gamma(s_Q^{DS}, a^Q, C) + \gamma(s_R^{DS}, a^R, C) \\
&\overset{\text{(by ind. hypothesis)}}{=} \gamma(s_Q^{SOS}, a^Q, C) + \gamma(s_R^{SOS}, a^R, C) \\
&\overset{\text{(by SOS semantics)}}{=} \gamma(s_P^{SOS}, a, C)
\end{aligned}$$

Here $\gamma(s_P^{DS}, a^Q, C)$ denotes the cumulation of those a-transitions which are due to an a-transition in subprocess Q (and similarly for R). We have used the induction hypothesis, namely that $s_Q^{DS} \sim_M s_Q^{SOS}$ and $s_R^{DS} \sim_M s_R^{SOS}$.

4. $P = Q|[L]|R$: We prove $s_P^{DS} \sim_M s_P^{SOS}$ by showing that for every action a and for every equivalence class $C_{k,l}^P$ we have $\gamma(s_P^{DS}, a, C_{k,l}^P) = \gamma(s_P^{SOS}, a, C_{k,l}^P)$ (note that the equivalence relation is now defined on the union of the state spaces $S = S_P^{DS} \cup S_P^{SOS}$ where $S_P^{DS} = S_Q^{DS} \times S_R^{DS}$ and $S_P^{SOS} = S_Q^{SOS} \times S_R^{SOS}$). We need to distinguish two cases:
(a) $a \notin L$, i.e. a is a non-synchronising action.

$$\begin{aligned}
\gamma(s_P^{DS}, a, C_{k,l}^P) &\overset{\text{(by construction)}}{=} \gamma(s_P^{DS}, a^Q, C_{k,l}^P) + \gamma(s_P^{DS}, a^R, C_{k,l}^P) \\
&\overset{\text{(by constr. and Lemma 1)}}{=} \gamma(s_Q^{DS}, a^Q, C_k^Q) + \gamma(s_R^{DS}, a^R, C_l^R) \\
&\overset{\text{(by ind. hypothesis)}}{=} \gamma(s_Q^{SOS}, a^Q, C_k^Q) + \gamma(s_R^{SOS}, a^R, C_l^R) \\
&\overset{\text{(by SOS sem. and Lemma 1)}}{=} \gamma(s_P^{SOS}, a^Q, C_{k,l}^P) + \gamma(s_P^{SOS}, a^R, C_{k,l}^P)
\end{aligned}$$

$$\stackrel{\text{(by SOS semantics)}}{=} \quad \gamma(s_P^{SOS}, a, C_{k,l}^P)$$

Here $\gamma(., a^Q, .)$ ($\gamma(., a^R, .)$) denotes those transitions which are due to an a-transition of process Q (R). Note that (assuming that $s_P^{DS} \in C_{i,j}^P$), in this non-synchronising case, target class $C_{k,l}^P$ is either $C_{i,l}^P$ or $C_{k,j}^P$ since only one partner makes a move, and thus one of the summed cumulative rates in the above sequence of equations is always equal to zero.

(b) $a \in L$, i.e. a is a synchronising action. Details omitted for lack of space.

5.-6. The proof for recursion and hiding proceeds essentially along the same line, by establishing the equivalence of the cumulative rates. Details are omitted.

6 Towards Minimal Semantics

We have shown that our MTBDD semantics is correct, but so far we have not made any considerations on its minimality. It would, of course, be desirable that the MTBDD constructed by our denotational semantics encodes an SLTS which is minimal with respect to Markovian bisimulation, where minimality means that every class of bisimilar states is represented by a single macro state. Trivially, this goal could be achieved by performing bisimulation minimisation [23,10,18] after every construction step, thereby ensuring that all intermediate representations are minimal. BDD-based bisimulation algorithms are available [4,16], they follow the usual iterative refinement scheme, but such a strategy is impracticable since the overhead for running the bisimulation algorithm would be prohibitive.

Ideally, we wish to perform bisimulation on-the-fly, keeping the encoded state space minimal at every step of the construction, by exploiting information about the operator at hand and the structure of the operand processes. For that purpose we investigate a set of heuristic algorithms. As a simple example, we now briefly discuss such an algorithm for the case of recursion. Fig. 4 shows an algorithm which merges known equivalence classes of process Q in order to obtain equivalence classes for $P = \text{rec}X : Q$. Starting from the newly formed class C'_{ini} which contains both X and the initial state, the algorithm looks for pairs of predecessor classes (denoted $Pred(C'_{ini})$) which can be merged. If two classes are merged, the search also considers the further predecessor classes in a chained fashion, and this aim-driven procedure makes the algorithm quite efficient. In the example shown in Fig. 5, the predecessor of X and the predecessor of the initial state are merged first, and in the subsequent step the two dark states are merged. This is a case where our heuristic algorithm finds the largest Markovian bisimulation, leading to a minimal state space. However, there exist situations like the one shown in Fig. 6 where the algorithm of Fig. 4 does not find the coarsest partition, since it is not possible to merge the two dark states without merging the three lightly shaded states at the same time. Altogether, one can say that the merging of two classes at a time is not sufficient, as our algorithm is correct but not complete.

(1) $Part := Part^Q$
(2) $C'_{ini} := C_{ini} \cup C_X$ /* the initial class and the class containing state X are merged */
(3) $Part := Part \backslash \{C_{ini}, C_X\} \cup \{C'_{ini}\}$ /* the partition is updated */
(4) $Mergers := \{C'_{ini}\}$
(5) **while** $Mergers \neq \emptyset$ **do**
(6)   *choose* $C_{mrg} \in Mergers$
(7)   **forall** $C_i, C_j \in Pred(C_{mrg})$ **do** /* consider pairs of predecessor classes of C_{mrg} */
(8)   **if** $\forall a : \gamma(C_i, a, C_i \cup C_j) = \gamma(C_j, a, C_i \cup C_j)$ **then** /* compare mutual rates */
(9)   **if** $\forall C_k : \forall a : \gamma(C_i, a, C_k) = \gamma(C_j, a, C_k)$ **then** /* compare rates to third party */
(10) $C'_i := C_i \cup C_j$ /* two classes are merged */
(11) $Part := Part \backslash \{C_i, C_j\} \cup \{C'_i\}$ /* the partition is updated */
(12) $Mergers := Mergers \cup \{C'_i\}$ /* a new merger is added */
(13)   **endif**
(14)   **endif**
(15)   **endfor**
(16) $Mergers := Mergers \setminus \{C_{mrg}\}$ /* the processed merger is removed */
(17) **endwhile**
(18) **return** $Part$

Fig. 4. Determining equivalence classes of $recX : Q$ from the classes of Q

For prefix, finding the optimal partition is trivial: If Q contains a state s' with a single outgoing transition $s' \xrightarrow{a,\lambda} s_Q^{DS}$ then s' is equivalent to the new initial state of $P = (a, \lambda); Q$. For choice, as already observed in [24], the key to minimality lies in the ability to detect common behaviour within the operands Q and R. This can be achieved by identifying and comparing the strongly connected components (SCC) of Q and R. SCCs can be determined symbolically in an efficient way [29][11]. For parallel composition, the resulting SLTS is not minimal if the two partners contain identical behaviour which leads to symmetries in the state space (but symmetry is not a necessary precondition for non-minimality). [24] describes state space reduction for replicated processes. Although this can yield a large reduction of the state space, it is shown in [3] that the resulting SLTS is not necessarily minimal. In fact, it is minimal only if all states of the replicated process are "relatively prime", which condition is difficult to verify in practice[12]. For hiding, the situation is essentially the same as for recursion: We have an algorithm which calculates equivalence classes of (hide b in Q) from the equivalence classes of Q, but again the resulting partition is not necessarily optimal.

---

[11] In addition, since only reachable behaviour should be represented, in case the initial state s_Q^{DS} (s_R^{DS}) is unreachable after the application of the choice operator, transitions emanating from this state can be deleted.

[12] Furthermore, since some states of the combined process may be unreachable due to synchronisation conditions, (symbolic) reachability analysis may be necessary in order to determine the set of reachable states.

7 Conclusion and Future Work

Conclusions: We presented a straight-forward approach to deriving a symbolic representation of the transition system underlying a given SPA term. The major contribution of this work is to provide a fully MTBDD-based semantics for a stochastic process algebra, in order to avoid the state space explosion problem by exploiting compositionality. Furthermore we discussed some heuristic algorithms that help to reduce the encoded state space at each stage of its construction.

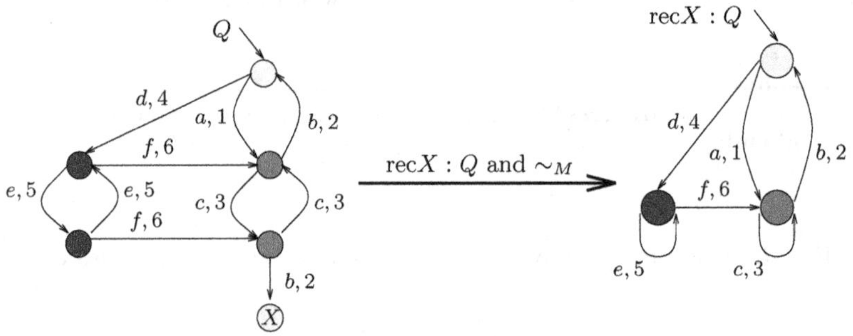

Fig. 5. Example where the algorithm of Fig. 4 finds the largest bisimulation

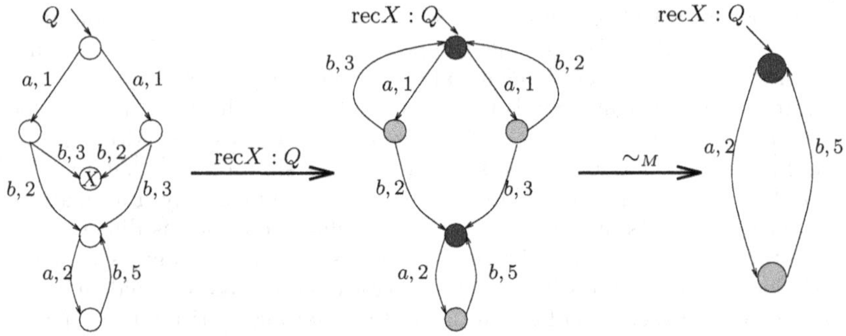

Fig. 6. Example (after [3]) that demonstrates non-optimality of the algorithm in Fig. 4

Future Work: Next, we plan to implement our denotational semantics, in order to compare the size of the resulting MTBDDs to that of existing MTBDD generation methods. In the implementation, some optimisations are possible (e.g.

reuse of the encodings of states that became unreachable). It is planned that our implentation works with an extended version of R-TIPP, for instance allowing parallel composition not only at the top level (as long as it is not in the scope of a recursion operator), and adding an immediate prefix. Finally, we intend to further investigate the questions addressed in Sect. 6, e.g.: Is it possible to classify the situations in which our heuristic algorithms lead to optimal/non-optimal results? What is the benefit of using our algorithms compared to standard bisimulation algorithms? Is it even possible to extend our algorithms in such a way that they guarantee minimality?

Acknowledgements

The authors would like to thank the anonymous referees whose comments helped to fix some technical flaws and improve the presentation of the paper.

References

1. R.I. Bahar, E.A. Frohm, C.M. Gaona, G.D. Hachtel, E. Macii, A. Pardo, and F. Somenzi. Algebraic Decision Diagrams and their Applications. *Formal Methods in System Design*, 10(2/3):171–206, April/May 1997.
2. M. Bernardo and R. Gorrieri. A Tutorial on EMPA: A Theory of Concurrent Processes with Nondeterminism, Priorities, Probabilities and Time. *Theoretical Computer Science*, 202:1–54, 1998.
3. H. Bohnenkamp. Kompositionelle Semantiken stochastischer Prozeßalgebren zur Erzeugung reduzierter Transitionssysteme. Master's thesis, University of Erlangen–Nürnberg, Informatik 7, 1995 (in German).
4. A. Bouali and R. de Simone. Symbolic Bisimulation Minimisation. In *Computer Aided Verification*, pages 96–108, 1992. LNCS 663.
5. R.E. Bryant. Graph-based Algorithms for Boolean Function Manipulation. *IEEE Transactions on Computers*, C-35(8):677–691, August 1986.
6. P. Buchholz. Exact and Ordinary Lumpability in Finite Markov Chains. *Journal of Applied Probability*, (31):59–75, 1994.
7. L. de Alfaro, M. Kwiatkowska, G. Norman, D. Parker, and R. Segala. Symbolic Model Checking for Probabilistic Processes using MTBDDs and the Kronecker Representation. In S. Graf and M. Schwartzbach, editors, *TACAS 2000, LNCS 1785*, pages 395–410, Berlin, 2000.
8. A. Dsouza and B. Bloom. Generating BDD models for process algebra terms. In *Computer Aided Verification*, pages 16–30, 1995. LNCS 939.
9. R. Enders, T. Filkorn, and D. Taubner. Generating BDDs for symbolic model checking in CCS. *Distributed Computing*, (6):155–164, 1993.
10. J.C. Fernandez. An Implementation of an Efficient Algorithm for Bisimulation Equivalence. *Science of Computer Programming*, 13:219–236, 1989.
11. M. Fujita, P. McGeer, and J.C.-Y. Yang. Multi-terminal Binary Decision Diagrams: An efficient data structure for matrix representation. *Formal Methods in System Design*, 10(2/3):149–169, April/May 1997.
12. N. Götz. Stochastische Prozeßalgebren – Integration von funktionalem Entwurf und Leistungsbewertung Verteilter Systeme. Ph.D. thesis, Universität Erlangen–Nürnberg, 1994 (in German).

13. H. Hermanns, U. Herzog, and V. Mertsiotakis. Stochastic Process Algebras — Between LOTOS and Markov Chains. *Computer Networks and ISDN (CNIS)*, 30(9-10):901–924, 1998.
14. H. Hermanns, J. Meyer-Kayser, and M. Siegle. Multi Terminal Binary Decision Diagrams to Represent and Analyse Continuous Time Markov Chains. In B. Plateau, W.J. Stewart, and M. Silva, editors, *3rd Int. Workshop on the Numerical Solution of Markov Chains*, pages 188–207. Prensas Universitarias de Zaragoza, 1999.
15. H. Hermanns and M. Rettelbach. Syntax, Semantics, Equivalences, and Axioms for MTIPP. In *Proc. of PAPM'94*, pages 71–88. Arbeitsberichte des IMMD 27 (4), Universität Erlangen-Nürnberg, 1994.
16. H. Hermanns and M. Siegle. Bisimulation Algorithms for Stochastic Process Algebras and their BDD-based Implementation. In J.-P. Katoen, editor, *ARTS'99, 5th Int. AMAST Workshop on Real-Time and Probabilistic Systems*, pages 144–264. Springer, LNCS 1601, 1999.
17. J. Hillston. *A Compositional Approach to Performance Modelling.* Cambridge University Press, 1996.
18. P. Kanellakis and S. Smolka. CCS Expressions, Finite State Processes, and Three Problems of Equivalence. *Information and Computation*, 86:43–68, 1990.
19. J.-P. Katoen, M. Kwiatkowska, G. Norman, and D. Parker. Faster and Symbolic CTMC Model Checking. In *PAPM-PROBMIV 2001*, pages 23–38. Springer, LNCS 2165, 2001.
20. M. Kuntz and M. Siegle. Deriving symbolic representations from stochastic process algebras. Tech. Rep. Informatik 7 03/02, Universität Erlangen-Nürnberg, 2002.
21. M. Kwiatkowska, G. Norman, and D. Parker. PRISM: Probabilistic Symbolic Model Checker. In P. Kemper, editor, *MMB-PNPM-PAPM-PROBMIV Tool Proceedings*, pages 7–12. Univ. Dortmund, Informatik IV, Bericht 760/2001, 2001.
22. R. Milner. *Communication and Concurrency.* Prentice Hall, London, 1989.
23. R. Paige and R. Tarjan. Three Partition Refinement Algorithms. *SIAM Journal of Computing*, 16(6):973–989, 1987.
24. M. Rettelbach and M. Siegle. Compositional Minimal Semantics for the Stochastic Process Algebra TIPP. In *Proc. of PAPM'94*, pages 89–106. Arbeitsberichte des IMMD 27 (4), Universität Erlangen-Nürnberg, 1994.
25. M. Siegle. Compact representation of large performability models based on extended BDDs. In *Fourth Int. Workshop on Performability Modeling of Computer and Communication Systems (PMCCS4)*, pages 77–80, Williamsburg, Sept. 1998.
26. M. Siegle. Compositional Representation and Reduction of Stochastic Labelled Transition Systems based on Decision Node BDDs. In D. Baum, N. Müller, and R. Rödler, editors, *MMB'99*, pages 173–185, Trier, September 1999. VDE Verlag.
27. M. Siegle. Advances in model representation. In L. de Alfaro and S. Gilmore, editors, *Process Algebra and Probabilistic Methods, Joint Int. Workshop PAPM-PROBMIV 2001*, pages 1–22. Springer, LNCS 2165, September 2001.
28. R. Sisto. A method to build symbolic representations of LOTOS specifications. In *Protocol Specification, Testing and Verification*, pages 323–338, 1995.
29. A. Xie and P.A. Beerel. Implicit Enumeration of Strongly Connected Components. In *ICCAD'99*, pages 37–40. IEEE, 1999.

A Generalization of Equational Proof Theory?

Olivier Bournez

INRIA
615 rue du Jardin Botanique, BP 101
54602 Villers-lès-Nancy Cedex, Nancy, France
Olivier.Bournez@loria.fr

1 Motivation

Rule based languages focussed in last decade on the use of term rewriting as a modeling tool [5]. To extend modeling capabilities of these languages, we explored recently the possibility of making the rule applications subject to probabilistic choices [2]. Dealing with rewriting with probabilistic firing of rules leads to numerous problems about the understanding of the underlying theoretical notions and results. In [2], we started to discuss what could be the generalizations of the classical notions in rewriting community for abstract reduction systems [1]. The next natural step is to understand if a generalization of equational proof theory for probabilistic systems exists. Our first attempt to do so yielded the theory which is presented in this abstract and which is actually closer to fuzzy logic than probabilities [3].

2 Valued Equational Logic

Let $*$ be some fixed continuous t-norm [3]: $*$ is a continuous operation $[0,1]^2 \to [0,1]$ which is commutative, associative, non-decreasing in both arguments and having 1 as zero element.

A valued identity $s \approx_p t$, is given by $s, t \in \mathcal{T}(\Sigma, X)$, and some cost $p \in [0,1]$, where $\mathcal{T}(\Sigma, X)$ is the set of terms over some signature Σ and some disjoint infinite set of variables V.

Let E be a finite set of valued identities.

Definition 1 (Valued Equational Logic). *Valued equational logic is the logic obtained using the following inference rules:*

$$\frac{(s \approx_q t) \in E}{E \vdash s \approx_q t} \qquad \frac{E \vdash s \approx_q t \;\; E \vdash t \approx_r u}{E \vdash s \approx_{q*r} u}$$

$$\frac{}{E \vdash s \approx_1 s} \qquad \frac{E \vdash s \approx_q t \quad \sigma \textit{ substitution}}{E \vdash \sigma(s) \approx_q \sigma(t)}$$

$$\frac{E \vdash s \approx_q t}{E \vdash t \approx_q s} \qquad \frac{E \vdash s_1 \approx_{q_1} t_1 \;\; E \vdash s_n \approx_{q_n} t_n}{E \vdash f(s_1, \dots, s_n) \approx_{q_1 * q_2 \cdots * q_n} f(t_1, \dots, t_n)}$$

Let $s, t \in \mathcal{T}(\Sigma, X)$ be two terms. The *provability degree* of $s \approx t$ is defined as $|s \approx t| = \sup\{p | E \vdash s \approx_p t\}$.

H. Hermanns and R. Segala (Eds.): PAPM-PROBMIV 2002, LNCS 2399, pp. 207–208, 2002.

3 Valued Σ-Algebras

A valued relation, i.e. a function $R : A \times A \to [0,1]$, is a $*$-equivalence relation if it satisfies $R(a,a) = 1$, $R(a,b) = R(b,a)$, $R(a,b) * R(b,c) \leq R(a,c)$ for all $a, b, c \in A$.

Definition 2 (Valued Σ-Algebra). *A valued Σ-algebra $\mathcal{A}$ consists of a carrier set A, a function $f^A : A^n \to A$ for each function symbol $f \in \Sigma$ of arity n, and a $*$-congruence $=^A$ on A, that is to say a valued $*$-equivalence relation $=^A$ on A with $=^A (a_1, b_1) * \ldots * =^A (a_n, b_n) \leq =^A (f^A(a_1, \ldots, a_n), f^A(b_1, \ldots, b_n))$ for all symbol $f \in \Sigma$ of arity n, $a_1, \ldots, a_n, b_1, \ldots, b_n \in A$.*

Definition 3 (Model). *A valued Σ-algebra $\mathcal{A}$ is a model of a set E of valued identities, denoted by $\mathcal{A} \models E$, iff every valued identity of E holds in $\mathcal{A}$: a valued identity $s \approx_p t$ holds in $\mathcal{A}$, denoted by $\mathcal{A} \models s \approx_p t$, if for all Σ-homomorphism $\phi : \mathcal{T}(\Sigma, X) \to \mathcal{A}$, we have $p \leq =^A (\phi(s), \phi(t))$, where a Σ-homomorphism $\phi : \mathcal{T}(\Sigma, X) \to \mathcal{A}$ is a mapping with $=^A (\phi(f(a_1, \ldots, a_n)), f^A(\phi(a_1), \ldots, \phi(a_n))) = 1$.*

Let $s, t \in \mathcal{T}(\Sigma, X)$ be two Σ-terms. The *truth degree* of $s \approx t$ is defined as $\|s \approx t\| = \inf\{p | \mathcal{A} \models s \approx_p t$ for some valued Σ-algebra $\mathcal{A}$ with $\mathcal{A} \models E\}$.

4 Main Properties

Valued equational logic is sound and complete.

Theorem 1 (Soundness and Completeness). *For any finite set E of valued identities, for any $s, t \in \mathcal{T}(\Sigma, X)$, $|s \approx t| = \|s \approx t\|$.*

Moreover, valued equational logic has initial models.

Theorem 2 (Existence of Initial Models). *For any finite set E of valued identities, there exists a valued Σ-algebra $\mathcal{I}_E$, with for any $s, t \in \mathcal{T}(\Sigma, X)$, $\mathcal{I}_E \models s \approx_p t$ iff $E \models s \approx_p t$.*

References

1. Franz Baader and Tobias Nipkow. *Term Rewriting and all That*. Cambridge University Press, 1998.
2. Olivier Bournez and Claude Kirchner. Probabilistic rewrite strategies. applications to elan. In *Rewriting Techniques and Applications (RTA'2002)*, Lecture Notes in Computer Science. Springer, 2002.
3. Petr Hájek and Lluis Godo. Deductive systems of fuzzy logic (a tutorial). *Tatra Mountains Mathematical Publications*, volume 13, pages 35–66. 1997.
4. Jan Willem Klop. Term rewriting systems. In S. Abramsky, D. M. Gabbay, and T. S. E. Maibaum, editors, *Handbook of Logic in Computer Science*, volume 2, chapter 1, pages 1–117. Oxford University Press, Oxford, 1992.
5. Narciso Martí-Oliet and José Meseguer. Rewriting logic: Roadmap and bibliography. *Theoretical Computer Science*, 2002. To appear.

An Integrated Approach for the Specification and Analysis of Stochastic Real-Time Systems
(Short Abstract)

Mario Bravetti

Università di Bologna, Dipartimento di Scienze dell'Informazione
Mura Anteo Zamboni 7, 40127 Bologna, Italy
bravetti@cs.unibo.it

Two approaches for expressing and analyzing time properties of systems have been developed which are based on formal description paradigms: (*i*) the real-time approach, mainly concerned with the expression of time constraints and the verification of exact time properties, and (*ii*) the probabilistic-time approach, mainly concerned with the probabilistic quantification of durations of system activities via exponential distributions and the evaluation of system performance.

The different aspects of time expressed by the real-time and probabilistic-time approaches can be seen as being *orthogonal.* According to the probabilistic-time approach the possible values for the duration of an activity are quantified through probabilistic (exponential) distributions, but no time constraint is expressible: all duration values are possible with probability greater than zero. According to the real-time approach some interval of time is definable for doing something, but the actual time the system spends in-between interval bounds is expressed non-deterministically. A specification paradigm capable of expressing both aspects of time should be able of expressing both time constraints and a probabilistic quantification for the possible durations which satisfy such constraints. We can obtain such an expressive power by considering models capable of expressing *general probability distributions* for the duration of activities. In this way time constraints are expressible via probability distributions that associate probability (density) greater than zero only to time values that are possible according to the constraints. Technically, the set of possible time values for the duration of an activity is given by the *support* of the associated duration distribution. This idea of deriving real-time constraints from distribution supports, that we have introduced in [2], was subsequently applied also in [3] and [4].

Representing real-time and probabilistic-time in a single specification paradigm allows us to model a concurrent system more precisely by expressing and analyzing the relationship between the two aspects of time. Moreover, the capability of expressing general distributions gives the possibility of producing much more realistic specifications of systems. System activities which have an uncertain duration can be represented by more adequate distributions than exponential ones (e.g. Gaussian or experimentally determined distributions).

Process algebra with generally distributed time offer the possibility of an integrated approach for the modeling and analysis of stochastic real-time concurrent/distributed systems [1]. In particular, specifications made with the calculus of Interactive Generalized Semi-Markov Processes (IGSMPs) [1] can be directly

H. Hermanns and R. Segala (Eds.): PAPM-PROBMIV 2002, LNCS 2399, pp. 209–210, 2002.

analyzed through standard discrete event simulation and by means of the techniques in [1]: minimization via bisimulation based congruence abstracting from internal system activities and derivation of the performance model (a GSMP).

Besides performing combined analysis, we have now introduced formal techniques for compositionally deriving, from an IGSMP specification: (*i*) a pure probabilistic-time (Markovian) specification in the form of a term of the calculus of *Interactive Weighted Markov Chains* (IWMCs) – an extension of Interactive Markov Chains (IMCs) with *weight* transitions expressing probabilistic choices –, by approximating general distributions with *phase-type* distributions, i.e. combinations of exponential distributions (as a consequence all duration values for delays get probability greater than 0 and the information about time constraints, related to the real-time behavior of the system, is lost); and (*ii*) a pure real-time specification in the form of a parallel composition of *Interactive Timed Automata* (ITA) – a variant of Timed Automata where action executions, events enabled on the basis of clock constraints and clock reset events are expressed by means of separate transitions –, by considering the support of general distributions and by turning probabilistic choices into non-deterministic choices (as a consequence the information related to the probabilistic-time behavior of the system is lost).

Deriving a pure Markovian representation (the IWMC) and a pure real-time representation (the ITA) is very important from a practical viewpoint in that it gives the possibility of reusing existing techniques and tools already developed for performance evaluation and model-checking of non-probabilistic real-time properties. Moreover, the advantage of deriving an IWMC and an ITA from the same initial IGSMP specification (w.r.t. generating them independently) is that they are guaranteed to be consistent, in that they represent different aspects of the same initial system specification. Both mappings can be found in [1].

The mapping into IWMCs is significant in that: (*i*) it shows process algebra to provide exactly the machinery necessary for approximating GSMPs with CTMCs through phase-type distributions, and (*ii*) it confirms ST semantics to be the adequate semantics for generally distributed time (as claimed in [1]) in that approximation of activity durations with phase-types is a form of action refinement. Moreover, w.r.t. the mapping in [4], which solves a problem in [3], the mapping into ITA does not cause an exponential growth of the state space.

References

1. M. Bravetti, *"Specification and Analysis of Stochastic Real-Time Systems"*, Ph.D. Thesis, University of Bologna (Italy), February 2002. Available at `http://www.cs.unibo.it/~bravetti`
2. M. Bravetti, *"Towards the Integration of Real-Time and Probabilistic-Time Process Algebras"*, in Proc. of the *3rd European Research Seminar on Advances in Distributed Systems (ERSADS'99)*, Madeira Island (Portugal), April 1999
3. J. Bryans, J. Derrick, *"Stochastic Specification and Verification"*, In Proc. of the *3rd Irish Workshop in Formal Methods*, Elect. Workshops in Comp., July 1999
4. P.R. D'Argenio, *"A Compositional Translation of Stochastic Automata into Timed Automata"*, Technical Report CTIT 00-08, Univ. Twente, May 2000

Probabilistic Abstract Interpretation and Statistical Testing
(Extended Abstract)

Alessandra Di Pierro[1] and Herbert Wiklicky[2]

[1] Dipartimento di Informatica, Universitá di Pisa, Italy
[2] Department of Computing, Imperial College, London, UK

Although generally too weak to guarantee correctness, software testing is an indispensable technique for the validation of software systems. It can be used for example to assess how good software is, to find faults and thus improve the software, or for measuring the quality and reliability of reactive systems.

One important phase of the testing process is *test selection* in which test scenarios are selected according to some given criteria. While in general test selection can involve quite elaborate considerations, for simple programs like controllers, embedded systems, etc. *statistical testing*, i.e. the random selection of test data is generally regarded as a feasible approach.

The concrete task we consider here is to determine the probability that a reactive system's response falls within a certain set of acceptable outputs, i.e. that it passes a certain test. We consider a test function $t : X \mapsto \mathbb{B}$ from the input space X of a program c onto $\mathbb{B} = \{0, 1\}$ which is 1 when c passes the test and 0 otherwise. The quality of the system is thus described by the expectation value $\mathbf{E}(t)$, i.e. the probability of a correct output.

As pointed out before testing of a concrete system may not always be feasible and thus tests can be performed instead on an abstract system. In [5] the abstract system is obtained via *abstract interpretation* [1]. In this classical case, where the semantics of the concrete and abstract system is defined via order-theoretic structures, the abstract semantics is guaranteed to be *safe* if concrete and abstract semantics are related via a *Galois connection*:

Definition 1. *Let $\mathcal{C} = (\mathcal{C}, \leq_\mathcal{C})$ and $\mathcal{D} = (\mathcal{D}, \leq_\mathcal{D})$ be two partially ordered set. If there are two functions $\alpha : \mathcal{C} \mapsto \mathcal{D}$ and $\gamma : \mathcal{D} \mapsto \mathcal{C}$ such that for all $c \in \mathcal{C}$ and all $d \in \mathcal{D}$: $c \leq_\mathcal{C} \gamma(d)$ iff $\alpha(c) \leq_\mathcal{D} d$, then $(\mathcal{C}, \alpha, \gamma, \mathcal{D})$ forms a* Galois connection.

Using classical abstract interpretation to define the abstract testing scenario, we obtain an abstract testing function T which is a *safe* abstraction of concrete one, i.e. $t(x) \leq T(x)$ for all $x \in X$. We can conclude therefore that: $\mathbf{E}(t) \leq \mathbf{E}(T)$, but we have no information about how much $\mathbf{E}(t)$ and $\mathbf{E}(T)$ actually differ.

We can measure the difference between $\mathbf{E}(t)$ and $\mathbf{E}(T)$ if we consider an abstract testing scenario obtained by *probabilistic abstract interpretation* which was introduced by the authors in [3,4] and which re-casts the classical framework in a probabilistic setting: The standard order-theoretic semantics is replaced by a probabilistic one based on linear spaces. The main properties of a Galois connection can be translated into an inner product or Hilbert space setting [2].

H. Hermanns and R. Segala (Eds.): PAPM-PROBMIV 2002, LNCS 2399, pp. 211–212, 2002.

The central element of this translation is the notion of an *orthogonal projection* $\pi_{\boldsymbol{\tau}}$ into the image of a linear map $\boldsymbol{\tau}$.

Definition 2. *Let $\mathcal{C}$ and $\mathcal{D}$ be two finite-dimensional vector spaces and $\boldsymbol{\alpha} : \mathcal{C} \mapsto \mathcal{D}$ a linear map between them. A linear map $\boldsymbol{\alpha}^\dagger = \boldsymbol{\gamma} : \mathcal{D} \mapsto \mathcal{C}$ is the (unique)* Moore-Penrose pseudo-inverse *of $\boldsymbol{\alpha}$ iff $\boldsymbol{\alpha} \circ \boldsymbol{\gamma} = \pi_{\boldsymbol{\alpha}}$, and $\boldsymbol{\gamma} \circ \boldsymbol{\alpha} = \pi_{\boldsymbol{\gamma}}$.*

Using a pair of Moore-Penrose pseudo-inverses $(\boldsymbol{\alpha}, \boldsymbol{\gamma})$ in place of a Galois connection (α, γ) we can define a probabilistic abstract interpretation. While classical abstract interpretation results in *safe* approximations the main aim of probabilistic abstract interpretations is to construct *close* approximations; in fact the notion of a so called "least square approximation" is closely related to the concept of the Moore-Penrose pseudo-inverse [2].

An important aspect of the probabilistic abstract interpretation methodology is the possibility of measuring the precision of the abstraction in numerical terms. In the general case of a generic semantic function $f : \mathcal{A} \mapsto \mathcal{B}$ it seems reasonable to measure the precision of a probabilistic abstract interpretation defined by $(\mathcal{A}^\#, \mathcal{B}^\#, f^\#)$ via the norm of the "pseudo-quotient" [4]: $\Delta = (\boldsymbol{\alpha}' \circ f) \circ (f^\# \circ \boldsymbol{\alpha})^\dagger$.

In the case of an abstract test $\boldsymbol{T}$ which is obtained by *lifting* a classical abstract interpretation to a probabilistic abstract interpretation [3] we can relate $\mathbf{E}(t)$ and $\mathbf{E}(T)$ quantitatively via the pseudo-quotient Δ.

Theorem 1. *Given a test function $t : X \mapsto \mathbb{B}$ for a finite domain X and an abstract test function $T : Y \mapsto \mathbb{B}$ obtained via the abstraction function $\alpha : X \mapsto Y$, let $\mathbf{t}$ and $\boldsymbol{T}$ be the lifting of t and T respectively. Then the pseudo-quotient is of the form:*

$$\Delta = \boldsymbol{t} \circ (\boldsymbol{T} \circ \boldsymbol{\alpha})^\dagger = \begin{pmatrix} \frac{\mathbf{E}(t)}{\mathbf{E}(T)} & 1 - \frac{\mathbf{E}(t)}{\mathbf{E}(T)} \\ 0 & 1 \end{pmatrix}$$

where the expectation values $\mathbf{E}(t) = \mathbf{E}(t)_{\boldsymbol{d}}$ and $\mathbf{E}(T) = \mathbf{E}(T)_{\boldsymbol{\alpha}(\boldsymbol{d})}$ are with respect to the uniform distribution $\mathbf{d}$ on X and its abstraction $\boldsymbol{\alpha}(\mathbf{d})$ respectively.

Given the pseudo-quotient Δ we can state for example that $\mathbf{E}(t) = \mathbf{E}(T)\Delta_{11}$.

References

1. P. Cousot and R. Cousot. Abstract Interpretation: A unified lattice model for static analysis of programs by construction or approximation of fixpoints. In *Proceedings of POPL'77*, pages 238–252, Los Angeles, 1977.
2. F. Deutsch. *Bet Approximation in Inner Product Spaces*, volume 7 of *CMS Books in Mathematics*. Springer Verlag, New York — Berlin, 2001.
3. A. Di Pierro and H. Wiklicky. Concurrent Constraint Programming: Towards Probabilistic Abstract Interpretation. In M. Gabbrielli and F. Pfenning, editors, *Proceedings of PPDP 2000*, pages 127–138, Monteéal, Canada, 2000.
4. A. Di Pierro and H. Wiklicky. Measuring the precision of abstract interpretations. In *Proceedings of LOPSTR 2000*, volume 2042 of *Lecture Notes in Computer Science*, pages 147–164, Berlin – New York, 2001. Springer Verlag.
5. D. Monniaux. An abstract Monte-Carlo method for the analysis of probabilistic programs. In *Proceedings of POPL 2001*, pages 93–101, London, 2001.

Approximate Verification of Probabilistic Systems

Richard Lassaigne[1] and Sylvain Peyronnet[2]

[1] Equipe de Logique, U. Paris VII, France
lassaign@logique.jussieu.fr
[2] LRI, U. Paris XI, France
syp@lri.fr

General methods have been proposed [2,4] for the model checking of probabilistic systems, where the verification of a probabilistic statement is reduced to the solution of a linear system over the system's state space. To overcome the state space explosion problem, some probabilistic model checkers, such as PRISM [3], use MTBDDs. We propose a different solution, in which we use a Monte-Carlo algorithm [6] to approximate $Prob[\psi]$, the probability that a temporal formula is true. We show how to obtain a randomized estimator of $Prob[\psi]$ for a fragment of LTL formulas. This fragment is sufficient to express interesting properties such as reachability and liveness.

We consider a subset of LTL formulas which have the property: truth at depth k implies truth in the entire model. The *essentially positive fragment* (EPF) of LTL is the set of formulas constructed from atomic formulas, their negations, closed under $\vee$, $\wedge$ and the temporal operators **X**, **U**. If ψ is any formula of the EPF fragment, we can use a BMC-like framework [1] to verify whether ψ is true on a path σ of depth k. The monotonicity of the property defined by an EPF formula yields the following result: for any formula of the essentially positive fragment of LTL and D$< b \leq$ L, there exists k such that if $Prob_k[\psi] \geq b$, then $Prob[\psi] \geq b$, where $Prob_k[\psi] \geq b$ is the probability over Kripke paths of depth k.

We show that we can approximate the probability $p{=}Prob_k[\psi]$ with a simple randomized algorithm. We generate random paths in the probabilistic space underlying the Kripke structure of depth k and compute the number A of paths on which the given formula is true. In order to approximate p with *approximation ratio* ε and *confidence ratio* δ, we use a sample of size $N = O5\frac{1}{b} \cdot \frac{1}{\varepsilon^2} \cdot \log \frac{1}{\delta}0$. To verify a statement $Prob_k[\psi]{\geq}b$, we test whether $5A/N0{>}b \cdot 5L{-}\varepsilon 0$. Then if $Prob_k[\psi]{\geq}b$, the probability that the algorithm accepts is greater than $5L{-}\delta 0$, where the probability is taken over the random choices of the algorithm. The lower bound is obtained by using Chernoff bound [7] on the tail of the distribution of a sum of independent random variables.

Our method proceeds in two steps: first we determine a lower bound c for p by binary search and successive applications of the algorithm described above, then we decide the property $Prob_k[\psi]{\geq}b$ by applying the algorithm once more. The method provides a framework for verifying probabilistic statements for EPF formulas. To approximately verify that $Prob[\psi]{\geq}b$, where ψ is an EPF formula, we check whether $Prob_k[\psi]{\geq}b$, for increasing values of k. If $Prob_k[\psi]{\geq}b$ is true then the monotonicity property guarantees that $Prob[\psi]{\geq}b$. Otherwise, we increment the value of k within a certain bound, for example the diameter of the system for reachability formulas, to conclude that $Prob[\psi]{\ngeq}b$.

We compare the performance of our method to PRISM's. The results indicate that large systems can be approximately verified in seconds, using very little memory. Ex-

H. Hermanns and R. Segala (Eds.): PAPM-PROBMIV 2002, LNCS 2399, pp. 213–214, 2002.

periments were performed on a 750 MHz SUN ultraSPARC III workstation with 2 GB memory. We tested our method on fully probabilistic versions of Pnueli and Zuck's protocol for the dining philosophers problem [8] and of Itai and Rodeh's synchronous leader election protocol [5]. We checked the following properties:

$$Prob[\bigvee_{i=1}^{n} hungry5i0 \Longrightarrow \ddot{}\ (\bigvee_{i=1}^{n} eat5i0)] \geq \text{L}$$
$$\text{and} \quad Prob[\ddot{}\ (leader\_election)] \geq \text{L}$$

The tables below show our results for ε=LD^{-2} and δ=LD^{-10}. We give path depth, time (in seconds) of our method and for PRISM.

phil.	depth	time	PRISM (time)	PRISM (states)
3	20	13.62	0.19	770
5	23	24.67	0.615	64858
10	33	70.32	13.059	4.21×10^{9}
15	42	146.22	68.926	2.73×10^{14}
20	51	261.43	167.201	1.77×10^{19}
25	58	412.06	3237.143	1.14×10^{24}
30	66	614.49	out of mem.	out of mem.
50	95	2020.79	out of mem.	out of mem.
100	148	11475.28	out of mem.	out of mem.

proc.	K	depth	time	PRISM (time)	PRISM (states)
5	4	72	24.82	4.254	5122
5	8	30	14.87	418.966	163842
5	16	18	11.48	out of mem.	out of mem.
6	4	140	42.55	15.931	20524
6	6	56	27.57	694.948	233340
6	8	35	22.36	out of mem.	out of mem.
8	2	1350	1035.09	0.851	1803
8	4	135	103.36	423.507	458847
10	8	121	81.34	out of mem.	out of mem.
10	16	111	49.9	out of mem.	out of mem.

Our experimental results indicate that the depth has the same order of magnitude as the diameter of the system. However, in practice, we cut off the construction of a path as soon as the formula is verified, thus reducing the depth of most of the paths. The main problem remaining is to determine, in the general case, an upper bound for the depth of the paths.

This work is a first step in applying approximation methods to probabilistic model checking. Our experiments point to an essential advantage of the method: the use of very little memory. In practice, this means that we are able to verify larger models than PRISM, allowing verification to be carried out when classical methods fail.

References

1. A. Biere, A. Cimatti, E. Clarke, and Y. Zhu. Symbolic model checking without BDD's. *Proc. of 5th TACAS* , LNCS 1573:193–207, 1999.
2. C. Courcoubetis and M. Yannakakis. The complexity of probabilistic verification. *Journal of the ACM*, 42(4):857–907, 1995.
3. L. de Alfaro, M. Kwiatkowska, G. Norman, D. Parker, and R. Segala. Symbolic model checking of concurrent probabilistic processes using MTBDDs and the Kronecker representation. In *Proc. of TACAS*, LNCS 1785, 2000.
4. H.Hansson and B. Jonsson. A logic for reasoning about time and reliability. *Formal Aspects of Computing*, 6:512–535, 1994.
5. A. Itai and M. Rodeh. Symmetry breaking in distributed networks. *Information and Computation*, 88(1):60–87, 1990.
6. R.M. Karp and M. Luby. Monte-Carlo algorithms for enumeration and reliability problems. *In Proceedings of the 24th FOCS*, 56–64, 1983.
7. R. Motwani and P. Raghavan. *Randomized Algorithms*. Cambridge University Press, 1995.
8. A. Pnueli and L. Zuck. Verification of multiprocess probabilistic protocols. *Distributed Computing*, pages 1:53–72, 1986.

Author Index

Lecture Notes in Computer Science

For information about Vols. 1–2302
please contact your bookseller or Springer-Verlag

Vol. 2303: M. Nielsen, U. Engberg (Eds.), Foundations of Software Science and Computation Structures. Proceedings, 2002. XIII, 435 pages. 2002.

Vol. 2304: R.N. Horspool (Ed.), Compiler Construction. Proceedings, 2002. XI, 343 pages. 2002.

Vol. 2305: D. Le Métayer (Ed.), Programming Languages and Systems. Proceedings, 2002. XII, 331 pages. 2002.

Vol. 2306: R.-D. Kutsche, H. Weber (Eds.), Fundamental Approaches to Software Engineering. Proceedings, 2002. XIII, 341 pages. 2002.

Vol. 2307: C. Zhang, S. Zhang, Association Rule Mining. XII, 238 pages. 2002. (Subseries LNAI).

Vol. 2308: I.P. Vlahavas, C.D. Spyropoulos (Eds.), Methods and Applications of Artificial Intelligence. Proceedings, 2002. XIV, 514 pages. 2002. (Subseries LNAI).

Vol. 2309: A. Armando (Ed.), Frontiers of Combining Systems. Proceedings, 2002. VIII, 255 pages. 2002. (Subseries LNAI).

Vol. 2310: P. Collet, C. Fonlupt, J.-K. Hao, E. Lutton, M. Schoenauer (Eds.), Artificial Evolution. Proceedings, 2001. XI, 375 pages. 2002.

Vol. 2311: D. Bustard, W. Liu, R. Sterritt (Eds.), Soft-Ware 2002: Computing in an Imperfect World. Proceedings, 2002. XI, 359 pages. 2002.

Vol. 2312: T. Arts, M. Mohnen (Eds.), Implementation of Functional Languages. Proceedings, 2001. VII, 187 pages. 2002.

Vol. 2313: C.A. Coello Coello, A. de Albornoz, L.E. Sucar, O.Cairó Battistutti (Eds.), MICAI 2002: Advances in Artificial Intelligence. Proceedings, 2002. XIII, 548 pages. 2002. (Subseries LNAI).

Vol. 2314: S.-K. Chang, Z. Chen, S.-Y. Lee (Eds.), Recent Advances in Visual Information Systems. Proceedings, 2002. XI, 323 pages. 2002.

Vol. 2315: F. Arhab, C. Talcott (Eds.), Coordination Models and Languages. Proceedings, 2002. XI, 406 pages. 2002.

Vol. 2316: J. Domingo-Ferrer (Ed.), Inference Control in Statistical Databases. VIII, 231 pages. 2002.

Vol. 2317: M. Hegarty, B. Meyer, N. Hari Narayanan (Eds.), Diagrammatic Representation and Inference. Proceedings, 2002. XIV, 362 pages. 2002. (Subseries LNAI).

Vol. 2318: D. Bošnački, S. Leue (Eds.), Model Checking Software. Proceedings, 2002. X, 259 pages. 2002.

Vol. 2319: C. Gacek (Ed.), Software Reuse: Methods, Techniques, and Tools. Proceedings, 2002. XI, 353 pages. 2002.

Vol.2320: T. Sander (Ed.), Security and Privacy in Digital Rights Management. Proceedings, 2001. X, 245 pages. 2002.

Vol. 2321: P.L. Lanzi, W. Stolzmann, S.W. Wilson (Eds.), Advances in Learning Classifier Systems. Proceedings, 2002. VIII, 231 pages. 2002. (Subseries LNAI).

Vol. 2322: V. Mařík, O. Štěpánková, H. Krautwurmová, M. Luck (Eds.), Multi-Agent Systems and Applications II. Proceedings, 2001. XII, 377 pages. 2002. (Subseries LNAI).

Vol. 2323: À. Frohner (Ed.), Object-Oriented Technology. Proceedings, 2001. IX, 225 pages. 2002.

Vol. 2324: T. Field, P.G. Harrison, J. Bradley, U. Harder (Eds.), Computer Performance Evaluation. Proceedings, 2002. XI, 349 pages. 2002.

Vol 2326: D. Grigoras, A. Nicolau, B. Toursel, B. Folliot (Eds.), Advanced Environments, Tools, and Applications for Cluster Computing. Proceedings, 2001. XIII, 321 pages. 2002.

Vol. 2327: H.P. Zima, K. Joe, M. Sato, Y. Seo, M. Shimasaki (Eds.), High Performance Computing. Proceedings, 2002. XV, 564 pages. 2002.

Vol. 2328: R. Wyrzykowski, J. Dongarra, M. Paprzycki, J. Waśniewski (Eds.), Parallel Processing and Applied Mathematics. Proceedings, 2001. XIX, 915 pages. 2002.

Vol. 2329: P.M.A. Sloot, C.J.K. Tan, J.J. Dongarra, A.G. Hoekstra (Eds.), Computational Science – ICCS 2002. Proceedings, Part I. XLI, 1095 pages. 2002.

Vol. 2330: P.M.A. Sloot, C.J.K. Tan, J.J. Dongarra, A.G. Hoekstra (Eds.), Computational Science – ICCS 2002. Proceedings, Part II. XLI, 1115 pages. 2002.

Vol. 2331: P.M.A. Sloot, C.J.K. Tan, J.J. Dongarra, A.G. Hoekstra (Eds.), Computational Science – ICCS 2002. Proceedings, Part III. XLI, 1227 pages. 2002.

Vol. 2332: L. Knudsen (Ed.), Advances in Cryptology – EUROCRYPT 2002. Proceedings, 2002. XII, 547 pages. 2002.

Vol. 2333: J.-J.Ch. Meyer, M. Tambe (Eds.), Intelligent Agents VIII. Revised Papers, 2001. XI, 461 pages. 2001. (Subseries LNAI).

Vol. 2334: G. Carle, M. Zitterbart (Eds.), Protocols for High Speed Networks. Proceedings, 2002. X, 267 pages. 2002.

Vol. 2335: M. Butler, L. Petre, K. Sere (Eds.), Integrated Formal Methods. Proceedings, 2002. X, 401 pages. 2002.

Vol. 2336: M.-S. Chen, P.S. Yu, B. Liu (Eds.), Advances in Knowledge Discovery and Data Mining. Proceedings, 2002. XIII, 568 pages. 2002. (Subseries LNAI).

Vol. 2337: W.J. Cook, A.S. Schulz (Eds.), Integer Programming and Combinatorial Optimization. Proceedings, 2002. XI, 487 pages. 2002.

Vol. 2338: R. Cohen, B. Spencer (Eds.), Advances in Artificial Intelligence. Proceedings, 2002. XII, 373 pages. 2002. (Subseries LNAI).

Vol. 2340: N. Jonoska, N.C. Seeman (Eds.), DNA Computing. Proceedings, 2001. XI, 392 pages. 2002.

Vol. 2342: I. Horrocks, J. Hendler (Eds.), The Semantic Web – ISCW 2002. Proceedings, 2002. XVI, 476 pages. 2002.

Vol. 2345: E. Gregori, M. Conti, A.T. Campbell, G. Omidyar, M. Zukerman (Eds.), NETWORKING 2002. Proceedings, 2002. XXVI, 1256 pages. 2002.

Vol. 2346: H. Unger, T. Böhme, A. Mikler (Eds.), Innovative Internet Computing Systems. Proceedings, 2002. VIII, 251 pages. 2002.

Vol. 2347: P. De Bra, P. Brusilovsky, R. Conejo (Eds.), Adaptive Hypermedia and Adaptive Web-Based Systems. Proceedings, 2002. XV, 615 pages. 2002.

Vol. 2348: A. Banks Pidduck, J. Mylopoulos, C.C. Woo, M. Tamer Ozsu (Eds.), Advanced Information Systems Engineering. Proceedings, 2002. XIV, 799 pages. 2002.

Vol. 2349: J. Kontio, R. Conradi (Eds.), Software Quality – ECSQ 2002. Proceedings, 2002. XIV, 363 pages. 2002.

Vol. 2350: A. Heyden, G. Sparr, M. Nielsen, P. Johansen (Eds.), Computer Vision – ECCV 2002. Proceedings, Part I. XXVIII, 817 pages. 2002.

Vol. 2351: A. Heyden, G. Sparr, M. Nielsen, P. Johansen (Eds.), Computer Vision – ECCV 2002. Proceedings, Part II. XXVIII, 903 pages. 2002.

Vol. 2352: A. Heyden, G. Sparr, M. Nielsen, P. Johansen (Eds.), Computer Vision – ECCV 2002. Proceedings, Part III. XXVIII, 919 pages. 2002.

Vol. 2353: A. Heyden, G. Sparr, M. Nielsen, P. Johansen (Eds.), Computer Vision – ECCV 2002. Proceedings, Part IV. XXVIII, 841 pages. 2002.

Vol. 2355: M. Matsui (Ed.), Fast Software Encryption. Proceedings, 2001. VIII, 169 pages. 2001.

Vol. 2358: T. Hendtlass, M. Ali (Eds.), Developments in Applied Artificial Intelligence. Proceedings, 2002 XIII, 833 pages. 2002. (Subseries LNAI).

Vol. 2359: M. Tistarelli, J. Bigun, A.K. Jain (Eds.), Biometric Authentication. Proceedings, 2002. X, 197 pages. 2002.

Vol. 2360: J. Esparza, C. Lakos (Eds.), Application and Theory of Petri Nets 2002. Proceedings, 2002. X, 445 pages. 2002.

Vol. 2361: J. Blieberger, A. Strohmeier (Eds.), Reliable Software Technologies – Ada-Europe 2002. Proceedings, 2002 XIII, 367 pages. 2002.

Vol. 2363: S.A. Cerri, G. Gouardères, F. Paraguaçu (Eds.), Intelligent Tutoring Systems. Proceedings, 2002. XXVIII, 1016 pages. 2002.

Vol. 2364: F. Roli, J. Kittler (Eds.), Multiple Classifier Systems. Proceedings, 2002. XI, 337 pages. 2002.

Vol. 2366: M.-S. Hacid, Z.W. Raś, D.A. Zighed, Y. Kodratoff (Eds.), Foundations of Intelligent Systems. Proceedings, 2002. XII, 614 pages. 2002. (Subseries LNAI).

Vol. 2367: J. Fagerholm, J. Haataja, J. Järvinen, M. Lyly. P. Råback, V. Savolainen (Eds.), Applied Parallel Computing. Proceedings, 2002. XIV, 612 pages. 2002.

Vol. 2368: M. Penttonen, E. Meineche Schmidt (Eds.), Algorithm Theory – SWAT 2002. Proceedings, 2002. XIV, 450 pages. 2002.

Vol. 2369: C. Fieker, D.R. Kohel (Eds.), Algebraic Number Theory. Proceedings, 2002. IX, 517 pages. 2002.

Vol. 2370: J. Bishop (Ed.), Component Deployment. Proceedings, 2002. XII, 269 pages. 2002.

Vol. 2372: A. Pettorossi (Ed.), Logic Based Program Synthesis and Transformation. Proceedings, 2001. VIII, 267 pages. 2002.

Vol. 2373: A. Apostolico, M. Takeda (Eds.), Combinatorial Pattern Matching. Proceedings, 2002. VIII, 289 pages. 2002.

Vol. 2374: B. Magnusson (Ed.), ECOOP 2002 – Object-Oriented Programming. XI, 637 pages. 2002.

Vol. 2375: J. Kivinen, R.H. Sloan (Eds.), Computational Learning Theory. Proceedings, 2002. XI, 397 pages. 2002. (Subseries LNAI).

Vol. 2378: S. Tison (Ed.), Rewriting Techniques and Applications. Proceedings, 2002. XI, 387 pages. 2002.

Vol. 2380: P. Widmayer, F. Triguero, R. Morales, M. Hennessy, S. Eidenbenz, R. Conejo (Eds.), Automata, Languages and Programming. Proceedings, 2002. XXI, 1069 pages. 2002.

Vol. 2382: A. Halevy, A. Gal (Eds.), Next Generation Information Technologies and Systems. Proceedings, 2002. VIII, 169 pages. 2002.

Vol. 2383: M.S. Lew, N. Sebe, J.P. Eakins (Eds.), Image and Video Retrieval. Proceedings, 2002. XII, 388 pages. 2002.

Vol. 2384: L. Batten, J. Seberry (Eds.), Information Security and Privacy. Proceedings, 2002. XII, 514 pages. 2002.

Vol. 2385: J. Calmet, B. Benhamou, O. Caprotti, L. Henocque, V. Sorge (Eds.), Artificial Intelligence, Automated Reasoning, and Symbolic Computation. Proceedings, 2002. XI, 343 pages. 2002. (Subseries LNAI).

Vol. 2386: E.A. Boiten, B. Möller (Eds.), Mathematics of Program Construction. Proceedings, 2002. X, 263 pages. 2002.

Vol. 2389: E. Ranchhod, N.J. Mamede (Eds.), Advances in Natural Language Processing. Proceedings, 2002. XII, 275 pages. 2002. (Subseries LNAI).

Vol. 2391: L.-H. Eriksson, P.A. Lindsay (Eds.), FME 2002: Formal Methods – Getting IT Right. Proceedings, 2002. XI, 625 pages. 2002.

Vol. 2392: A. Voronkov (Ed.), Automated Deduction – CADE-18. Proceedings, 2002. XII, 534 pages. 2002. (Subseries LNAI).

Vol. 2393: U. Priss, D. Corbett, G. Angelova (Eds.), Conceptual Structures: Integration and Interfaces. Proceedings, 2002. XI, 397 pages. 2002. (Subseries LNAI).

Vol. 2398: K. Miesenberger, J. Klaus, W. Zagler (Eds.), Computers Helping People with Special Needs. Proceedings, 2002. XXII, 794 pages. 2002.

Vol. 2399: H. Hermanns, R. Segala (Eds.), Process Algebra and Probabilistic Methods. Proceedings, 2002. X, 215 pages. 2002.

Vol. 2405: B. Eaglestone, S. North, A. Poulovassilis (Eds.), Advances in Databases. Proceedings, 2002. XII, 199 pages. 2002.